Strain Effect in Semiconductors

Yongke Sun • Scott E. Thompson • Toshikazu Nishida

Strain Effect in Semiconductors

Theory and Device Applications

Springer

Yongke Sun
SanDisk Corporation
601 McCarthy Boulevard
Milpitas, CA 95035
USA
Yongke.Sun@sandisk.com

Scott E. Thompson
University of Florida
Department Electrical & Computer Engineering
Gainesville, FL 32611
535, Engineering Bldg.
USA
thompson@ece.ufl.edu

Toshikazu Nishida
University of Florida
Department Electrical & Computer Engineering
223 Benton Hall
Gainesville FL 32611-6200
USA
nishida@ufl.edu

ISBN 978-1-4419-0551-2 e-ISBN 978-1-4419-0552-9
DOI 10.1007/978-1-4419-0552-9
Springer New York Dordrecht Heidelberg London

Library of Congress Control Number: 2009938434

Printed on acid-free paper

Springer is part of Springer Science+Business Media (www.springer.com)

Contents

Preface

Strain is an old concept in semiconductor physics. However, strain applied in Si logic technology is a relatively new response to the diminishing returns of pure geometric scaling. Process-induced strain was the first additive feature enhancement introduced into planar Si MOSFET transistors by Intel in 2002 which heralded a new age of feature enhanced CMOS scaling. Prior to strain in logic technologies, Si and Ge piezoresistive strain sensors were initiated much earlier, circa 1957, to respond to variable strain. Strain has also been used to enhance optoelectronic devices such as quantum well lasers incorporated via lattice-mismatched heterostructures. Before the advent of strain enhanced MOSFETs, there were already many, though scattered, research reports on strain effects in semiconductors. However, there had not been a strong driving force for strain studies in semiconductors until it began to play a major role in the mainstream VLSI semiconductor industry. Now in almost every semiconductor workshop, strain is induced by various means to boost device performance. Device and process engineers apply advantageous strain to improve electronic product performance and power at low additive cost to meet the demand of consumers.

There are excellent books on strain physics, such as Symmetry and Strain-induced Effects in Semiconductor by Pikus and Bir, and also many books on device physics, such as Fundamentals of Solid-State Electronics by Sah and Physics of Semiconductor Devices by Sze, as well as numerous papers published on the topic of strained Si, Ge, and other semiconductors, but there is a lack of a single text that combines both strain and device physics. Therefore, drawing from our experience both in the semiconductor industry and in the academic field, we have attempted to summarize in this book some of the latest efforts to reveal the physics underlying the advantages that strain has brought forth as well as its applications in devices, and perhaps help guide the development of strained semiconductor devices. Thus in this book, we have included much of our own research, and have collected many valuable achievements and ideas by the research community. However, due to space constraints, we note that unfortunately only representative papers and not all key papers have been cited in this work.

This book is designed for two levels of readers. For readers such as students and applications engineers who seek a qualitative discussion, we provide a qualitative overview at the beginning of every chapter. For advanced graduate students and research and development engineers with a background in semiconductor physics who wish to dig deeper, the second part of each chapter provides a more systematic and mathematically involved treatment of the subject. We hope this book provides useful insight into the common physics of strain effects in semiconductors that serve as the foundation for the varied strained semiconductor device applications for both sets of readers.

1

Overview: The Age of Strained Devices

1.1 ORIGIN OF THE STRAINED-SI TECHNOLOGY

One of the predecessors of strained Si to enhance MOSFET performance is the research that showed enhanced electron mobilities in n-type (100) $Si/Si_{1-x}Ge_x$ multilayer heterostructures and hole mobilities in p-type (100) $Si/i\text{-}Si_{1-x}Ge_x/Si$ double-heterostructures in the early 1980s (Manasevit et al, 1982; R.People et al, 1984). Strain caused by the lattice mismatch was suspected as one of the factors for the mobility enhancement. The physical mechanism for the enhancement can be traced back to the theoretical formulation of deformation potentials by Bardeen and Shockley (Bardeen and Shockley, 1950; Shockley and Bardeen, 1950) in 1950 and the experimental measurements of the piezoresistance effect, a change in resistance with mechanical stress, by Smith (Smith, 1954).

In an era of rapidly changing technology, strain is a relatively old topic in semiconductor physics, yet its tangible effects on band structure and carrier transport have spurred a renewed interest in strained semiconductor physics. To model lattice scattering, deformation potential theory was developed by Bardeen and Shockley to characterize the band energy shift with strain caused by phonons (Bardeen and Shockley, 1950; Shockley and Bardeen, 1950). Herring and Vogt (Herring and Vogt, 1956) then extended deformation potentials to model transport in strained semiconductors. Deformation potential theory is still the primary method to model the band shift and warping in energy band calculations (Oberhuber et al, 1998; Fischetti and Laux, 1996).

1.2 STRAIN IN SEMICONDUCTOR DEVICES

While strain physics is fundamental, the source of strain is technology and device dependent. For example, strain can result from phonon-induced lattice vibrations in homogeneous semiconductors, lattice-mismatched film growth

Y. Sun et al., *Strain Effect in Semiconductors: Theory and Device Applications*, DOI 10.1007/978-1-4419-0552-9_1,

in epitaxial heterostructures, intrinsic stress in deposited thin films, and applied external stress. Prior to the development of heteroepitaxy and chemical vapor deposition, variable strain transducers were developed to exploit the piezoresistive effect in Si and Ge to construct strain gauges and stress transducers (Mason and Thurston, 1957; Burns, 1957) that responded to different values of strain or stress. With the advent of micromachining, more elaborate piezoresistive transducers have been fabricated using microelectromechanical systems (MEMS) technology. Simultaneously, integrated circuits were invented and evolved exponentially in density and performance along the path portended by Moore's law through improvements in lithography and microelectronics fabrication technologies until various obstacles began to loom. Finally, continual geometric scaling of the metal-oxide-semiconductor field-effect transistor (MOSFET) channel length, gate dielectric thickness, and junction depth led to increasing off-state channel leakage, gate leakage, source–drain resistance, and short-channel effects. Performance improvement at each technology node by simple geometrical scaling became more problematic and costly until the end of simple scaling for CMOS was predicted.

1.2.1 Conventional Simple Scaling

The end of simple scaling for a solid state device technology is not new. Scaling of silicon bipolar junction transistors (BJT) ended in the 90s for various reasons including voltage, base width, and power density limits. However, the unrelenting scaling of the competing complementary MOS (CMOS) as another factor cannot be overestimated. By the mid-90s, the performance of 0.1-μm CMOS devices measured by the unity current gain frequency, f_T, was comparable to the highest reported values for BJTs, but at lower power and cost (Taur et al, 1997). Now more than a decade later, conventional CMOS has reached its simple scaling limits. However, unlike the 90s, there is presently no new device to realistically compete with or potentially replace the industry work horse for VLSI applications. Carbon nanotubes and silicon nanowires are considered to be leading contenders but have yet to achieve commercial success in even a niche logic or memory market.

1.2.2 Feature-Enhanced CMOS

In order to meet customer needs for a continuation of Moore's law, feature enhancement instead of simple geometric scaling of the silicon CMOS platform is recognized as the necessary driver for the microelectronics industry. Key features include strain, metal gate, high-κ dielectric, nonplanar geometries, and heterogeneous semiconductor integration. The first key feature to enhance 90, 65, and 45-nm technology nodes is uniaxial process-induced stress (Chan et al, 2003; Murthy et al, 2003; Ghani et al, 2003; Yang et al, 2004) (Chidambaram et al, 2004; Chien-Hao et al, 2004; Liu et al, 2005). Successive features have been added to stress including metal gate and high-κ

dielectric (Mistry et al, 2007; Packan et al, 2008). Instead of increasing the geometrical scaling, future advancements are expected to be an increase in additive feature enhancements.

The development of the first commercial strain feature-enhanced silicon technology is reviewed in (Thompson et al, 2006a). An early question was on the source of the strain: lattice-mismatched epitaxial layers on a fully relaxed substrate or process-induced source/drain stress or intrinsic stress in deposited thin films. Following the promising $Si/Si_{1-x}Ge_x$ heteroepitaxy results, wafer based substrate strain was experimentally and theoretically studied by a large number of researchers (Rim et al 2003, and references therein). In the 90s, two process-induced strain sources were investigated, high stress capping layers deposited on MOSFETs (Shimizu et al, 2001; Ito et al, 2000) and embedded SiGe source and drain (Gannavaram et al, 2000) although the SiGe source and drain was originally proposed for higher boron activation and reduced external resistance. The embedded SiGe literature prompted Intel to evaluate the technology, which resulted in larger than expected device performance enhancement, which after considerable internal debate was later attributed to compressive channel stress (Thompson et al, 2002). However, neither biaxial nor uniaxial stress was immediately adopted in CMOS logic technologies for several reasons (Thompson et al, 2006a). For biaxial stress, issues included defects in the substrate and performance loss at high vertical electric fields (Fischetti et al, 2003). Process-induced uniaxial channel stress was not initially adopted since different stress types (compressive and tensile for n and p-channel respectively) were needed.

After careful analysis of the hole mobility enhancement at high vertical electric fields and the potential for continued effectiveness at nanoscale dimensions, process-induced uniaxial strain was adopted over biaxial stress. Uniaxial stress provided significantly larger hole mobility enhancement at both low strain and high vertical electric field (Thompson et al, 2004b). Since high strain can lead to strain relaxation via defect formation, large mobility enhancement at low strain is critical for yield. Uniaxial stress also provided larger drive current improvement for nanoscale short channel devices since the enhanced electron and hole mobility arises mostly from reduced conductivity effective mass instead of primarily from reduced scattering for biaxial stress. Another important consideration was the strain effect on the threshold voltage. Process-induced uniaxial stress resulted in an approximately five times smaller n-channel threshold voltage shift. The smaller threshold voltage shift was manifested in a smaller penalty for threshold voltage shift retargeting by adjusting the channel doping. Alternately, the larger threshold voltage shift for the substrate-induced biaxial tensile stress causes approximately half of the stress-enhanced electron mobility to be lost (Fossum and Weimin, 2003). Based on the merits of uniaxial stress and the necessity for opposite stress types for n- and p-channel MOSFETs, two process flows were developed that independently targeted the stress magnitude and direction (Thompson et al, 2006b). The first involved embedded and raised SiGe in the p-channel source and drain and a tensile capping layer on the n-channel device. The second used dual stress liners: compressive and tensile silicon nitride (SiN) for p- and n-channel devices, respectively. As a feature enhancement for CMOS, process-induced stress is employed in nearly all high-performance logic technologies at the 90, 65, and 45-nm technology nodes for both microprocessor and consumer products (Zhang et al, 2005; Bai et al, 2004; Yang et al, 2004; Chan et al, 2003; Thompson et al, 2004a; Qiqing et al, 2005; Ghani et al, 2003; Pidin et al, 2004; Liu et al, 2005; Mistry et al, 2007; Packan et al, 2008).

1.2.3 Variable Strain Sensors

In contrast to the fixed strain incorporated in logic devices for a fixed or constant improvement in device performance, piezoresistive strain sensors respond to variable strain through a modulation in the device resistance. The underlying physics of performance improvement in logic devices and strain transduction in piezoresistive strain sensors is the same. While improvement of logic device performance requires an increase in mobility, which dictates the "sign" of the fixed strain, strain sensors respond to both negative (compressive) and positive (tensile) strains. Since the strain is fixed in logic devices, the linearity of mobility increase with strain is not an issue since the strain is theoretically frozen into the device by stressors incorporated into the device structure during the manufacturing process. In contrast, piezoresistive strain sensors are designed to transduce or detect varying strains by producing a proportional change in resistance. Hence, linear resistance change with strain is important to sense/transduce strains of varying amplitudes into an electrical signal without introducing distortion.

In contrast to discrete strain gauge sensors that are assumed to measure the local strain without significantly affecting the stiffness of the structure in question, integrated stress transducers are devices that integrate the piezoresistive strain gauge within a sensing structure. The combination of MEMS and semiconductor strain gauges has enabled the development of integrated stress transducers. A conventional discrete strain transducer is contrasted with a MEMS piezoresistive pressure stress transducer (microphone) and a fixed stress-enhanced p-channel MOSFET in Fig. 8.13 which is also reproduced as Fig. 1.1 here for an example.

Although it is possible to construct a discrete thin and compliant silicon strain gauge in the same manner as a metal film strain gauge, the vast silicon integrated circuit manufacturing knowledge base coupled with the fortuitous mechanical properties of silicon (Peterson, 1982) has enabled the fabrication of MEMS stress transducers that integrate silicon piezoresistors with a mechanical structure made of the same silicon material.

1.2.4 Strained Quantum Well Optoelectronics

Strain has been an inevitable part of modern heterostructure devices employed in advanced quantum well solid-state lasers and other optoelectronic devices.

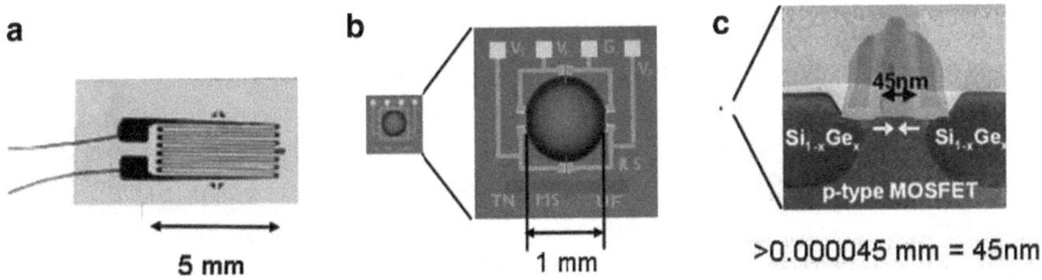

Fig. 1.1. Applications of strain and stress: (**a**) Discrete strain gauge (Omega.com, 2003–2009) (**b**) Si MEMS piezoresistive variable stress transducer with four integrated Si piezoresistors (Arnold et al, 2001) and (**c**) fixed stress enhanced transistor (Thompson et al, 2004a)

While a positive feature enhancement for CMOS, in optoelectronic devices strain was a undesirable by-product from lattice-mismatched semiconductor interfaces, a scourge to be managed in order to avoid quantum efficiency-killing nonradiative defects created by strain relaxation. However, due to improvements in semiconductor film growth technology, strain grading buffer layers, and scaling of device size, strain relaxation is better controlled. With better defect control, strain effects on the band structure offer great potential for enhancing the performance of optoelectronic devices such as solid-state lasers.

Because of the different operation mechanisms, the emphasis of strain effects on band structure is different for optoelectronic devices compared with CMOS. While strain feature enhancement for CMOS is tied primarily to its benefits for electron transport, strain effects on photon emission caused by radiative electron transitions are key for emissive optoelectronic devices. Photon emission is caused by radiative electron transitions in semiconductors such as electron–hole recombination. Since recombination involves two electronic states, the transition probability is determined by the electronic state properties such as energy and wave function. The collective processes of photons, i.e., light emission (as well as absorption), depend on the semiconductor band structure. Strain affects the band structure and thus also affects the performance of optoelectronic devices such as wavelength, gain, linewidth, and quantum efficiency. As will be seen, strain effects on the band structure that affect optoelectronic devices include shifts of the bandgap, changes of energy level density of states (DOS), and electronic wavefunction variation or mixing.

1.3 ORGANIZATION

This book is organized in three major parts. The overall arc of the book follows a trajectory beginning with strain fundamentals for a semiconductor at equilibrium through nonequilibrium transport to strain applications on semiconductor devices.

Part I, "Band Structures of Strained Semiconductors," first reviews stress and strain and crystal symmetry. Strain effects on semiconductors are then introduced by its effects on symmetry. The main physics of interest here that underpins all of the physics of strain is its effects on band structure, which is discussed next initially for an unstrained semiconductor crystal. Strain effects are introduced within two major band structure calculation frameworks, the tight-binding approach and the $\mathbf{k} \cdot \mathbf{p}$ approach. Since the discussion of band structure is originally for a bulk semiconductor, i.e., one that is unconstrained in dimension, the first part finishes with a discussion of the unique differences in the band structure and its strain effects for low-dimensional structures such as a two-dimensional (2D) electron gas such as that created in a MOSFET and a quantum well heterostructure and a one-dimensional nanowire.

Part II, "Transport Theory of Strained Semiconductors," discusses how changes in band structure coupled with changes in carrier scattering caused by strain affect carrier transport. Carrier transport is first reviewed beginning with the Drude model for electron transport in an unstrained semiconductor followed by a qualitative discussion of how strain affects each component of carrier transport. A key transport factor, scattering, is then reviewed by covering the primary scattering processes, lattice, phonon, piezoelectric, and impurity scattering, in a three-dimensional (3D) spatially unconfined structure. Strain effects on the bulk scattering rates are then discussed followed by a discussion of the scattering mechanism unique to spatially confined structures, surface roughness scattering. Finally, the strain effects on carrier transport are summarized in terms of the piezoresistance effect, electron and hole transport, and surface roughness scattering. Strain effects on the high lateral field and near-ballistic transport are also explored.

With the formalities of strain effects on equilibrium and nonequilibrium semiconductors largely discussed, Part III discusses applications of strain to semiconductor devices. Three categories of strain applications are included: (1) fixed strain feature enhancement of electron devices such as Si and SiGe planar and nonplanar MOSFETs, (2) variable strain transducers such as discrete strain gauges and integrated MEMS piezoresistive stress transducers, and (3) optoelectronic devices where strain is a by-product of the quantum well heterostructures employed and an effect to be managed as well as exploited.

Part I

Band Structures of Strained Semiconductors

2

Stress, Strain, Piezoresistivity, and Piezoelectricity

2.1 STRAIN TENSOR

Strain in crystals is created by deformation and is defined as relative lattice displacement. For simplicity, we use a 2D lattice model in Fig. 2.1 to illustrate this concept, but discuss the general concept in 3D cases. As shown in Fig. 2.1a, we may use two unit vectors $\hat{x}$, $\hat{y}$ to represent the unstrained lattice, and in a simple square lattice, they correspond to the lattice basis vectors. Under a small uniform deformation of the lattice, the two vectors are distorted in both orientation and length, which is shown in Fig. 2.1b. The new vectors $\hat{\mathbf{x}}'$ and $\hat{\mathbf{y}}'$ may be written in terms of the old vectors:

$$\hat{\mathbf{x}}' = (1+\varepsilon_{xx})\hat{x} + \varepsilon_{xy}\hat{y} + \varepsilon_{xz}\hat{z}, \tag{2.1}$$

$$\hat{\mathbf{y}}' = \varepsilon_{yx}\hat{x} + (1+\varepsilon_{yy})\hat{y} + \varepsilon_{yz}\hat{z}, \tag{2.2}$$

and in the 3D case, we also have

$$\hat{\mathbf{z}}' = \varepsilon_{zx}\hat{x} + \varepsilon_{zy}\hat{y} + (1+\varepsilon_{zz})\hat{z}. \tag{2.3}$$

The strain coefficients $\varepsilon_{\alpha\beta}$ define the deformation of the lattice and are dimensionless. The 3×3 matrix

$$\bar{\bar{\varepsilon}} = \begin{pmatrix} \varepsilon_{xx} & \varepsilon_{xy} & \varepsilon_{xz} \\ \varepsilon_{yx} & \varepsilon_{yy} & \varepsilon_{yz} \\ \varepsilon_{zx} & \varepsilon_{zy} & \varepsilon_{zz} \end{pmatrix} \tag{2.4}$$

is called the strain tensor. A tensor is a mathematical notation, usually represented by an array, to describe a linear relation between two physical quantities. A tensor can be a scalar, a vector, or a matrix. A scalar is a zero-rank tensor, a vector is a first-rank tensor, and a matrix is a second-rank tensor, and so on. The strain tensor is a second-rank tensor, which in this book is labeled with two bars over the head. However, in places without confusion, we usually neglect the bars. Suppose a lattice point is originally located

Y. Sun et al., *Strain Effect in Semiconductors: Theory and Device Applications*, DOI 10.1007/978-1-4419-0552-9_2,

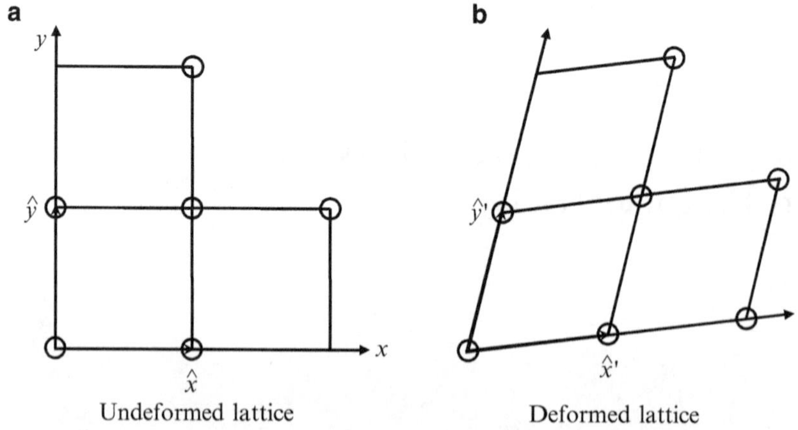

Fig. 2.1. Diagram for (**a**) an undeformed lattice and (**b**) a deformed lattice

at $\mathbf{r} = x\hat{x} + y\hat{y} + z\hat{z}$, then with a uniform deformation this point will be at $\mathbf{r}' = x\hat{\mathbf{x}}' + y\hat{\mathbf{y}}' + z\hat{\mathbf{z}}'$. For a general varying strain, the strain tensor may be written as

$$\varepsilon_{\alpha,\beta} = \frac{\partial u_\alpha}{\partial x_\beta}, \qquad u_\alpha = u_x, u_y, u_z, \; x_\beta = x, y, z, \tag{2.5}$$

where u_α is the displacement of lattice point under study along x_α. A strain tensor (2.4) is symmetric, i.e.,

$$\varepsilon_{\alpha\beta} = \varepsilon_{\beta\alpha} = \frac{1}{2}\left(\frac{\partial u_\alpha}{\partial x_\beta} + \frac{\partial u_\beta}{\partial x_\alpha}\right). \tag{2.6}$$

The antisymmetric part of tensor (2.4) represents a rotation of the entire body.

Usually people work with the other set of strain components, which are defined as

$$e_{xx} = \varepsilon_{xx}; \quad e_{yy} = \varepsilon_{yy}; \quad e_{zz} = \varepsilon_{zz}, \tag{2.7}$$

which describe infinitesimal distortions associated with a change in volume, and the other strain components e_{xy}, e_{yz}, and e_{zx} are defined in terms of changes of angle between the basis vectors. Neglecting the terms of order ε^2 in the small strain approximation, they are

$$\begin{aligned} e_{xy} &= \hat{\mathbf{x}}' \cdot \hat{\mathbf{y}}' = \varepsilon_{xy} + \varepsilon_{yx}, \\ e_{yz} &= \hat{\mathbf{y}}' \cdot \hat{\mathbf{z}}' = \varepsilon_{yz} + \varepsilon_{zy}, \\ e_{zx} &= \hat{\mathbf{z}}' \cdot \hat{\mathbf{x}}' = \varepsilon_{zx} + \varepsilon_{xz}. \end{aligned} \tag{2.8}$$

These six coefficients completely define the strain. We can write these six strain coefficients in the form of an array as $\mathbf{e} = \{e_{xx}, e_{yy}, e_{zz}, e_{yz}, e_{zx}, e_{xy}\}$. The introduction of this set of notation for the strain components is merely for the convenience of describing the relations between strain and the other strain-related physical quantities. The relation between two second-rank tensors must be described by a fourth-rank tensor, which is very complicated; while after transforming the second-rank tensors to first-rank, only a second-rank tensor is required.

The crystal dilation under deformation can be evaluated through calculating the volume defined by $\hat{\mathbf{x}}'$, $\hat{\mathbf{y}}'$, and $\hat{\mathbf{z}}'$,

$$V' = \hat{\mathbf{x}}' \cdot \hat{\mathbf{y}}' \times \hat{\mathbf{z}}' = 1 + e_{xx} + e_{yy} + e_{zz}, \tag{2.9}$$

and the dilation δ then is given by

$$\delta = \frac{\delta V}{V} = e_{xx} + e_{yy} + e_{zz}, \tag{2.10}$$

which is the trace of the strain tensor. The dilation is negative for hydrostatic pressure.

2.2 STRESS TENSOR

Crystal deformations can be induced by externally applied forces, or in other words, a solid resists deformations, thus deformations will generate forces. Stress is defined as the force in response to strain in a unit area. Stress has nine components and is a second-rank tensor, which we write as $\tau_{\alpha\beta}$, $\alpha, \beta = x, y, z$. On the surface of an infinitesimal volume cube, the stress distribution is illustrated in Fig. 2.2, where τ_{xx} represents a force applied in the x direction to a unit area of the plane whose outward-drawn normal lies in the x direction, and τ_{xy} represents a force applied in the x direction to a unit area of the

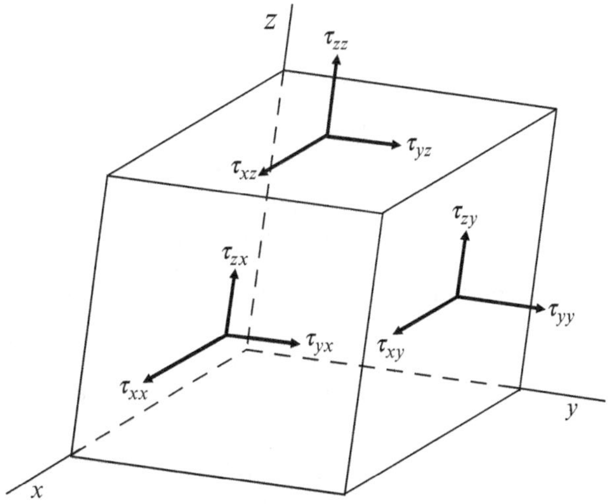

Fig. 2.2. Illustration for stress components on the surfaces of an infinitesimal cube

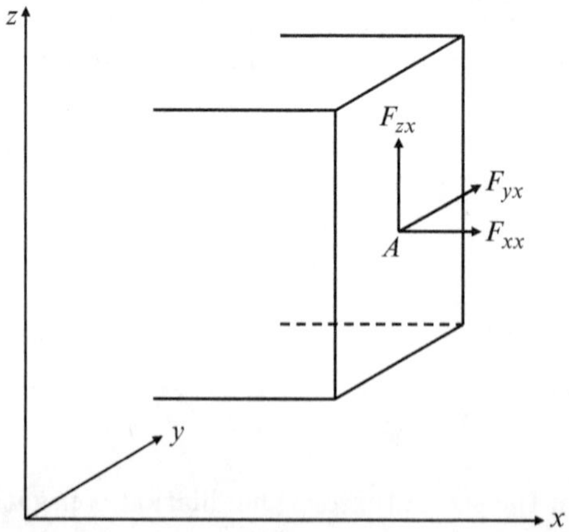

Fig. 2.3. Illustration of the forces applied on one surface with area of A of the cube shown in Fig. 2.2

plane whose outward-drawn normal lies in the y direction. The stress tensor is symmetric just as the strain tensor. The antisymmetric part of the stress tensor represents a torque, and in a state of equilibrium, all torques must vanish inside a solid.

The stress and force relation is better illustrated in Fig. 2.3 where we show a force applied on an infinitesimal plane whose normal is along x and has an area A. In such a case, we resolve the force into components along the coordinate axes, i.e., F_{xx}, F_{yx}, and F_{zx}. The stress components in this plane are

$$\tau_{xx} = \frac{F_{xx}}{A}, \ \tau_{yx} = \frac{F_{yx}}{A}, \ \tau_{zx} = \frac{F_{zx}}{A}. \tag{2.11}$$

We now study some simple stress cases to determine the stress tensors.

1. Hydrostatic pressure:
   Under a hydrostatic pressure P, all shear stress is zero. Stress along any principle direction is $-P$, namely,
   $$\tau = \begin{pmatrix} -P & 0 & 0 \\ 0 & -P & 0 \\ 0 & 0 & -P \end{pmatrix}. \tag{2.12}$$
   Here the sign convention is that tensile stress is positive and compressive stress is negative.
2. Uniaxial stress T along the [001] direction:
   For a uniaxial stress T along the [001] direction, all stress components but τ_{zz} are zero, and $\tau_{zz} = T$. So
   $$\tau = \begin{pmatrix} 0 & 0 & 0 \\ 0 & 0 & 0 \\ 0 & 0 & T \end{pmatrix}. \tag{2.13}$$

3. Uniaxial stress T along the [110] direction:
The case for a uniaxial stress along the [110] direction is a little more complicated. Generally when we talk about a stress T along the $\langle 110 \rangle$ direction, it refers to the force exerted along the $\langle 110 \rangle$ direction divided by the cross section of the (110) surface, but not necessarily equal to any of the stress tensor elements. To find the stress elements, we can use two methods. First is to resolve the force into three coordinate axes. For [110] uniaxial stress T as shown in Fig. 2.4a, the force along the [110] direction is $F = Ta^2$. Its component along [001] is zero. Along both x and y direction, the force is $F/\sqrt{2}$. However, the cross area for the force along [110] shown in Fig. 2.4a is a^2 and is $\sqrt{2}a^2$ for the forces along the x and y direction. Thus, the stress along both x and y is $F/2a^2 = T/2$. The shear stress on both [100] and [010] planes is also $T/2$. The second method to obtain the stress components is through the coordinate transformation method. Suppose in an unprimed coordinate system, stress T is along the x direction, and thus $\tau_{xx} = T$, and all the other stress components are zero. We can rotate the x and y axes 45° clockwise, and then an original [100] uniaxial stress that only has one nonvanishing component $\tau_{xx} = T$ now corresponds to a [110] uniaxial stress in a primed coordinate system, as shown in Fig. 2.4b. The stress elements in the primed coordinate system are given by the transformation,

$$\tau'_{ij} = \sum_{mn} \tau_{mn} \frac{\partial x'_i}{\partial x_m} \frac{\partial x'_j}{\partial x_n}, \tag{2.14}$$

where $\frac{\partial x'_i}{\partial x_m}$, etc. represent the directional cosines of the transformed axes made to the original axes. This equation results from the general tensor transformation of S to S' under an orthogonal coordinate transformation A,

$$S' = ASA^{\mathrm{T}}, \tag{2.15}$$

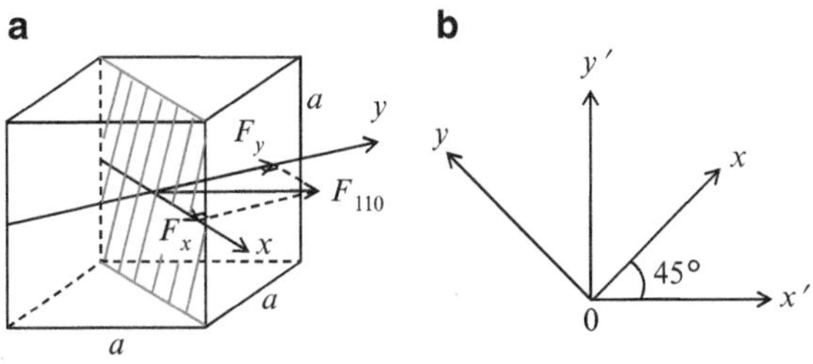

Fig. 2.4. (**a**) The decomposition of a force along the [110] direction along the x and y directions, and their stress relations. Please note that in this figure, the x and y directions are along the diagonals of the surfaces instead of along the edges. (**b**) The coordinate systems before and after a 45° rotation clockwise. The unprimed and primed systems are the coordinate systems before and after the rotation

where A^{T} is the transpose of matrix A. The stress tensor under the [110] uniaxial stress found using both methods is

$$\tau = \frac{\mathrm{T}}{2}\begin{pmatrix} 1\,1\,0 \\ 1\,1\,0 \\ 0\,0\,0 \end{pmatrix}. \tag{2.16}$$

Because a stress tensor is symmetric, similar to the strain tensor case, the six coefficients, τ_{xx}, τ_{yy}, τ_{zz}, τ_{yz}, τ_{zx}, and τ_{xy} completely define the stress. Similar to a strain tensor, a second-rank stress tensor can be reduced to a 1D array form.

2.3 ELASTIC COMPLIANCE AND STIFFNESS CONSTANTS

In the linear elastic theory, Hooke's law is justified and stress is proportional to strain

$$\tau_{ij} = \sum_{\alpha\beta} C_{ij\alpha\beta} e_{\alpha\beta}, \quad i, j, \alpha, \beta = x, y, z, \tag{2.17}$$

where the coefficients $C_{ij\alpha\beta}$ are called elastic stiffness constants. Elastic stiffness constants are a fourth-rank tensor. Because of the symmetry of both the strain tensor and the stress tensor, we have

$$C_{ij\alpha\beta} = C_{ji\alpha\beta} = C_{ij\beta\alpha}, \tag{2.18}$$

so we may write both strain and stress tensor as a six-component array as

$$\mathbf{e} = (e_{xx}, e_{yy}, e_{zz}, e_{yz}, e_{zx}, e_{xy}) \tag{2.19}$$

and

$$\tau = (\tau_{xx}, \tau_{yy}, \tau_{zz}, \tau_{yz}, \tau_{zx}, \tau_{xy}) \tag{2.20}$$

and reduce the elastic stiffness tensor to a 6×6 matrix

$$\tau_i = \sum_m C_{im} e_m. \tag{2.21}$$

This 6×6 matrix has a very simple form in cubic crystals due to the high symmetry. It has only three independent components and has the form

$$C_{ij} = \begin{pmatrix} C_{11} & C_{12} & C_{12} & 0 & 0 & 0 \\ C_{12} & C_{11} & C_{12} & 0 & 0 & 0 \\ C_{12} & C_{12} & C_{11} & 0 & 0 & 0 \\ 0 & 0 & 0 & C_{44} & 0 & 0 \\ 0 & 0 & 0 & 0 & C_{44} & 0 \\ 0 & 0 & 0 & 0 & 0 & C_{44} \end{pmatrix}. \tag{2.22}$$

This is easy to understand by inspecting (2.21) and considering the transformation of this equation under symmetry operations. First, the elastic stiffness tensor must be symmetric. Second, since for cubic crystals, the three axes are equivalent, therefore we must have $C_{11} = C_{22} = C_{33}$, and $C_{44} = C_{55} = C_{66}$. Third, a shear strain cannot cause a normal stress, so terms like $C_{14} = 0$. And a shear strain along one axis cannot induce forces causing a shear along another axis, so terms like $C_{45} = 0$. Finally in the view of a force along one axis, the other two axes are equivalent, and thus we have $C_{12} = C_{13}$, etc. These results can also be obtained by investigating the transformation of the components in (2.17) under symmetry operations using an equation similar to (2.14)

$$C'_{lk\gamma\delta} = \sum_{ij\alpha\beta} C_{ij\alpha\beta} \frac{\partial x'_l}{\partial x_i} \frac{\partial x'_k}{\partial x_j} \frac{\partial x'_\gamma}{\partial x_\alpha} \frac{\partial x'_\delta}{\partial x_\beta}. \tag{2.23}$$

For example, it is easy to verify that under a reflection and thus $x \to -x$, $C_{xyzz} = -C_{xyzz}$, so in the 6×6 matrix, $C_{63} = 0$.

In many cases it is convenient to work with the inverse of the elastic stiffness tensor, which is defined through the relation between strain and stress

$$\varepsilon_{\alpha\beta} = \sum_{ij} S_{\alpha\beta ij} \tau_{ij}. \tag{2.24}$$

The fourth-rank tensor $S_{\alpha\beta ij}$, called the compliance tensor, can also be reduced into a 6×6 matrix. Under cubic symmetry, it has the same form as the stiffness tensor

$$\begin{pmatrix} S_{11} & S_{12} & S_{12} & 0 & 0 & 0 \\ S_{12} & S_{11} & S_{12} & 0 & 0 & 0 \\ S_{12} & S_{12} & S_{11} & 0 & 0 & 0 \\ 0 & 0 & 0 & S_{44} & 0 & 0 \\ 0 & 0 & 0 & 0 & S_{44} & 0 \\ 0 & 0 & 0 & 0 & 0 & S_{44} \end{pmatrix}, \tag{2.25}$$

and the strain–stress relation can be written as

$$e_m = \sum_i S_{mi} \tau_i. \tag{2.26}$$

Because the elastic stiffness tensor and compliance tensor are inverse to each other, so it is easy to work out the relations between the components as

$$\begin{aligned} S_{11} &= \frac{C_{11} + C_{12}}{(C_{11} - C_{12})(C_{11} + 2C_{12})}, \\ S_{12} &= \frac{-C_{12}}{(C_{11} - C_{12})(C_{11} + 2C_{12})}, \\ S_{44} &= \frac{1}{C_{44}}. \end{aligned} \tag{2.27}$$

In mechanical engineering, Young's modulus Y and Poisson ration ν are commonly used. For a homogeneous, isotropic material, strain is related to stress through

$$\varepsilon_{xx} = \frac{1}{Y}(\tau_{xx} - \nu(\tau_{yy} + \tau_{zz})),$$
$$\varepsilon_{yy} = \frac{1}{Y}(\tau_{yy} - \nu(\tau_{zz} + \tau_{xx})),$$
$$\varepsilon_{zz} = \frac{1}{Y}(\tau_{zz} - \nu(\tau_{xx} + \tau_{yy})). \tag{2.28}$$

In cubic systems Young's modulus and Poisson ration ν are related to the compliance constants by

$$Y = \frac{1}{S_{11}}, \quad \nu = -\frac{S_{12}}{S_{11}}. \tag{2.29}$$

2.4 EXAMPLES OF STRESS–STRAIN RELATIONS

Now we use two examples to illustrate how to determine the strain tensor from stress using the relations we have discussed earlier.

1. Biaxial stress:
   A semiconductor layer pseudomorphically grown on a (001)-oriented lattice-mismatched substrate is schematically shown in Fig. 2.5. In this case, the top layer is biaxially strained, and the strain components e_{xx} and e_{yy} are

$$e_{xx} = e_{yy} = \frac{a_0 - a}{a}. \tag{2.30}$$

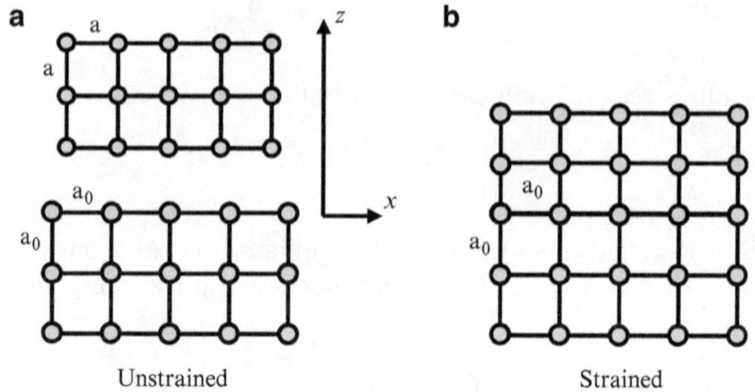

Fig. 2.5. Illustration of biaxial stress (strain). (**a**) Two different material layers have different lattice constant before growth; (**b**) After pseudomorphic film growth, the lattice constant of the top layer conforms to that of the bottom layer and is under biaxial stress (strain)

The strain is tensile in the x-y plane. To obtain the strain in the z direction, we use the strain–stress relation (2.26), i.e.,

$$\begin{bmatrix} e_{xx} \\ e_{yy} \\ e_{zz} \\ e_{zx} \\ e_{yz} \\ e_{xy} \end{bmatrix} = \begin{bmatrix} S_{11} & S_{12} & S_{12} & 0 & 0 & 0 \\ S_{12} & S_{11} & S_{12} & 0 & 0 & 0 \\ S_{12} & S_{12} & S_{11} & 0 & 0 & 0 \\ 0 & 0 & 0 & S_{44} & 0 & 0 \\ 0 & 0 & 0 & 0 & S_{44} & 0 \\ 0 & 0 & 0 & 0 & 0 & S_{44} \end{bmatrix} \begin{bmatrix} \tau_{xx} \\ \tau_{yy} \\ \tau_{zz} \\ \tau_{zx} \\ \tau_{yz} \\ \tau_{xy} \end{bmatrix}. \tag{2.31}$$

In the current case, $\tau_{xx} = \tau_{yy} = T$, $\tau_{zz} = 0$, and $\tau_{zx} = \tau_{yz} = \tau_{xy} = 0$. Therefore, we have

$$\begin{aligned} e_{xx} &= e_{yy} = (S_{11} + S_{12})T, \\ e_{zz} &= 2S_{12}T. \end{aligned} \tag{2.32}$$

Thus,

$$e_{zz} = \frac{2S_{12}}{S_{11} + S_{12}} e_{xx}. \tag{2.33}$$

Strain tensor in this case is

$$e = \begin{pmatrix} e_{xx} & 0 & 0 \\ 0 & e_{xx} & 0 \\ 0 & 0 & e_{zz} \end{pmatrix}. \tag{2.34}$$

2. [110] uniaxial stress:
The x-y plane of a cubic crystal under a [110] uniaxial stress is illustrated in Fig. 2.6. The stress tensor is already obtained in Eq. (2.16), i.e., $\tau_{xx} = \tau_{yy} = \tau_{xy} = T/2$, and $\tau_{zz} = \tau_{zx} = \tau_{yz} = 0$. Substituting into (2.31), we obtain

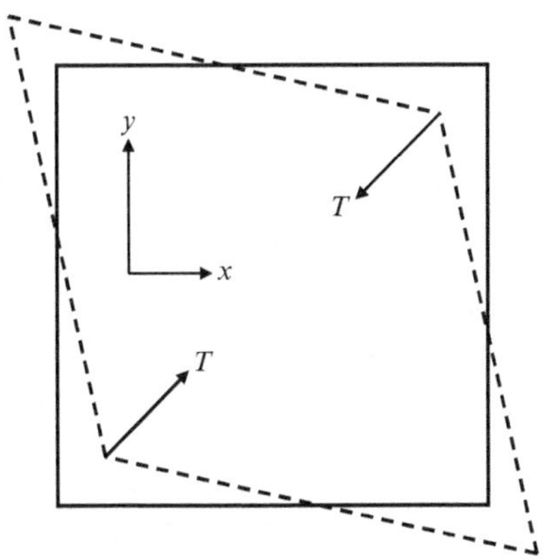

Fig. 2.6. Illustration of the [110] uniaxial compressive stress (strain)

$$
\begin{aligned}
e_{xx} &= e_{yy} = \frac{S_{11} + S_{12}}{2} T, \\
e_{xy} &= \frac{S_{44}}{2} T, \\
e_{zz} &= S_{12} T.
\end{aligned}
\tag{2.35}
$$

The strain tensor in this case then is

$$
\varepsilon = \begin{pmatrix} e_{xx} & e_{xy}/2 & 0 \\ e_{xy}/2 & e_{xx} & 0 \\ 0 & 0 & e_{zz} \end{pmatrix}. \tag{2.36}
$$

2.4.1 Hydrostatic and Shear Strain

An arbitrary strain tensor can be decomposed into three separate tensors as following:

$$
\begin{aligned}
\begin{pmatrix} \varepsilon_{xx} & \varepsilon_{xy} & \varepsilon_{xz} \\ \varepsilon_{yx} & \varepsilon_{yy} & \varepsilon_{yz} \\ \varepsilon_{zx} & \varepsilon_{zy} & \varepsilon_{zz} \end{pmatrix} &= \frac{1}{3} \begin{pmatrix} \varepsilon_{xx} + \varepsilon_{yy} + \varepsilon_{zz} & 0 & 0 \\ 0 & \varepsilon_{xx} + \varepsilon_{yy} + \varepsilon_{zz} & 0 \\ 0 & 0 & \varepsilon_{xx} + \varepsilon_{yy} + \varepsilon_{zz} \end{pmatrix} \\
&+ \frac{1}{3} \begin{pmatrix} 2\varepsilon_{xx} - (\varepsilon_{yy} + \varepsilon_{zz}) & 0 & 0 \\ 0 & 2\varepsilon_{yy} - (\varepsilon_{zz} + \varepsilon_{xx}) & 0 \\ 0 & 0 & 2\varepsilon_{zz} - (\varepsilon_{xx} + \varepsilon_{yy}) \end{pmatrix} \\
&+ \begin{pmatrix} 0 & \varepsilon_{xy} & \varepsilon_{xz} \\ \varepsilon_{yx} & 0 & \varepsilon_{yz} \\ \varepsilon_{zx} & \varepsilon_{zy} & 0 \end{pmatrix},
\end{aligned}
\tag{2.37}
$$

where the first constant tensor whose diagonal element is one-third of the trace of the original strain tensor accounts for the volume change [see (2.10)], and the latter two traceless tensors account for the shape change of an infinitesimal cube. Correspondingly, the first tensor describes the effect of a hydrostatic strain, and the latter two tensors describe the effect of shear strain. Among the two shear strain tensors, the diagonal one is related to the change of lengths along the three axes and the other one with diagonal elements being zero is related to the rotation of the axes of an infinitesimal cube. For a cubic crystal, the first type of shear occurs when a uniaxial stress is applied along any of the three $\langle 100 \rangle$ axes, and the second type of shear is nonzero only when stresses are applied along $\langle 110 \rangle$ or $\langle 111 \rangle$ axes. Obviously, for a cube under the hydrostatic strain, the shape does not change, while under an arbitrary first type of shear, the shape of the cube will become orthorhombic, and under an arbitrary second type of shear, the shape of the cube will become triclinic. A cubic crystal under biaxial stress becomes tetragonal, and it becomes orthorhombic under a uniaxial stress along $\langle 110 \rangle$.

For a first look, applying a compressive uniaxial stress along [001] and a biaxial tensile stress in the x-y plane to a cubic crystal seems identical. Indeed, if for both cases the stress is T, and we decompose the resulting strain tensor into the hydrostatic and shear parts, the shear strain coincides. However, the hydrostatic strain differs in sign and a factor of 2 in magnitude.

2.5 PIEZORESISTIVITY

Piezoresistivity is an effect of stress-induced resistivity change of a material. The piezoresistance coefficients (π coefficients) that relate the piezoresistivity and stress are defined by

$$\pi = \frac{\Delta R/R}{T}, \tag{2.38}$$

where R is the original resistance that is related to semiconductor sample dimension by $R = \rho\frac{l}{wh}$, ΔR signifies the change of resistance, and T is the applied mechanical stress. The ratio of ΔR to R can be expressed in terms of relative change of the sample length $\Delta l/l$, width $\Delta w/w$, height $\Delta h/h$, and resistivity $\Delta \rho/\rho$ as

$$\frac{\Delta R}{R} = \frac{\Delta l}{l} - \frac{\Delta w}{w} - \frac{\Delta h}{h} + \frac{\Delta \rho}{\rho}, \tag{2.39}$$

where resistivity ρ is inversely proportional to the conductivity. The first three terms of the RHS of (2.39) depict the geometrical change of the sample under stress, and the last term $\Delta\rho/\rho$ is the resistivity dependence on stress. For most semiconductors, the stress-induced resistivity change is several orders of magnitude larger than the geometrical change-induced resistance change, so the resistivity change by stress is the determinant factor of the piezorestivity.

In general conditions, resistivity $\rho = 1/\sigma$ is a second-rand tensor, and stress T is also a second-rank tensor. The resistivity change, $\Delta\rho_{ij}$, is connected to stress by a fourth-rank tensor π_{ijkl}, the piezoresistance tensor. Under arbitrary stress in linear response regime,

$$\frac{\Delta\rho_{ij}}{\rho} = -\frac{\Delta\sigma_{ij}}{\sigma} = \sum_{k,l} \pi_{ijkl}\tau_{kl}, \tag{2.40}$$

where summation is over x, y, and z.

Following the same discussion for compliance and stiffness tensor, and writing $\Delta\rho_{ij}$ to a vector form $\Delta\rho_i$, where $i = 1, 2, \ldots, 6$, as we did for stress and strain, we can rewrite (2.40) as

$$\frac{\Delta\rho_i}{\rho} = \sum_{k=1}^{6} \pi_{ik}\tau_k, \tag{2.41}$$

where π_{ik} is a 6×6 matrix. For cubic structures, it has only three independent elements due to the cubic symmetry,

$$\pi_{ik} = \begin{pmatrix} \pi_{11} & \pi_{12} & \pi_{12} & 0 & 0 & 0 \\ \pi_{12} & \pi_{11} & \pi_{12} & 0 & 0 & 0 \\ \pi_{12} & \pi_{12} & \pi_{11} & 0 & 0 & 0 \\ 0 & 0 & 0 & \pi_{44} & 0 & 0 \\ 0 & 0 & 0 & 0 & \pi_{44} & 0 \\ 0 & 0 & 0 & 0 & 0 & \pi_{44} \end{pmatrix}. \quad (2.42)$$

Among the three independent π-coefficients, π_{11} depicts the piezoresistive effect along one principal crystal axis for stress along this principal crystal axis (longitudinal piezoresistive effect), π_{12} depicts the piezoresistive effect along one principal crystal axis for stress directed along one perpendicular crystal axis (transverse piezoresistive effect), and π_{44} describes the piezoresistive effect on an out-of-plane electric field by the change of the in-plane current induced by in-plane shear stress.

The detailed discussion of semiconductor piezoresistivity will be covered in Chap. 5.

2.6 PIEZOELECTRICITY

Different from the piezoresistive effect, the piezoelectric effect arises from stress-induced charge polarization in a crystal that lacks a center of inversion. Thus, the piezoelectric effect does not exist in Si, Ge, etc. elementary semiconductors. The zinc-blende semiconductors are the simplest crystals with this property. The polarization is related to stress through the piezoelectric tensor $\bar{\bar{e}}$,

$$\mathbf{P} = [e]\mathbf{e}_{\mathbf{strain}}, \quad (2.43)$$

where $\mathbf{P}$ is the polarization vector and $\mathbf{e}_{\mathbf{strain}}$ is the strain written as a six-component vector. Thus the piezoelectric tensor is a 3×6 matrix. For zinc-blende semiconductors, the piezoelectric tensor only has one nonvanishing tensor element, e_{14}, and the polarization induced by strain is then given by

$$\begin{pmatrix} P_x \\ P_y \\ P_z \end{pmatrix} = \begin{pmatrix} 0 & 0 & 0 & e_{14} & 0 & 0 \\ 0 & 0 & 0 & 0 & e_{14} & 0 \\ 0 & 0 & 0 & 0 & 0 & e_{14} \end{pmatrix} \begin{pmatrix} e_{xx} \\ e_{yy} \\ e_{zz} \\ e_{yz} \\ e_{zx} \\ e_{xy} \end{pmatrix}. \quad (2.44)$$

Because of the special form of the piezoelectric tensor, only the shear strain generates the piezoelectricity. For zinc-blende semiconductors such as GaAs grown on (001) direction, the biaxial strain does not generate piezoelectricity.

The piezoelectric effect is largest along the $\langle 111 \rangle$ axes, since the anions and cations are stacked in the (111) planes, thus strain creates relative displacement between them.

The piezoelectric constants of GaAs were measured and theoretically calculated (Adachi, 1994). The commonly adopted value is

$$e_{14} = -0.16 \ \ \mathrm{C/m^2}. \tag{2.45}$$

On the other hand, the applied electric field across the piezoelectric material can generate strain. The piezoelectric strain tensor $\bar{\bar{d}}$ has the same form as the piezoelectric tensor and also has only one nonvanishing component, d_{14}, for zinc-blende semiconductors. It is related to e_{14} by

$$d_{14} = S_{44} e_{14}. \tag{2.46}$$

The commonly adopted value for d_{14} for GaAs is -2.7×10^{-12} m/V.

The sign of e_{14} or d_{14} is negative for III–V semiconductors. If the crystal is expanded along the $\langle 111 \rangle$ direction, the A-faces (cation faces) becomes negatively charged. This is different from the II–V semiconductors, where e_{14} is positive.

For the other semiconductors lacking inversion symmetry, the piezoelectric tensor may have more than one nonvanishing component. In wurtzite semiconductors such as GaN, there are three nonvanishing components, e_{13}, e_{33}, and e_{15}. Piezoelectric effect may play an important role in semiconductor transport. In an AlGaN/GaN heterostructure, the spontaneous polarization and the piezoelectric effect can induce large density of electrons even when there is no doping (Bernardini and Fiorentini, 1997; Jogai, 1998; Sacconi et al, 2001). In GaAs/InGaAs superlattices grown in the $\langle 111 \rangle$ direction, piezoelecticity induced band bending can greatly change the potential profile, and thus alter the charge distribution and transport properties (Smith and Mailhiot, 1988; Kim, 2001).

3

Strain and Semiconductor Crystal Symmetry

3.1 INTRODUCTION

One common question asked about strain on semiconductor band structures is: why is there band splitting with strain? Essentially, the answer to this question can only be sought from the semiconductor crystal symmetry considerations. Band degeneracies are defined by semiconductor crystal symmetry. When strain reduces the original symmetry, some degeneracies are lifted, and thus we see band splitting.

Symmetry considerations are intuitively simple yet very complicated when treated in a systematical way. The mathematical tool to treat symmetry is group theory. In this chapter, we are going to investigate the semiconductor crystal symmetry and its effects on the semiconductor band structures, then study strain-altered symmetry-induced band splitting and warping. For simplicity, we will avoid use of the abstract concepts of group theory, but rather will treat the relation of symmetry and properties of band structures using simple observations and examples.

We will divide the discussion in this chapter into two main parts. The first part is for readers who want a simple qualitative picture of symmetry and its effects on band structures. Through the qualitative discussion alone, one can understand how strain changes the band structure by reducing the crystal symmetry. The second part is aimed for readers with an interest in more detailed yet more abstract conceptions and reasonings. Thus, this chapter is organized as follows: in Sect. 3.2, a qualitative discussion is first given for symmetry and its associated band properties; strain-induced symmetry reduction and band property alteration are illustrated using simple diagrams for qualitative understanding. Then in Sect. 3.3 and the following sections, detailed discussions of crystal symmetries are given. Symmetry-determined band properties are studied in a more theoretical point of view. In the last section, a summary is given, which discusses the merits and limitations of symmetry considerations in terms of the current strain research.

Y. Sun et al., *Strain Effect in Semiconductors: Theory and Device Applications*, DOI 10.1007/978-1-4419-0552-9_3,

3.2 SYMMETRY AND STRAIN: OVERVIEW

In this section, a general and broad view of crystal symmetry is given. Symmetry effects on energy bands are discussed in an illustrative manner. Some terms, for instance, the reciprocal lattice, Brillouin zone, etc., which are more mathematically involved will be further discussed in the later sections.

3.2.1 Examples of Crystal Lattices

Crystals are formed by stacking atoms in 3D space. The stacking obeys certain rules. For example, if we assume the atoms to be spheres, the atoms stacking in a plane naturally form a hexagon if they are packed closely, just as illustrated in Fig. 3.1a by atoms labeled as "A". When it comes to the stacking along

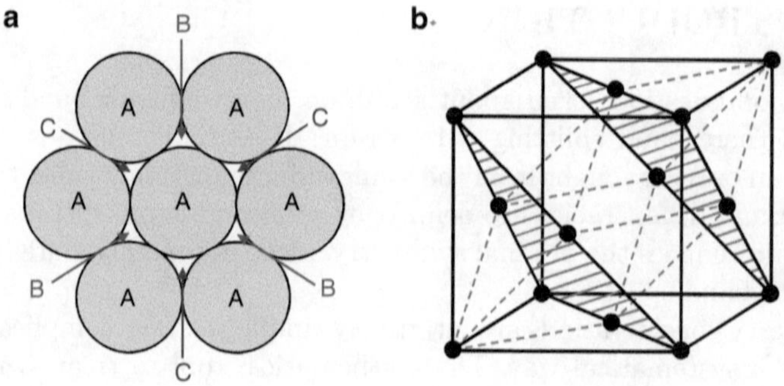

Fig. 3.1. (**a**) Atom stacking sequence for close pack structures. "A", "B" and "C" are possible sites for atoms to occupy. (**b**) The crystalline cell of a simple FCC lattice. Atoms are closely packed following "ABCABC..." stacking sequence along the body diagonals. Shaded planes are the (111) planes where atoms are closely packed, and the enclosed area by the dashed lines is the primitive cell

the perpendicular direction, there are two options for positions of the atoms to occupy in the next atom layer, labeled as "B" and "C" in Fig. 3.1a. If the stacking sequence is "ABCABC...," then the crystal formed has the cubic symmetry. The sequence "ABABAB..." has the hexagonal symmetry. The former stacking sequence constructs the face-centered cubic (FCC) crystal lattice, and the latter constructs the hexagonal crystal lattice. In Fig 3.1(b), a crystalline cell of an FCC lattice is shown. It is a cube with atoms at every vertex and also an atom at the center of each cube surface. The atoms on the vertices are in the layer "A" if we assume the shadowed layers are "B" and "C." We can see that the closely packed planes are the (111) planes if we follow the convention to define the edges of the crystalline cell along the $\langle 100 \rangle$ directions. In this example, every atom is equivalent no matter for FCC or hexagonal lattice, and the crystal can be considered formed by units stacked periodically in space. We have to distinguish the atomic sites from the lattice points. A lattice is a geometrical array of points generated by a set of discrete translation operations. A crystal is made up of a basis of one or more atoms which is repeated at each lattice point. The minimum cell corresponding to a single lattice point of a crystal structure with translational symmetry is called the primitive cell. A crystal can be viewed as constructed of replicas of the primitive cell arrayed in precise spatial order in three dimensions. The primitive cell for the FCC lattice is shown in Fig. 3.1b enclosed by the dashed lines. If a primitive cell of a lattice contains only one atom, it is a simple lattice. The lattice points can be chosen to coincide with the atomic sites. But most semiconductors we encounter with, e.g., Si, Ge, and GaAs, have complex lattices, where one lattice point corresponds to two atoms, which differ by neighboring or electronic configuration, and are not equivalent geometrically or electronically. For example in Si, whose atomic stacking sequence along the $\langle 111 \rangle$ direction is shown in Fig. 3.2a, the close packed (111) plane is only periodic every two atomic layers (illustrated by open and solid circles), and the atoms on these two layers are configurationally different, so the Si crystal is actually formed by interpenetrating FCC lattices (each of them are identical), by shifting each other a distance of $a/4$ along the crystalline cell (shown in Fig. 3.2b) diagonal, where a is the length of the crystalline cell edge. In this case, we may also consider the two atoms circled in Fig. 3.2b as one geometrical point. Then these points comprise an FCC lattice. Zinc-blende semiconductors such as GaAs have similar crystal structures, except that the two interpenetrating FCC lattices are formed by different types of atoms. So, the close packed atom stacking forms the FCC lattice, which is found in semiconductors such as Si and GaAs, or the hexagonal lattice, which is found in wurtzite semiconductors such as GaN.

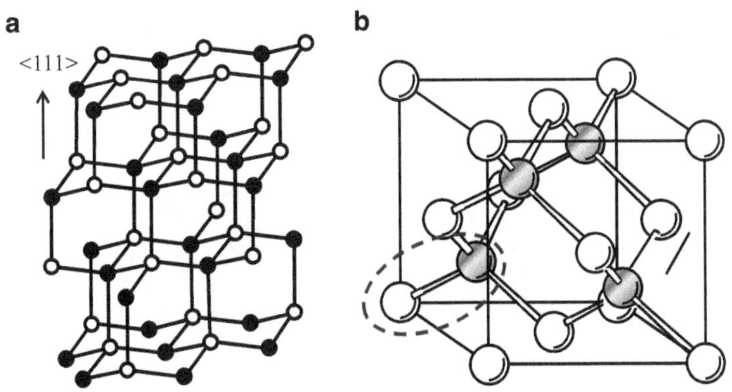

Fig. 3.2. (**a**) Diamond or zinc-blende structure. For diamond structures, the solid and open dots represent the same kind of atoms; for zinc-blende structures, the solid and open dots represent two difference atoms. (**b**) The crystalline cell of the diamond or zinc-blende structure

3.2.2 Crystal Symmetry

The stacking order of the atoms in the process of crystal formation determines the symmetry properties of the crystal. There are two systems of symmetry that exist in a semiconductor crystal. One we call the translational symmetry. It is another name for the periodicity of the atomic arrangement in solids. The other is the point symmetry. Next we use Fig. 3.3 which is a diagram for a 2D grid for a better illustration for these two types of symmetry. The grid points are evenly spaced, and if they are extended infinitely, they represent a 2D lattice. This 2D lattice is the same after being translated by na in either x or y direction, where n is an arbitrary integer and a is the lattice constant. This is a very simple example of translational symmetry. Point symmetry refers to the operations that keep the crystal unchanged, but with the exclusion of the translational symmetry. They include rotations, inversions, and reflections. Translational symmetry constrains the operations of the point symmetry, because the periodical lattice must be the same after rotation operations. So unlike a sphere that is symmetrical under a rotation with any angle about any axis through the center, only 1st-, 2nd-, 3rd-, 4th-, and 6th-order rotation axes can exist in a periodical lattice. The illustration of point symmetry is shown in Fig. 3.3b. About the axis perpendicular to the paper plane through point A, we obviously have a fourfold rotational symmetry, i.e., this grid keeps unchanged when we rotate it by 0°, 90°, 180°, and 270°. Each dashed line in the figure is a twofold rotation axis, i.e., the grid is the same if flipped about these dashed lines. Furthermore, the grid is also symmetrical by reflection about these dashed lines. In this special case, the reflection is equivalent to the 180° rotation. However in most cases, a reflection operation is not equivalent to a 180° rotation. The point A is also an inversion center. It means if a point with coordinate (x, y) changes into $(-x, -y)$, it is still a grid point, and under this inversion, the grid as a whole is unchanged. For an infinite grid, every grid point is equivalent to point A and has all the symmetry

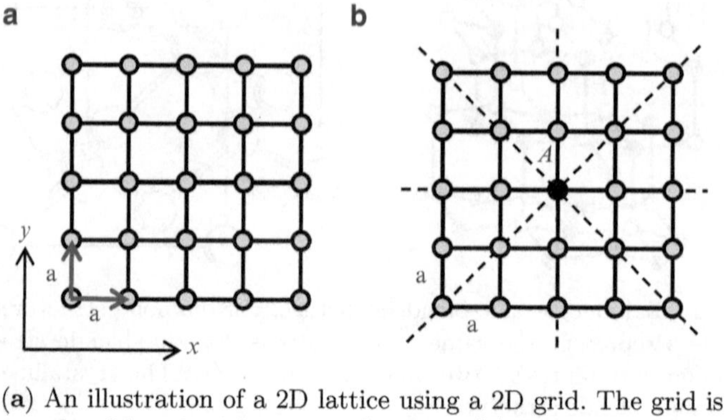

Fig. 3.3. (**a**) An illustration of a 2D lattice using a 2D grid. The grid is the same after being translated by a in either x or y direction. (**b**) Diagram of point symmetry for the 2D grid

properties of A. This example also illustrates the meaning of the term "point symmetry", because for each symmetry operation we mentioned above, the point A is fixed. A real crystal lattice can be considered a 3D grid. The point symmetry for an FCC lattice can be proven to have as many operations as a cube has. There are a total of 48 symmetry operations, including 3 fourfold rotational axes through the square surfaces, 4 threefold rotational axes through the body diagonals, and inversion about the cube center, etc. In some crystals, point symmetry is manifested by distinct geometrical shapes, such as in rock salt that is cubic, and quartz that is hexagonal, etc.

These symmetry properties are not only exemplified by appearance, but also have profound effects that determine the essential properties of every aspect of a crystal. As we studied in the last chapter, the independent elastic stiffness tensor constants are reduced from 21 in a general form (due to the symmetric form of the stiffness tensor) to only 3 under cubic symmetry. This is an example of symmetry effects on the mechanical properties of crystals. The semiconductor electronic properties, which are the center of this book, are also strongly dependent on crystal symmetry. This can be studied by investigating the semiconductor energy band symmetry.

3.2.3 Energy Band Symmetry

Solving the quantum mechanical problem and finding the electronic states and energies in solids is very difficult and time-consuming because of the complex nature of the electric interaction in solids. However, symmetry has fundamental effects on band structures, i.e., the E–k relations in the k-space. Some very important properties of the band structures can be acquired without going into complicated band calculations, and by studying the crystal symmetry alone.

Translational symmetry restricts the electronic wave functions in a crystal to be Bloch waves, which are plane waves modulated by periodic functions whose periodicity follows the crystal periodicity. This effect cannot be readily shown graphically, but it imposes a fundamental law on electronic states in solids. We will see this point in the later chapters when band calculation methods are introduced.

For a more straightforward and graphic illustration of symmetry restrictions on energy band structures, here we may consider only the crystal point symmetry. It is first important to understand that the band structure is constructed in the reciprocal space, which can be obtained by a Fourier transformation of the real crystal lattice. We will leave this transformation to later sections. The reciprocal space of a real lattice is also a lattice. The reciprocal lattice for the 2D lattice shown in Fig. 3.3a is also a 2D lattice with unit vector $2\pi/a$,and is shown in Fig. 3.4a. The reciprocal lattice for an FCC lattice is a body-centered cubic (BCC) lattice. The traditional illustration of an energy band structure is to plot the energy levels in the first Brillouin zone, which is geometrically a primitive cell of the reciprocal lattice, and conventionally defined as the

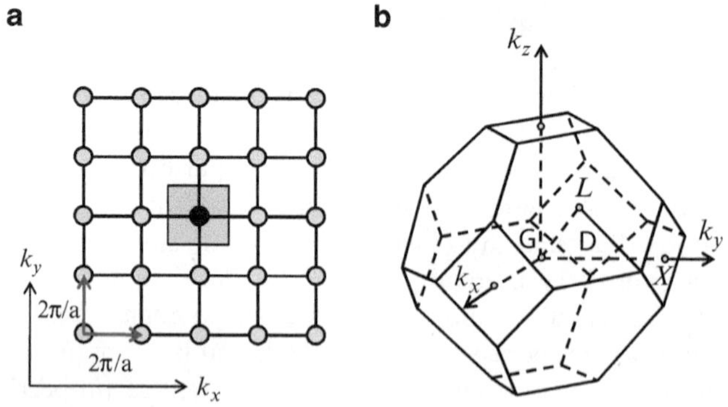

Fig. 3.4. The first Brillouin zone for (**a**) a 2D lattice, and (**b**) a FCC lattice, with a lattice constant a

space enclosed by planes perpendicular to and bisecting the lines connecting a reciprocal lattice point and its neighbors. The first Brillouin zone for the 2D lattice in Fig. 3.4a is shown as the shaded area. The Brillouin zone for a FCC lattice is shown in Fig. 3.4b. In this figure, the Γ point is the symbol for the center, the X point $(0, 2\pi/a, 0)$ labels the center of the square surfaces, the L point $2\pi/a(0.5, 0.5, 0.5)$ labels the center of the hexagonal surfaces. The line that connects the Γ point to the X point is labeled as Δ, and the direction from Γ to L is called Λ. The band structure, i.e., the E–k diagram, describes the relation of the electron energy versus wave vector $\mathbf{k}$ from the origin of the Brillouin zone to the zone edge. Why do we have to study the electron energy in k-space? Because an electron has 6 dimensions, e.g., 3 real space dimensions, and 3 momentum space dimensions. An electron moves in solids in a form of Bloch waves, and $\mathbf{k}$ is its wave vector. In the sense of effective mass theory, $\hbar\mathbf{k}$ is the momentum of the electron in solids, where $\hbar$ is the Plank constant.

Next we use the FCC lattice semiconductor as an example to discuss the band symmetry. At the Γ point, the band structure has the same point symmetry as the crystal has. In fact, the center of any type of Brillouin zone is labeled as Γ. Without consideration of spin, the highest order of the valence band degeneracy normally equals the number of coordinates transformed by the symmetry operations. This is due to the fact that most semiconductor valence bands have p-characteristics. Cubic symmetry results in a threefold degeneracy at the Γ point without inclusion of spin because of the equivalency of the x, y, z directions in the Brillouin zone, and the energy diagrams along these directions are the same. However, their corresponding wave functions are linearly independent. In the valence bands of semiconductors other than cubic semiconductors, the x, y, and z directions are not all equivalent, and this results in the existence of the effective crystal field. In this case, the degeneracy order of the valence bands is less than three. Spin has very special rotational symmetry, and with spin–orbit interaction, the valence bands of cubic

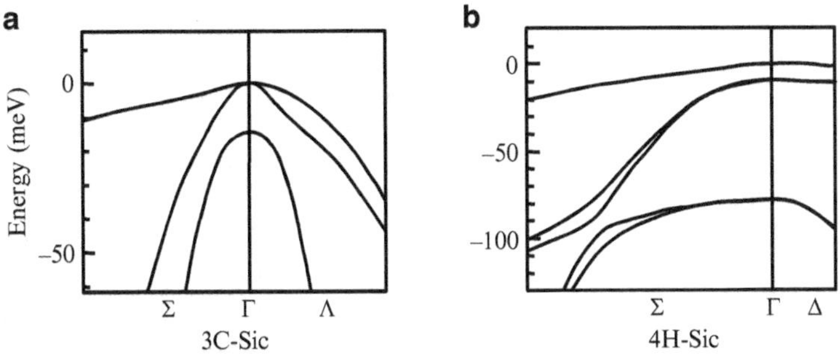

Fig. 3.5. The valence band structures of (**a**) zinc-blende 3C-SiC and (**b**) hexagonal 4H–SiC. From Persson and Lindefelt (Persson and Lindefelt, 1997)

semiconductors include doubly degenerate heavy-hole and light-hole bands and one split-off band resulting from spin–orbit interaction, while hexagonal semiconductors have three split valence bands with each doubly (spin) degenerate at the Γ point resulting from both the spin–orbit interaction and nonzero crystal field. SiC is a very good material to illustrate this point, because SiC has various crystal structures or polytypes. The different crystal symmetry of these polytypes results in different band structures. The 3C-SiC is zinc-blende while the 4H-SiC is hexagonal. Their valence band structures are shown in Fig. 3.5 (Persson and Lindefelt, 1997). Similar to hexagonal semiconductors, tetragonal (e.g., CaC_2, etc.) and orthorhombic semiconductors (e.g., CdSb, CaSi, etc.) also only have split valence bands at the Γ point. In short words, with spin–orbit interaction, the highest order of degeneracy is fourfold, and only cubic crystals possess this degeneracy. Semiconductors having lower symmetry always have lower degeneracy orders, except sometimes they may have accidental degeneracy, e.g., at the point of energy level crossing, which is not determined by symmetry.

Besides the degeneracy at the same k point, there exists another kind of degeneracy, i.e., the star degeneracy, where energies are the same at different k points that can be transformed into each other through some symmetry operations and become equivalent. For example, in the FCC semiconductor Brillouin zone shown in Fig. 3.4, any point on a Δ-axis can be transformed into the corresponding point located at another Δ-axis by rotation. But due to the cubic symmetry, the energies at these equivalent points must be the same, and these states are degenerate. When crystal symmetry changes, star degeneracy is also partially or fully lifted.

The dependence of the band degeneracy on crystal symmetry clearly states the origin of the band splitting with strain. Shear strain reduces the crystal symmetry; its defined band degeneracy is lifted, and thus the originally degenerate energy levels are split.

3.2.4 Strain Effects on Energy Bands

For cubic semiconductors under hydrostatic strain, the crystal symmetry is not altered. Thus, the hydrostatic strain has no effect on lifting the band degeneracy. If we concentrate on, for example, the Si conduction band valleys, or the valence band edge under hydrostatic strain, we will not see valley splitting or HH and LH band splitting. It looks like nothing happens. However, hydrostatic strain has an important effect on shifting the bandgaps. Just imagine in the process when two atoms are brought together to form a molecule as shown in Fig. 3.6a. The interatomic interaction increases when they approach each other, and thus the resulting bonding and antibonding energy states become further apart in energy. It is similar to what takes place when atoms in crystals are pulled apart or pushed together. So, normally when a compressive hydrostatic stress is applied, the semiconductor bandgap widens, and when a tensile hydrostatic stress is applied, the bandgap diminishes. Figure 3.6b shows this trend of bandgap dependence on hydrostatic strain for Si. Therefore, a hydrostatic strain changes the distance between bands.

Shear strain that reduces the cubic symmetry will cause the valence band degeneracy lifting for cubic semiconductors. A general stress will induce both hydrostatic and shear strain. We will use two technologically important stress types as examples: the biaxial stress and the $\langle 110 \rangle$ uniaxial stress. This time, only shear stress is considered. A cube under these two types of stress has the deformation as shown in Fig. 3.7. For biaxial stress, the shear strain elongates or shortens the crystalline cell edge along z with respect to those along x and y and thus changes the cubic symmetry into tetragonal symmetry. For uniaxial stress, the shear strain changes the upper and lower planes of the crystalline cell into rhombuses and thus changes the cubic symmetry into orthorhombic symmetry. Under either stress, the degenerate HH and LH valence bands of a cubic semiconductor split into separate HH and LH bands, as shown in Fig. 3.8.

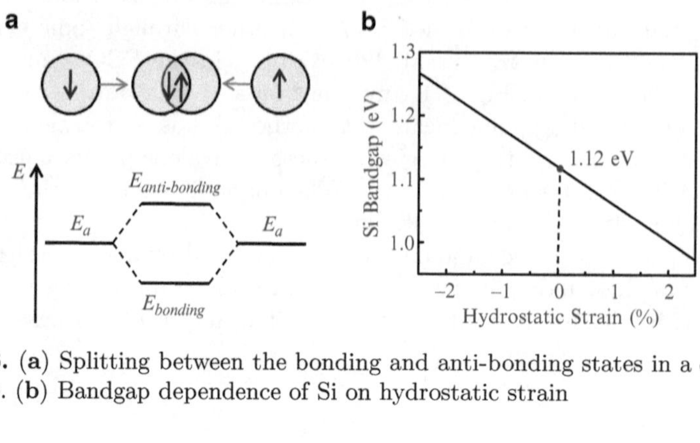

Fig. 3.6. (**a**) Splitting between the bonding and anti-bonding states in a diatomic molecule. (**b**) Bandgap dependence of Si on hydrostatic strain

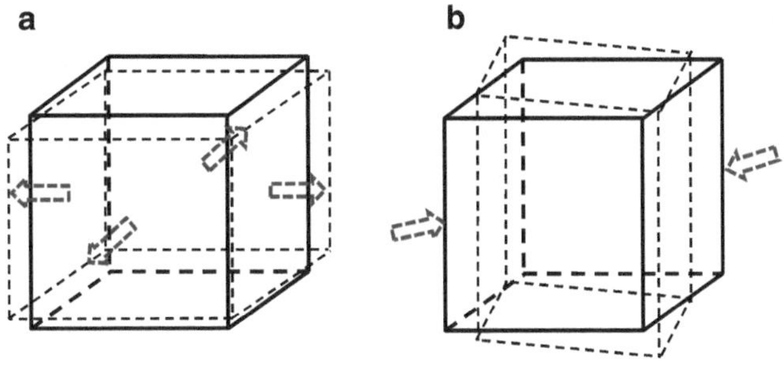

Fig. 3.7. The deformation of a cube under (**a**) biaxial tensile and (**b**) [110] uniaxial compressive stress

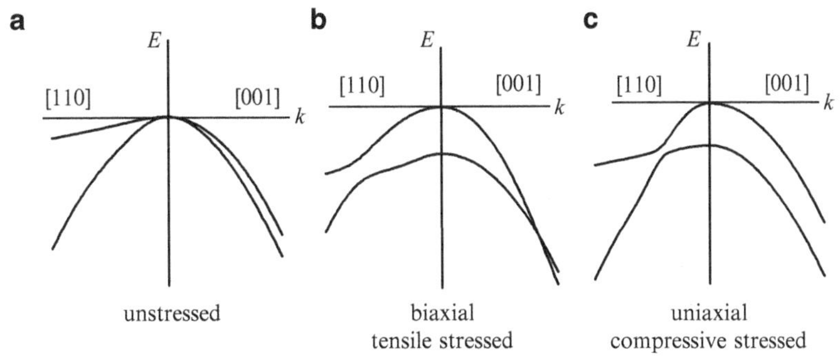

Fig. 3.8. The HH and LH bands for (**a**) unstressed, (**b**) biaxial tensile stressed, and (**c**) uniaxial compressive stressed Si along the [110] and [001] directions

Consider next the strain effects on the conduction bands. The conduction band edges are not like the valence bands, and different semiconductors may have them at different locations in the Brillouin zone, related to their different covalent and polar interactions. For example, the GaAs conduction band edge is located at the Γ point, with L valleys about 290 meV higher. The Si conduction bands are six Δ-valleys located along the $\langle 100 \rangle$ directions at about $0.85X$. Ge has its conduction band edges at the L points. Strain effects vary for different types of conduction valleys. For the Γ valley, since it is singly degenerate, there is no observable splitting. Si and Ge both have multiple conduction valleys, and have the star degeneracy. For Si the star degeneracy is sixfold and for Ge fourfold. One may wonder why the star degeneracy for Ge is only fourfold since there are 8 L points. The answer lies in the structure of the FCC Brillouin zone shown in Fig. 3.4. At each L point, only half of the valley lies within the Brillouin zone. In fact, the two L points at the opposite ends are equivalent. They are the same one. Similarly, if the Si conduction valleys were exactly at the X points, the star degeneracy would be only threefold. The trends of splitting of these valleys due to strain are easy to obtain. For example, both biaxial and [110] uniaxial stresses affect the four valleys along x and y in an identical way in Si, but affects the other two valleys along z differently. Thus, Si conduction valleys are split into the so-called Δ_2 and Δ_4 valleys. Biaxial stress affects the 4 L valleys of Ge identically, and thus they do not split. The [110] uniaxial stress distinguishes the L valleys according to their projection locations in the x-y plane (two along [110] and two along [$\bar{1}$10]), and thus the Ge L valleys split into two double-valley groups. Although there is no splitting for the GaAs Γ valley, it has to be noted that due to the energetic proximity between the Γ and L valley, strain can alter the electron transport by shifting the gap between these two valleys, since the electron occupation of the L valley cannot usually be neglected.

As a summary, the band splitting diagrams for Si, Ge, and GaAs due to the two types of stress are plotted in Fig. 3.9.

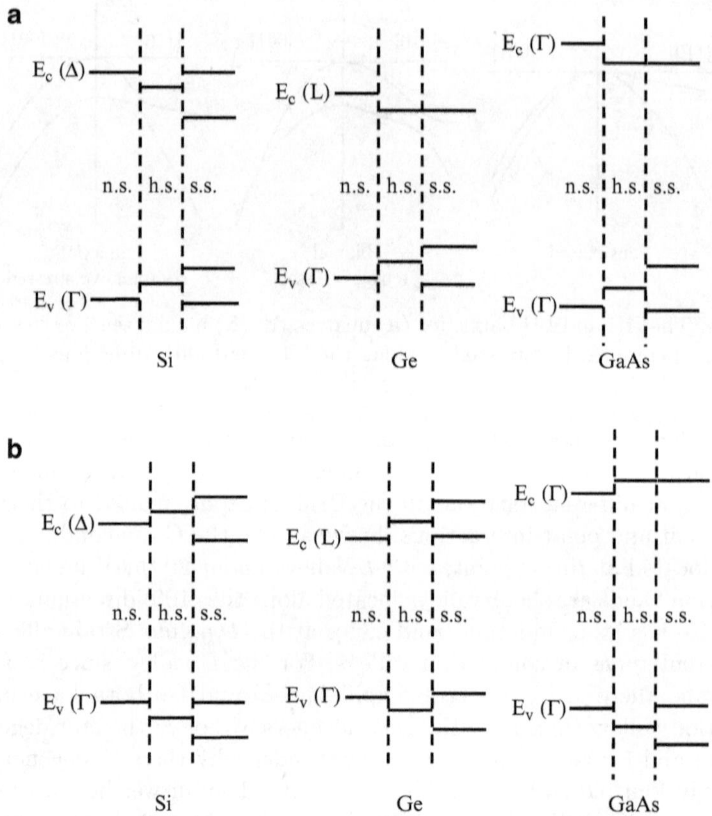

Fig. 3.9. Diagrams of band splitting of Si, Ge and GaAs under (**a**)in-plane biaxial tensile stress, and (**b**)⟨110⟩ uniaxial compressive stress, where "n.s." labels the unstrained band, "h.s." labels the band shifts with hydrostatic strain, and "s.s." labels the band splitting with shear strain

Band warping can also be understood by symmetry considerations. For Si conduction valleys, the Δ-axis, which connects the Γ point to the center of a square surface, has the same symmetry as a central axis that bisects the center of a square. The energy structure at this Δ-valley must follow this symmetry. Consequently, the energy structure for Si conduction valleys is an ellipsoid of revolution with transverse contours being circles. In fact, the ellipsoidal symmetry is not necessarily required but rather a consequence of no energetically nearby bands. If stress reduces the square symmetry of the perpendicular plane along one axis, the in-plane energy bands will no longer have the square symmetry. Then, the in-plane energy contours are no longer circular but warped. For example, uniaxial stress along [110] turns a square perpendicular to the [001] direction into a rhombus just as shown in Fig. 3.10a. The in-plane energy contour for Δ-valleys on the z-axis will become an ellipse with major and minor axes along the $\langle 110 \rangle$ directions, as shown in Fig. 3.10b.

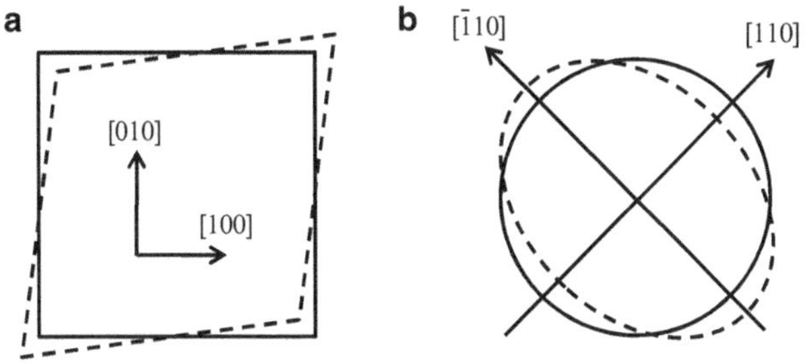

Fig. 3.10. (**a**) Deformation of a square under $\langle 110 \rangle$ uniaxial stress. (**b**) The corresponding energy contour of the Si out-of-plane valleys

Warping in the valence bands by stress follows the same arguments. Since the valence band edge is located at the Γ point, the energy surface shall always follow the strained macroscopic crystal symmetry. We illustrate this point in Fig. 3.11, where we show the deformation of a cube with no stress, biaxial tensile stress, and $\langle 110 \rangle$ uniaxial compressive stress, and the 3D equi-energy surfaces at 25 meV for the top valence band of GaAs accordingly. The two types of shapes lined-up vertically have exactly the same symmetry. First of all, the unstress energy surface is just like a cube. The symmetry of it is the same as a cube. Here, we select GaAs as the example instead of Si because the split-off energy in Si is much smaller so that the equi-energy surfaces for the HH band are very irregular and do not resemble a cube even though the symmetry is the same. The biaxial stress does not change the symmetry in the x-y plane, so the unstressed and biaxially stressed contours have the same symmetry in the x-y plane, but the nonequivalency between x, y, and z directions is evidently manifested by the energy surface as in Fig. 3.11b. The uniaxial stress has an even lower symmetry than the biaxial stress, and the energy surface is greatly warped, as can be seen from Fig. 3.11c.

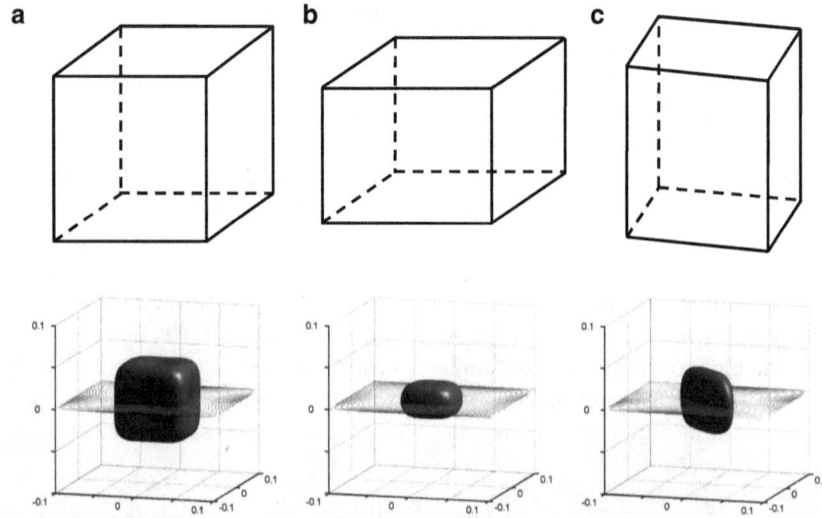

Fig. 3.11. Deformation of a cube, and the corresponding energy surfaces of the top valence band of GaAs at 25 meV under (**a**) no stress, (**b**) 1GPa biaxial tensile stress, and (**c**) 1GPa ⟨110⟩ uniaxial compressive stress

To be specific, the x-y plane only has the symmetry of a rhombus, and the equi-energy surface under this stress becomes an ellipsoid of revolution along the $[\bar{1}10]$ direction. In fact, the shapes of the energy surfaces at relatively high and low energies for stressed cases are different. Lower energy surfaces may possess higher symmetry (e.g., the ellipsoid of revolution in Fig. 3.11c) because strain splits the bands, and at the zone center, warping due to strong interband coupling becomes weak, so the top band becomes parabolic.

For strain engineering purpose, people need to choose an appropriate type of stress to create a proper symmetry reduction and thus to produce a desirable band splitting and warping.

3.3 SYMMETRY EFFECTS IN DETERMINING ELECTRONIC STATES

Systemic and rigid study of symmetry needs mathematical reasoning rather than apriori assertions as we did in the earlier section. Group theory is the tool for this task. The order of band degeneracies, the properties of electronic wave functions, strain perturbation-induced characteristic change of the band structures, etc., can be obtained to a great detail through group theory by considering the semiconductor symmetry without requiring concrete knowledge about the semiconductor itself, such as material composition, electronic bonding, electron affinity, and so on. However, group theory is complicated and abstract, and it is beyond our scope in this book. Interested readers are referred to other textbooks. In this section, we are trying to find the relation between the energy bands and symmetry from the quantum mechanical point of view, to have a better understanding of the consequences of reduction of symmetry induced by strain. Some conceptions that are mathematically involved are also introduced here for readers who want to explore more into abstract deductions.

In calculation of band structures, the core task is to solve the single electron Schrödinger equation

$$H\psi(\mathbf{r}) = \left(\frac{p^2}{2m} + V(\mathbf{r})\right)\psi(\mathbf{r}) = E\psi(\mathbf{r}), \tag{3.1}$$

where $V(\mathbf{r})$ is the effective crystal potential. Quantum mechanical systems are different because they have different forms of effective potentials, $V(\mathbf{r})$. Quantum mechanical laws dictate that the solutions to the Schrödinger equation (3.1) have symmetry properties defined by $V(\mathbf{r})$. Therefore, the symmetry of crystal electronic band structures is essentially determined by the crystal symmetry, which sets the symmetry for the crystal potential $V(\mathbf{r})$.

3.3.1 Translational Symmetry and Reciprocal Space

First of all, all crystals posses the translational symmetry, which can be represented by crystal lattices, as mentioned in the earlier section. Crystal potential $V(\mathbf{r})$ in the lattice is a periodic function of space. Lattice periodicity requires that every point in the lattice can be reached from an origin by a vector that is a multiple of some basis vectors. In Fig. 3.3a, any grid point can be reached by $a(n_1\hat{x} + n_2\hat{y})$, where $\hat{x}$ and $\hat{y}$ are unit vectors along the x and y directions, respectively, and n_1 and n_2 are integers. In the 3D case, the lattice can be represented by three basis vectors $\mathbf{a_1}$, $\mathbf{a_2}$, and $\mathbf{a_3}$, and an arbitrary lattice point may be written as

$$\mathbf{R} = n_1\mathbf{a_1} + n_2\mathbf{a_2} + n_3\mathbf{a_3}, \quad n_1, n_2, n_3 = \text{integer} \tag{3.2}$$

with the origin selected on any lattice point. Therefore, the crystal potential satisfies:

$$V(\mathbf{r} + \mathbf{R}) = V(\mathbf{r}). \tag{3.3}$$

The parallelepiped formed by the three basis vectors is called the primitive cell of the crystal. Usually the basis vectors $\mathbf{a_1}$, $\mathbf{a_2}$, and $\mathbf{a_3}$ might not be orthogonal to each other.

For a periodic function, $V(\mathbf{r})$ can be expanded in a Fourier series. First let us consider a one-dimensional periodic function $V(x)$ with periodicity of a,

$$V(x) = V(x + na), \quad n = \text{integer}. \tag{3.4}$$

The corresponding Fourier series then is

$$V(x) = \sum_n V_n \exp\left(\frac{2\pi i n}{a}x\right), \tag{3.5}$$

with

$$V_n = \frac{1}{a} \int_{-a/2}^{a/2} V(x) \exp\left(-\frac{2\pi \mathrm{i} n}{a} x\right) \mathrm{d}x. \quad (3.6)$$

It is easily proven that the Fourier series on the right hand side of (3.5) is identical to the original periodic function V(x), invariant by translation with a displacement $x_m = ma$, i.e., $V(x) = V(x + x_m)$, where m is an arbitrary integer. In analogy, $V(\mathbf{r})$ in the 3D case can be written as

$$V(\mathbf{r}) = \sum_{\mathbf{G}} V_{\mathbf{G}} \exp(\mathrm{i}\mathbf{G} \cdot \mathbf{r}). \quad (3.7)$$

The vector **G** is the usual wave vector for the Fourier transformation. However, to preserve the translational invariance of $V(\mathbf{r})$, the value of **G** must fulfil certain conditions, i.e., if we substitute **r** with $\mathbf{r}+\mathbf{R} = \mathbf{r}+n_1\mathbf{a_1}+n_2\mathbf{a_2}+n_3\mathbf{a_3}$ into the right hand side of (3.7), the exponential term $\exp[\mathrm{i}\mathbf{G} \cdot (\mathbf{r} + \mathbf{R})]$ must have the same value as $\exp(\mathrm{i}\mathbf{G} \cdot \mathbf{r})$. This requires that $\mathbf{G} \cdot \mathbf{R} = 2m\pi$, where m is an integer. Since **R** has only discrete values, then **G** also has only discrete values. Translations of a lattice point by **R** comprise the real lattice. Similarly, all **G** also comprise a lattice, the reciprocal lattice. Without further ado, we write down the conventional definition of the basis vectors of the reciprocal space,

$$\mathbf{g_1} = 2\pi \frac{\mathbf{a_2} \times \mathbf{a_3}}{\mathbf{a_1} \cdot (\mathbf{a_2} \times \mathbf{a_3})}, \ \mathbf{g_1} = 2\pi \frac{\mathbf{a_3} \times \mathbf{a_1}}{\mathbf{a_1} \cdot (\mathbf{a_2} \times \mathbf{a_3})}, \ \mathbf{g_1} = 2\pi \frac{\mathbf{a_1} \times \mathbf{a_2}}{\mathbf{a_1} \cdot (\mathbf{a_2} \times \mathbf{a_3})}. \quad (3.8)$$

The product $|\mathbf{a_1} \cdot (\mathbf{a_2} \times \mathbf{a_3})|$ is the volume of a real lattice space primitive cell. It can be readily verified that the product between an arbitrary reciprocal lattice vector **G**, where

$$\mathbf{G} = l_1\mathbf{g_1} + l_2\mathbf{g_2} + l_3\mathbf{g_3}, \quad l_1, l_2, l_3 = \text{integer} \quad (3.9)$$

and the vector **R**, $\mathbf{G} \cdot \mathbf{R}$, equals a multiple of 2π. The Brillouin zone is the smallest polyhedron centered at one reciprocal lattice point and enclosed by perpendicular bisectors of reciprocal lattice vectors.

Consider an FCC lattice with the cubic cell shown in Fig. 3.1a with side length a; its basis vectors are

$$\mathbf{a_1} = \frac{a}{2}(\hat{y} + \hat{z}), \ \mathbf{a_2} = \frac{a}{2}(\hat{z} + \hat{x}), \ \mathbf{a_3} = \frac{a}{2}(\hat{x} + \hat{y}). \quad (3.10)$$

Applying Eq. (3.8), the basis vectors for the reciprocal lattice are

$$\mathbf{g_1} = \frac{2\pi}{a}(\hat{y} + \hat{z} - \hat{x}), \ \mathbf{g_2} = \frac{2\pi}{a}(\hat{z} + \hat{x} - \hat{y}), \ \mathbf{g_3} = \frac{2\pi}{a}(\hat{x} + \hat{y} - \hat{z}). \quad (3.11)$$

These are precisely the basis vectors for a BCC lattice with side length $4\pi/a$.

3.3.2 Bloch Theorem

We only mentioned in the earlier section that electrons in semiconductor crystals move in the form of Bloch waves. This is a fundamental property of electrons in solids and deserves a proof. The translational symmetry of crystals, represented by (3.3), has a strict constraint to the eigenfunctions of (3.1). The electronic wavefunction can be expanded using plane waves in a general form:

$$\psi(\mathbf{r}) = \sum_{\mathbf{k}} C_{\mathbf{k}} e^{i\mathbf{k}\cdot\mathbf{r}}, \tag{3.12}$$

where $\mathbf{k}$ is a point in the reciprocal lattice compatible with the periodic boundary conditions. Expand the crystal potential $V(\mathbf{r})$ as in (3.7), and substitute (3.7) and (3.12) into (3.1) so we obtain

$$\sum_{\mathbf{k}} \frac{\hbar^2 k^2}{2m} C_{\mathbf{k}} e^{i\mathbf{k}\cdot\mathbf{r}} + \sum_{\mathbf{k}'\mathbf{G}} C_{\mathbf{k}'} V_{\mathbf{G}} e^{i(\mathbf{k}'+\mathbf{G})\cdot\mathbf{r}} = E \sum_{\mathbf{k}} C_{\mathbf{k}} e^{i\mathbf{k}\cdot\mathbf{r}}. \tag{3.13}$$

After renaming the summation indices this becomes

$$\sum_{\mathbf{k}} e^{i\mathbf{k}\cdot\mathbf{r}} \left[\left(\frac{\hbar^2 k^2}{2m} - E \right) C_{\mathbf{k}} + \sum_{\mathbf{G}} V_{\mathbf{G}} C_{\mathbf{k}-\mathbf{G}} \right] = 0. \tag{3.14}$$

For this equation to be valid at every point $\mathbf{r}$, the expression in the brackets, which is independent of $\mathbf{r}$, must vanish for every $\mathbf{k}$, i.e.,

$$\left(\frac{\hbar^2 k^2}{2m} - E \right) C_{\mathbf{k}} + \sum_{\mathbf{G}} V_{\mathbf{G}} C_{\mathbf{k}-\mathbf{G}} = 0. \tag{3.15}$$

This is a set of coupled equations in the reciprocal space for the expansion coefficients of $\psi(\mathbf{r})$ in (3.12). Here, $C_{\mathbf{k}}$, and $C_{\mathbf{k}-\mathbf{G}}$ are Fourier coefficients, whose k values differ from one another by a reciprocal lattice vector $\mathbf{G}$. Thus, $C_{\mathbf{k}}$ is only coupled to $C_{\mathbf{k}-\mathbf{G}}$, $C_{\mathbf{k}-\mathbf{G}'}$, $C_{\mathbf{k}-\mathbf{G}''}$,.... Under this condition, a solution to (3.1) is a superposition of plane waves whose wave vectors $\mathbf{k}$ differ only by reciprocal lattice vectors $\mathbf{G}$. Thus, the electronic wavefunctions and eigenvalues can be indexed by the vector of $\mathbf{k}$ in the first Brillouin zone, and

$$\begin{aligned} \psi_{\mathbf{k}}(\mathbf{r}) &= \sum_{\mathbf{G}} C_{\mathbf{k}-\mathbf{G}} e^{i(\mathbf{k}-\mathbf{G})\cdot\mathbf{r}} \\ &= \sum_{\mathbf{G}} C_{\mathbf{k}-\mathbf{G}} e^{-i\mathbf{G}\cdot\mathbf{r}} e^{i\mathbf{k}\cdot\mathbf{r}} \\ &= u_{\mathbf{k}}(\mathbf{r}) e^{i\mathbf{k}\cdot\mathbf{r}}. \end{aligned} \tag{3.16}$$

This is called the Bloch theorem, and $\psi_{\mathbf{k}}(\mathbf{r})$ is called the Bloch function where

$$u_{\mathbf{k}}(\mathbf{r}) = \sum_{\mathbf{G}} C_{\mathbf{k}-\mathbf{G}} e^{-i\mathbf{G}\cdot\mathbf{r}} \tag{3.17}$$

has the same periodicity of the lattice,

$$u_{\mathbf{k}}(\mathbf{r}) = u_{\mathbf{k}}(\mathbf{r}+\mathbf{R}). \tag{3.18}$$

Therefore, $u_{\mathbf{k}}(\mathbf{r})$ is a function determined by the material properties and repeated at every primitive cell. The electronic states in solids are these periodic functions modulated by plane waves indexed by **k**. Electronic wave functions in any crystals obey Bloch theorem because periodicity is a common property for all types of crystals. When there is strain and subsequent symmetry reduction, the periodic part of the Bloch function, $u_{\mathbf{k}}(\mathbf{r})$, changes. From the perturbation point of view, the new periodic functions are the superposition of the old functions. That is, band mixing occurs.

Bloch functions are periodic functions of the reciprocal lattice:

$$\begin{aligned}
\psi_{\mathbf{k}+\mathbf{G}}(\mathbf{r}) &= \sum_{\mathbf{G}'} C_{\mathbf{k}+\mathbf{G}-\mathbf{G}'} e^{-i\mathbf{G}'\cdot\mathbf{r}} e^{i(\mathbf{k}+\mathbf{G})\cdot\mathbf{r}} \\
&= \left(\sum_{\mathbf{G}''} C_{\mathbf{k}-\mathbf{G}''} e^{-i\mathbf{G}''\cdot\mathbf{r}} \right) e^{i\mathbf{k}\cdot\mathbf{r}} \\
&= \psi_{\mathbf{k}}(\mathbf{r}).
\end{aligned} \tag{3.19}$$

From this characteristic of the Bloch functions, we can understand why the two L points at the opposite ends in the Brillouin zone shown in Fig. 3.4b are actually one point, because firstly by symmetry they have the same energy and secondly from (3.11), they differ by a reciprocal vector $\frac{2\pi}{a}(1,1,1) = \mathbf{g}_1 + \mathbf{g}_2 + \mathbf{g}_3$, and thus they have the same wave function. The reasoning for two opposite X points being the same one follows the same argument.

3.3.3 Point Symmetry Effects on Electronic States

Point symmetry-determined electronic properties are illustratively discussed in Sect. 3.2. We may look this problem from some new perspectives. Overall speaking, point symmetry has three distinct effects on crystal electronic states.

The first is state degeneracy. The band degeneracy can be naively understood from the process of formation of crystals by atoms. In an atom, the electric potential is spherically symmetrical, and an electronic state with angular momentum l has an order of degeneracy $2l+1$. When atoms are brought together, the electronic states from different atoms interact with each other to form energy bands. But when the atoms are arrayed together by a special order, they also experience an addition potential from the arrangement of the other atoms. Since the maximum symmetry of this atomic array is the cubic symmetry under which the maximum degeneracy order is 3 (we will see this point later), so the atomic states with $l \geq 2$ are all split, while the levels with $l = 3/2$ (p-state coupled with spin) or $l = 1$ (p-state) do not split. Thus, in any type of crystals, the highest degeneracy order is 3 without inclusion of spin and is 4 with spin being taken into consideration.

The energy states of a crystal can be classified by its symmetry groups. The crystal Hamiltonian commutes with all symmetry operation, i.e., $HR = RH$, where R is any symmetry operation. In fact, all the eigenstates of crystal Hamiltonian can be selected as the common eigenstates of both the Hamiltonian and symmetry operations. If states $|i\rangle$ are the eigenstates of R, then

$$\sum_k \langle i|R|k\rangle\langle k|H|j\rangle = \sum_k \langle i|H|k\rangle\langle k|R|j\rangle, \tag{3.20}$$

where $|k\rangle\langle k|$ is the identity operator. Because

$$\langle i|R|k\rangle = \delta_{ik}, \tag{3.21}$$

it follows from (3.20) that

$$(R_{ii} - R_{jj})H_{ij} = 0. \tag{3.22}$$

This indicates that $H_{ij} = 0$ when R_{ii} and R_{jj} are different eigenvalues of R. Therefore, in the common eigenstates of H and R, the Hamiltonian matrix is block diagonal. We can always find a linear combination of the eigenstates of R as the basis in which the symmetry operation R is most block diagonalized (under this basis, the representation of R becomes irreducible), and at the same time the Hamiltonian H is also a block matrix in the same format as R. The dimension of the block matrix gives the order of state degeneracy of the Hamiltonian.

The second is the symmetry of the electronic wave functions. As discussed above, the eigenstates of the Hamiltonian can be classified according to symmetry. For example, for a system with mirror symmetry σ, the eigenstates for σ follow

$$\sigma\psi = C\psi, \ \sigma(\sigma\psi) = \psi = C^2\psi, \tag{3.23}$$

so

$$C^2 = 1, \quad C = \pm 1. \tag{3.24}$$

This means the eigenstates of the mirror symmetry are either even or odd with respect to the mirror symmetry operation. Thus, the eigenstates of H may also possess this symmetry. This is an example involving one dimensional transformation of symmetry. For systems such as the cubic or tetrahedral crystals, the C_3 rotation through the [111] direction involves all three coordinates. If ψ is one eigenstate of the system, after rotation the system is unchanged, and the state ψ is transformed to the other function $C_3\psi$, which is also an eigenstate of the system, due to the commutation between H and C_3. Normally since the transformation involves all three coordinates, $C_3\psi$ is linearly independent of ψ, and so is $C_3^2\psi$. These three functions are related by the threefold rotation. These three functions are the eigenstates belonging to one eigenvalue, so this energy state is threefold degenerate. Generally, if the highest number of coordinates transformed by symmetry operations in one system is n, then this system may have n-fold degeneracy.

The third is the symmetry of the energy dispersion constrained by the crystal point symmetry. First the entire energy band as a whole has the point symmetry of the crystal lattice. It means that the energy dispersion has the same point symmetry as the crystal lattice about the Γ point. Energy dispersions of cubic crystals such as Si and GaAs are the same along k_z, k_y, and k_z directions, while they may differ for hexagonal crystals such GaN. The symmetry of the k point is determined by the little group of k, which is determined by the crystal lattice symmetry accordingly. The little group of Γ is isomorphic to the point group of the crystal. Apart from the Γ point, a general k point only has the symmetry operation of identity, so the symmetry of energy dispersion around this point is not symmetrical. But along some high symmetry axes such as the Δ axis or at some high symmetry points such as the X point in the cubic crystals, because the members of the symmetry groups are beyond the identity operation, the symmetry of the energy dispersion is more complicated. Through the symmetry of these symmetry lines or points, the general energy dispersion, i.e., the $E-k$ relation along these lines or around these points, can be constructed, and the symmetry properties of the equi-energy surface can be intuitively obtained. One example may be the band structure of the Si conduction band valleys. Since the Si conduction band valleys are located at the Δ axes, and the Δ axes have fourfold rotation symmetry, then the energy dispersion near the valley edge may be written as

$$E = \frac{\hbar^2(k_1^2+k_2^2)}{2m_t} + \frac{\hbar^2 k_3^2}{2m_l}, \tag{3.25}$$

where k_1, k_2 are the k vectors orthogonal to, and k_3 is along one Δ axis, and m_t and m_l are the transverse and longitudinal effective masses that enter the dispersion relation as parameters.

In many cases, given the crystal symmetry properties, one can immediately obtain the information of the band degeneracy, phonon dispersion along symmetry axes, selection rules for phonon scattering or photon scattering, etc. All band structure computation methods, such as tight-binding, pseudopotential, $\mathbf{k}\cdot\mathbf{p}$ method and so on use symmetry to greatly simplify the task of solving the electronic energies and wave functions. In some cases, the exact expressions of the electronic wave functions are not needed. The symmetry properties alone of these wave functions are sufficient in the construction of Hamiltonian matrix. With some experimentally acquired parameters, the energy dispersion and density of states can be obtained, and the electron transitions such as phonon and photon scattering rate can be accurately accounted for.

3.4 SEMICONDUCTOR CRYSTAL CLASSES AND SYSTEMS

Because symmetry is such an important and convenient tool to obtain the information of a crystal, it is natural to classify crystals by their symmetry properties. By their point symmetry, crystals are classified into crystal classes, each of which has one distinct symmetry group. We can also classify the crystal lattices by their point symmetry. Crystals under this classification comprise the crystal systems. This section very briefly introduces the 32 crystal classes and 7 crystal systems. Actually, based on the crystal symmetry groups and their crystal lattices, there exist 230 space groups. Crystals that belong to the same crystal class and the same crystal system may belong to different space groups, e.g., diamond and NaCl, and have distinct properties in some aspects. Detailed introduction of the space groups is beyond the scope of this book, and thus we will neglect their introduction.

3.4.1 Crystal Classes and Systems

The crystal point symmetry operations include rotational axes, mirror planes, inversion centers, and rotation-inversion axes, etc. However, mirror planes can be proven to be equivalent to twofold rotation plus inversion, and the rotation-inversion is just the combined operation of proper rotation (pure rotation about an axis) and inversion. So all point symmetry operations can be considered as proper rotations with or without inversion.

The different combinations of these rotational symmetries plus the inversion constitute the basis to classify the crystals into crystal classes (point groups). There exist total 32 different crystal classes, i.e., the crystals can exhibit 32 different types of macroscopic symmetry. In crystal group theory, these 32 crystal classes are labeled with 32 different point groups. These groups are simply introduced in the following.

The simplest point group only contains one element, namely, the identity element, labeled as C_1. It represents the crystals without any symmetry.

The point groups containing only one rotational axis are called cyclic groups and are labeled as C_n. As we discussed before, there are totally 4 of this type of groups; $n = 2, 3, 4$, and 6. So C_n represents an n-fold rotation axis.

The point groups containing one n-fold rotation axis and n twofold axes orthogonal to the n-fold axis are labeled as D_n. Similar to C_n, there are four of them.

Including the inversion center and some mirror planes, the above point groups can become some new point groups.

For example, including the inversion center to C_1, the new point group is C_i. Including a mirror plane to C_1, the obtained point group is labeled as C_s.

Adding a mirror plane which is perpendicular to the nth-order rotational axis, the C_n group becomes group C_{nh}. Adding a mirror plane that is perpendicular to the n-fold rotation axis, the D_n group becomes group D_{nh}. Adding to the D_n group mirror planes that contain the nth-order axis and bisecting the two twofold axes forms the point group D_{nd}, and $n = 2$ or 3.

The point groups containing only rotation-inversion axes are labeled as S_n, among which $S_1 = C_i$, $S_2 = C_s$, $S_3 = C_{3h}$, so there are only S_4 and S_6.

Now we have discussed above 27 point groups, which contains at most one n-fold rotation axis with $n \geq 3$. The other five point groups with more than one high order rotation axes are related to the symmetry operations of a cube or a tetrahedron. A cube has a total of 48 symmetry operations and form a point group called the octahedron group, which is labeled as O_h. A tetrahedron has a total of 24 symmetry operations and forms a point group called the tetrahedron group, which is labeled as T_d. The other three point groups are: the 24 proper rotation operations form a group labeled as O; the 12 proper rotation operations forms a group labeled as T; adding an inversion center to group T, the new point group is labeled as T_h.

The symbols conventionally used in labeling the point groups are summarized in Table 3.1.

Table 3.1. The 32 crystal classes and their group symbols

Symbol	meaning
C_n	n-fold rotational axis (n =1, 2, 3, 4, 6)
S_n	n-fold rotation-inversion axis (n =1, 2, 3, 4, 6)
D_n	n 2-fold rotational axes orthogonal to a nth-order axis
$C_i = S_1$	Inversion center
$C_s = S_2$	One single mirror plane
h	"Horizontal," mirror planes perpendicular to the rotation axis
v	"Vertical," mirror planes parallel to the main rotation axis
d	"Diagonal," mirror planes parallel to the main rotation axis and bisecting the 2-fold axes
T	"Tetra-," four 3-fold and three 2-fold rotation axes as in a tetrahedron
O	"Octa-," four 3-fold and three 4-fold rotation axes as in an octahedron

The crystal lattices are classified by the crystal point symmetry into seven crystal systems. From low to high symmetry, they are triclinic, monoclinic, orthorhombic, rhombohedral, tetragonal, hexagonal, and cubic systems. Triclinic system corresponds to point group C_1 and C_i and possesses the lowest macroscopic symmetry. This type of lattice is composed of by three arbitrary vector bases without any requirement to the unit vector length and orthogonality. The other extremum is the cubic crystal system that possesses the highest macroscopic symmetry and corresponds to point groups T, T_d, T_h, O, and O_h. The seven crystal systems and the crystal classes that belong to each of them are shown in Table 3.2. Different crystals can be evidently presented by their crystalline cells, which have the symmetries the crystal possesses.

Because crystal systems are classified by the crystal lattices, so one crystal system may contain more than one crystal classes, which have the same type of crystal lattice, but due to the different primitive structures, they differ in

Table 3.2. The seven crystal systems and the corresponding crystal classes

Crystal systems	Basis vectors (crystal axes)	Crystal classes
Triclinic	Arbitrary $\mathbf{a}_1$, $\mathbf{a}_2$, $\mathbf{a}_3$	C_1, C_i
Monoclinic	$a_1 \neq a_2 \neq a_3$, $\alpha = \gamma = 90°$, $\beta \neq 90°$	C_2, C_s, C_{2h}
Orthorhombic	$a_1 \neq a_2 \neq a_3$, $\alpha = \beta = \gamma = 90°$	D_2, C_{2v}, D_{2h}
Tetragonal	$a_1 = a_2 \neq a_3$, $\alpha = \beta = \gamma = 90°$	C_4, C_{4h}, C_{4v}, D_4, D_{4h}, S_4, D_{2d}
Rhombohedral	$a_1 = a_2 = a_3$, $\alpha = \beta = \gamma \neq 90°$	C_3, C_{3v}, S_6, D_3, D_{3d}
Hexagonal	$a_1 = a_2 \neq a_3$, $\alpha = \beta = 90°$, $\gamma = 120°$	C_6, C_{6h}, D_6, C_{6v}, D_{6h}, C_{3h}, D_{3h}
Cubic	$a_1 = a_2 = a_3$, $\alpha = \beta = \gamma = 90°$	T, T_d, T_h, O, O_h

point symmetries. One example is that both Si and GaAs belong to the cubic crystal system, but Si has the O_h symmetry and GaAs has the T_d symmetry, since in the primitive cells of both Si and GaAs, there are two atoms, but for Si, these two atoms are of the same kind, while for GaAs, these two atoms are of the different kind, thus there is no inversion symmetry in GaAs. The Brillouin zone of each lattice type possesses the same point symmetry. Because the band structure is constructed in the Brillouin zone, the band structures of one lattice type has the same symmetry operations.

3.4.2 Cubic Semiconductors

Technologically attractive semiconductors include Si, Ge, GaAs, which all belong to the cubic crystal system, and GaN that belongs to the hexagonal crystal system. In this particular section, we will concentrate on the diamond structure group IV elementary semiconductors such as Si and Ge and zinc-blende III–V compound semiconductor such as GaAs, which are currently attractive in CMOS technology and briefly introduce their symmetry properties.

Both diamond structure and zinc-blende semiconductors have the FCC lattice structure, one of the typical close packing structures. Associated to every lattice point there are two atoms that are displaced relatively to each other by one quarter of the body diagonal along the $\langle 111 \rangle$ direction. In the diamond or zinc-blende semiconductors, each atom is surrounded by four nearest neighbors forming a tetrahedron. In zinc-blende semiconductors, the types of the atom and its neighbors are different.

The point group of the zinc-blende structure is the tetrahedral group T_d. There are a total of 24 symmetry operations, namely:

The identity E
The rotations by π about the axes x, y, z $(3C_4^2)$
The rotations by $\pm 2\pi/3$ about the four body diagonals $(8C_3)$
The rotation-inversions by $\pm\pi/2$ about the axes x, y, z $(6S_4)$
The rotation-inversions by π about the bisectrices xy, yz, zx [$6C_s$, or six mirror planes with respect to the (110), $(1\bar{1}0)$, (101), $(10\bar{1})$, (011), and $(01\bar{1})$ planes]

These are the symmetry operations for a tetrahedron. In a tetrahedron, there is no center of inversion. While the diamond structures are composed of the identical atoms, so there also exists the operations of inversion. The crystal is invariant under inversion with respect to the midpoint of two nearest atoms. Because of the existence of the inversion center, there are 48 symmetry operations of the diamond structure. If choosing one atom as the origin, to have the crystal coincided with each other after inversion, the inverted crystal must also be translated by one quarter of the body diagonal along the $\langle 111 \rangle$ direction. Except this translation, the 48 operations are the 24 operations of a tetrahedron, together with the 24 operations followed by the inversion operation. Thus, the point group of the diamond structure is the symmetry group of a cube, or a octahedron, labeled as O_h.

However, the primitive cell structure does not present in lattices and reciprocal lattices. Thus, the Brillouin zone of the cubic crystals, no matter they have diamond or zinc-blende structures, has the cubic symmetry. The band structure as an entire body possesses all these symmetry operations about the point Γ. The first Brillouin zone of all cubic crystals can be divided into 48 equivalent wedges. For example, $k_x \geq k_y \geq k_z$ is one of them. The integrations over the entire Brillouin zone can then be operated on one wedge only then multiplied by 48.

To illustrate the symmetry-determined band structure properties, consider first the empty lattice model or nearly free electron band structure model for the diamond structure that is shown in Fig. 3.12. In this model, only the uniform part of the periodic potential is considered, and the electronic states are classified solely by symmetry properties of the lattice. All bands are parabolic and several bands are degenerate at the Γ point because of the uniform potential. In this model, there is no real difference between diamond and zinc-blende structures because the perturbation of the potentials by the atoms is not considered. The degeneracy at the Γ point is labeled by various symbols such as the Γ with a subscript. The perturbation of the potentials by atoms, i.e., the nonuniform part of $V(\mathbf{r})$ will split up some band gaps between the separately labeled bands. The valence bands include the singly degenerate bottom band labeled as Γ_1 and the triply degenerate band labeled as $\Gamma_{25}^{'}$. Taking into account the spin degeneracy, the Γ_1 and the $\Gamma_{25}^{'}$ band can hold $8N$ electrons, where N is the number of the lattice sites in the crystal. This is exactly the total number of the valence electrons in the crystal, so the valence bands are fully filled, and the bands above the gap are totally empty at zero temperature. The triple degeneracy of the valence band $\Gamma_{25}^{'}$ comes from the equivalency of the three orthogonal directions in the space due to the cubic symmetry.

The other points other than the Γ point have less symmetries. The little group of the points on the Δ axes of diamond structures is C_{4v}. That is to say, the band structure around a Δ point has a fourfold rotation axis with four mirror planes that contain this axis. Similar to the Δ axes, the little group of the points on the Λ axes is C_{3v}, containing a threefold rotation axis and three parallel mirror planes. The little group for point X is D_{4h} and for point L is D_{3d}.

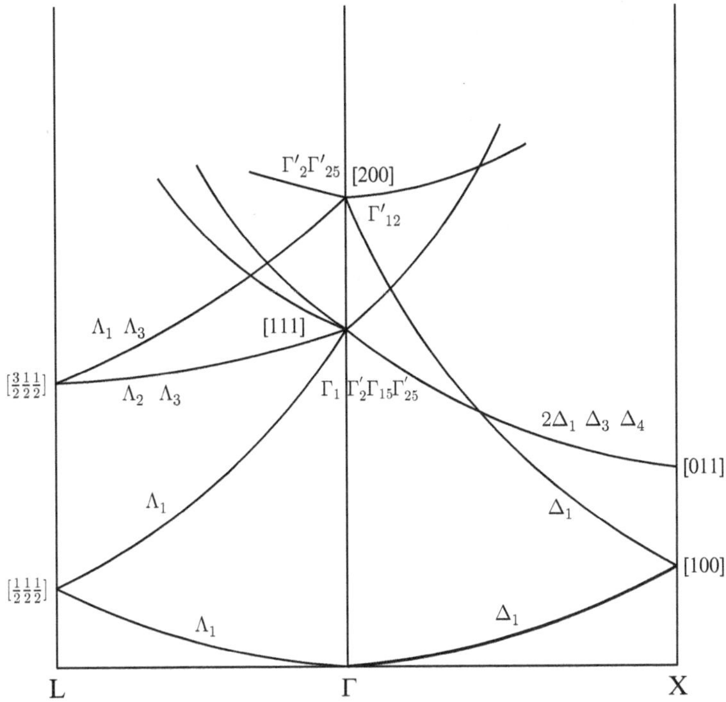

Fig. 3.12. The band structure of diamond structure semiconductors under nearly-free electron approximation

The band structures of the diamond and zinc-blende structure semiconductors are overall very similar especially for the valence bands even considering the realistic form of $V(\mathbf{r})$. Unlike the diamond structure semiconductors, energy levels of zinc-blende semiconductors at the X points are generally not degenerate, due to the lack of inversion symmetry.

3.5 STRAIN EFFECTS ON ELECTRONIC BAND STRUCTURES

Crystals in the same class or in the same system have some essential common properties. From this perspective, it is meaningful to see how a crystal evolves in crystal classes and systems when strain alters its symmetry.

3.5.1 Evolution of Crystal Systems with Strain

Of all crystal systems, cubic and hexagonal crystals possess the highest symmetry. All the other crystal systems can be derived from these two crystal systems by successively lowering the symmetry by shear strain. But note that hydrostatic strain can change the shape of other crystal systems than cubic crystals, e.g., the hexagonal crystals where ratio of the c/a (the lattice spacing along the c direction to the in-plane lattice spacing) will change with the same dilation of c and a, whereas normally this does not alter the crystal symmetry. Shear strain has less symmetry than the crystal and thus reduces the crystal symmetry.

A cubic crystalline cell will change into other shapes illustrated in Fig. 3.13. If a cubic crystal (Fig. 3.13a) is dilated (or compressed) along one of its 4-fold axes, it becomes tetragonal (Fig. 3.13b). The crystalline cube becomes a right parallelepiped on a square base. An original O_h crystal point group of Si or Ge turns into a point group D_{4h}. In a tetragonal crystalline cell, the rotation symmetry of the axis perpendicular to the square base is fourfold, with four twofold axes along the base plane edges and the diagonals. Meantime, the base is a mirror plane. The number of symmetry operations for this system is reduced by a factor of 3 compared to the cubic case. The crystalline cell of a tetragonal crystal may be converted to that of the orthorhombic system through two ways. One is through dilation (or compression) on one of the lateral faces (Fig. 3.13c), which reduces the tetragonal crystalline cell into a right parallelepiped on a rectangle base with mutually perpendicular edges. The second way is by shearing the base plane of the tetragonal crystalline cell, thus altering the angle between the edges of the base, which is originally a right angle (Fig. 3.13d). This results in a rectangular parallelepiped on a rhombic base. Either way, the symmetry of the crystalline cell is D_{2h}. The original fourfold axis of the tetragonal cell becomes twofold, and only two along the base plane diagonals of the four twofold axes remain. The number of symmetry operations is reduced to a mere 8. If the angle of the rectangle base or the angle between the diagonals of the rhombic base is changed, the monoclinic system that is invariant under C_{2h} is obtained (Fig. 3.13e). If the twofold axis is removed, a monoclinic system then is reduced to a triclinic system which is invariant under S_2 (Fig. 3.13h). This is one way how a cubic crystal system evolves to a triclinic system with application of strain. We can also first dilate a cubic along one of its body diagonals, which results in a transition from point group O_h to D_{3d}. This deformation makes the cube into a rhombohedron (Fig. 3.13f). The number of symmetry operations is 12, reduced by a factor 4 compared to a cube. From a rhombohedral system, a triclinic system can also be obtained by successive shearing deformation (Fig. 3.13g, h).

According to the strain–stress relation we have discussed in Chap. 2, applying a biaxial stress to a cubic system reduces it to a tetragonal system, applying a uniaxial stress along $\langle 110 \rangle$ to a cubic system converts it to a orthorhombic system, and applying a uniaxial stress along $\langle 111 \rangle$ changes a cubic system to a trigonal system.

From a hexagonal system, a rhombohedral system cannot be obtained through infinitesimal transformations of the lattice vectors. However, a deformation along the horizontal twofold axes converts the symmetry of D_{6h} to D_{2h}, and the hexagonal system becomes a base-centered orthorhombic system.

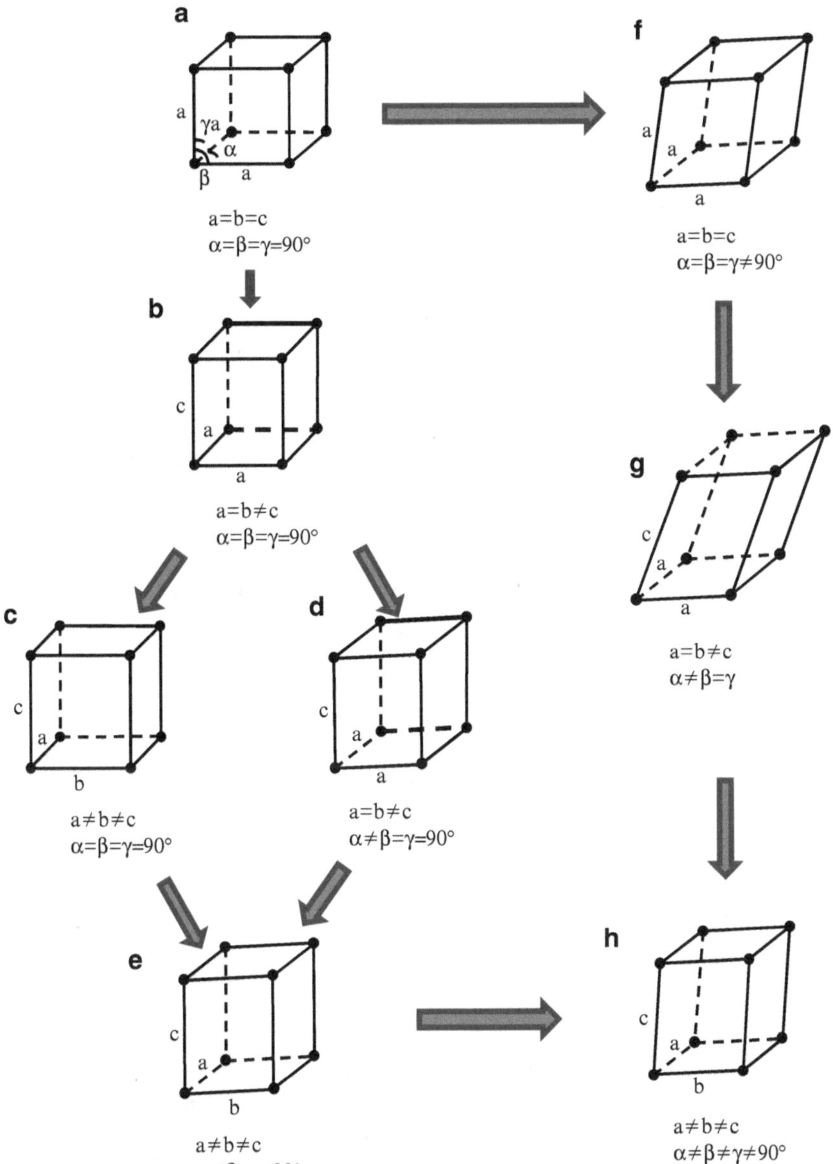

Fig. 3.13. Diagram for the evolution of a cubic crystal system under successive strain. (**a**) Cubic; (**b**) Tetragonal; (**c**, **d**) Orthorhombic; (**e**) Monoclinic; (**f**) Triclinic; (**g**) Rhombohedral; (**h**) Triclinic

3.5.2 Strained Band Structures

Crystal system is a good classification of band structure symmetries, since crystal system is defined by the lattice symmetry, while the band structure is constructed on the reciprocal lattice. Although Si and GaAs have different point symmetry, they have the same lattice symmetry, and thus the band structure has the same symmetry, the cubic symmetry. The valence bands of cubic semiconductors include doubly degenerate heavy-hole and light-hole bands with symmetry Γ_8 and one split-off band with symmetry Γ_7 resulting from spin–orbit interaction, while hexagonal semiconductors have three split valence bands with symmetry $\Gamma_9 + \Gamma_7 + \Gamma_7$ resulting from both the spin–orbit interaction and nonzero crystal field. The valence bands of tetragonal and orthorhombic semiconductors are essentially similar to those of the hexagonal semiconductors. A sheared crystal evolves into the other crystal system as discussed earlier. The band structure of the sheared crystal has the band degeneracy of the new system. This is the fundamental reason of degeneracy lifting.

Band warping by strain is merely due to the fact the band structure of a strained crystal has to comply with the altered crystal symmetry. This point was discussed to a great detail in Sect. 3.2. Here, we want to look again at the valence band warping for cubic semiconductors from the perspective of band mixing caused by symmetry reduction. In cubic semiconductors, the valence bands along $\langle 100\rangle$ are classified by angular momentum. Angular momentum is related to rotational invariance. The eigenstates at the valence band edge in cubic crystals can also be chosen as the eigenstates of L_z (or M), the z component of angular momentum L, with value of 1, due to the fourfold rotational symmetry. We may analogize the electronic eigenstates to the spherical harmonics:

$$|L\ \ M\rangle = \begin{cases} |1\ \ -1\rangle = (x - iy)/\sqrt{2}, \\ |1\ \ 0\rangle = z, \\ |1\ \ 1\rangle = (x + iy)/\sqrt{2}, \end{cases} \tag{3.26}$$

where x, y, and z may be replaced by functions that transform like x, y, and z under cubic symmetry operations for valence band eigenstates. When coupled with electron spin ($S = 1/2$), the eigenstates at the Γ point split into two groups classified by $J = 3/2$ and $J = 1/2$. HH ($J = 3/2$, $M_J = \pm 3/2$) and LH ($J = 3/2$, $M_J = \pm 1/2$) bands are degenerate at the Γ point and are not mixed with each other due to symmetry. Biaxial stress can change this rotational symmetry only in two orthogonal directions, while leaving the z-direction unchanged. Thus, along the z-direction, the mixing between HH and LH is negligible. However the $\langle 110\rangle$ uniaxial stress changes the rotational symmetry along the z-direction to twofold. This rotational symmetry cannot distinguish angular momentum differing by two, and thus will couple the LH (e.g., $M_J = -1/2$) and HH (e.g., $\mathrm{M}_J = +3/2$) states. This leads to the following results: a) under biaxial stress, even though degeneracy is lifted and states rise or descend in energy, band warping along $\langle 001\rangle$ is not significant and b) under uniaxial stress, in addition to the degeneracy lifting, mixing between the HH and LH bands significantly warps the bands. Symmetry breaking is also the reason for a phenomenon called "anti-crossing," which often takes place in quantum well valence band structures. Anticrossing means two energy levels that tend to cross near the Γ point curve away and form inflexions when approaching each other. This can be understood by the wave function mixing of these two states. We may consider the energies of the two states as the eigenenergies of a 2×2 matrix, which is diagonalized when there is no mixing, while it has nonzero off-diagonal elements and thus cannot have the same energy when there is wave function mixing. In strained semiconductor valence bands, anticrossing can also take place due to the band mixing caused by symmetry reduction, e.g., along the [110] direction for uniaxial stress semiconductors.

3.6 SUMMARY OF SYMMETRY, AND ITS LIMITATION

Crystal symmetry alone gives us a lot of information that is essential for determining the energy band properties among which the most important is the band degeneracy. Apart from the accidental degeneracy, band degeneracy is totally determined by the crystal symmetry. Band degeneracy lifting, including degeneracy at one single k point and the star degeneracy, plays a critical role when studying strain effects on electronic properties. Also, the symmetry lowering by strain also warps the unperturbed energy bands. The warping can be roughly understood from the symmetry constraints on the band structure. However, considerations from symmetry alone do not give the band details, e.g., the band energies and curvatures, etc, because symmetry only explores the geometrical form of the inter-atomic interaction, but not the real electric potential distribution. Even when we can write down a general form of the energy dispersion relation as in Eq. 3.25, symmetry cannot provide the information of the parameters [e.g., the effective masses in Eq. 3.25]. Although we have the information of how degeneracy lowers by a specific type of strain, the details of how the energy bands change with strain, e.g., identifying which state rises or descends under a certain type of stress, cannot be obtained. If we want to obtain a realistic band structure and acquire detailed qualitative or quantitative description of strain effects, we need to employ some band structure calculation methods. In the next chapter, two extensively used band structure computation formalisms, the tight-binding and $\mathbf{k} \cdot \mathbf{p}$ method, are introduced. Using them, we will be able to predict, e.g., how the Δ valleys split, or how the HH and LH band warps in strained-Si.

4

Band Structures of Strained Semiconductors

4.1 INTRODUCTION

In Chap. 3, a qualitative picture of band splitting and warping under stress, e.g., as shown in Figs. 3.9–3.11, was obtained merely from symmetry considerations. But for application purpose in strain engineering, further band structure details with strain are required. Otherwise, it is not possible to determine which type of stress is advantageous for a specific application goal. That is, symmetry alone cannot provide enough information that we have to acquire before we can implement the desirable stress. For example, $\langle 110 \rangle$ uniaxial stress, no matter tensile or compressive, will result in the same symmetry lowering to a cubic crystal. As in Fig. 3.10, the x-y plane energy contour, which has an ellipse shape resulting from the compressive $\langle 110 \rangle$ uniaxial stress, could have its major axis either in the [110] direction or in the $[\bar{1}10]$ direction, both complying with the uniaxial stress symmetry. Symmetry alone is not adequate to determine along which direction the ellipse major axis is oriented. However, this knowledge is critical for strain applications in n-type Si MOSFETs. Similar situations also exist in the valence bands. The band warping and splitting details, which are crucial for strain to enhance the p-type MOSFET performance, are not decisively determined by stress symmetry.

Therefore in this chapter, following the discussion of symmetry in the last chapter, we introduce band computation formalisms and study the band structures of strained semiconductors. There are already many solid state (Ashcroft and Mermin, 1976; Kettel, 1996) or semiconductor physics books (Yu and Cardona, 1996) covering the band structure theories. But for completeness and coherence of this book, we still elaborate a great detail on band calculation methods, especially on tight-binding (Slater and Koster, 1954; Chadi and Cohen, 1975) and $\mathbf{k} \cdot \mathbf{p}$ method (Kohn and Luttinger, 1954; Luttinger and Kohn, 1955; Luttinger, 1956), which respectively provide a great framework to investigate the strain effects, and on strained band structure calculations. For understanding of strain effects on band structures, as well as

Y. Sun et al., *Strain Effect in Semiconductors: Theory and Device Applications*, DOI 10.1007/978-1-4419-0552-9_4,

band formation from atomic orbitals, we first give a qualitative overview from tight-binding point of view in Sect. 4.2. The band calculation methods are introduced in later sections. First the tight-binding method is introduced, which visualizes the semiconductor bonding and strain effects. The concepts of the electronic wave function, and the relation between band structures with bond lengths and angles that are graphic and straightforward, are easy to comprehend under the tight-binding framework. Although it is possible to calculate the semiconductor band structure accurately using tight-binding method, the discussion of the tight-binding picture in this chapter is qualitatively based, since high-precision quantitative calculations based on tight-binding method require a large number of basis states, and the dimension of the Hamiltonian matrix needed to diagonalize to obtain the band structure is usually high. The consequences are long computer hours and difficulties to analyze the final results such as electronic wave function mixing. So following the introduction of the tight-binding method, we introduce another band calculation method, the $\mathbf{k} \cdot \mathbf{p}$ method, which usually has a much lower dimension Hamiltonian matrix to diagonalize and treats with high precision the strained bands at band edges where most carriers are located. Through $\mathbf{k} \cdot \mathbf{p}$ calculations, detailed band shifts and warping with strain are computed and presented finally. Readers can selectively read the materials of their interest.

As we mentioned in Chap. 3, every band structure calculation formalism employs crystal symmetry. Through the discussion in this chapter, we can also understand how symmetry properties affect the band structure and how they are implemented in the energy band calculation procedures.

4.2 STRAIN EFFECTS ON SEMICONDUCTOR BAND STRUCTURES: A QUALITATIVE OVERVIEW

In describing the electronic state in solids, there exist two seemingly contradictory limiting cases, the weak-binding and tight-binding approximation. In the weak-binding approximations, the interatomic interaction is sufficiently small, and the periodic crystal function is weak. The electrons behave nearly free. Electrons in metals fall into this category. In the tight-binding approximation, the interatomic interaction is strong, and the electrons are more likely to bond to discrete atomic sites and retain much of their properties as in a free atom. Covalent bonding crystals fall into this category. The elementary semiconductors such as Si and Ge are purely covalently bonded; compound semiconductors such as GaAs and GaN are also covalently bonded. Thus, the tight-binding approximation is usually used to semiconductor band structure calculations and has achieved great success.

4.2.1 Tight-Binding Formation of Semiconductor Crystals

Before we look into the tight-binding formation of the covalent semiconductors, first let us recall some atomic physics. Among various elementary materials, inert gases are stablest, because their outermost electronic shell is fully filled. For an elementary or a compound material to stay stable, its outermost electronic shell needs to be completely filled, too, just like the inert gases. If the outermost electronic shell is $1s$, then it is completely filled by two electrons. For instance, this is the case for hydrogen molecules and helium. Other than $1s$ shell, the outermost electronic shells all need eight electrons to completely fill to stay stable. Atoms in covalent solids achieve this by sharing electrons with their neighbors, i.e., through the bonds. Hydrogen atom has one electron. It needs one more atom to contribute the other electron to complete the outermost shell. By this, a hydrogen molecule is formed through the hydrogen bond. For an elementary semiconductor such as Si, the electronic configuration is $1s^2 2s^2 2p^6 3s^2 3p^2$. The inner two full shells with configuration $1s^2 2s^2 2p^6$ are very stable and the electrons in them have very low energy. The inner shell electronic wave functions are strongly localized, and generally, it is considered that they are not energetically active and do not overlap with the wave functions in the neighboring atoms and so do not contribute to the energy band formation. The electrons in the outermost half shell with configuration $3s^2 3p^2$ are called valence electrons and are energetically active. In a free atom, these valence electrons occupy two $3s$ orbitals and two $3p$ orbitals as shown in the left of Fig. 4.1. Note that the $3p$ orbitals are 3-fold degenerate with three linearly independent p orbitals, p_x, p_y, and p_z, and can hold six electrons including two spin states. According to the bonding principle, one Si atom needs four more electrons to fully fill the outermost shell. A naive thought is that since the Si crystal is extended in the 3D space and has cubic symmetry, one symmetric arrangement of the bonds shall be that each Si atom has four bonds by sharing two electrons with its neighboring atoms, and the four bonds have a common angle between each other. This is exactly what occurs in Si. The four bonds in Si are tetrahedrally configured, and the common angle between any two bonds is $109°28'$. The tetrahedral atomic configuration is also optimal for other group IV elementary semiconductors such as diamond and Ge, and III–V and II–VI zinc-blende compound semiconductors. In the compound semiconductor cases, although with some extent of electronic polarization for each atom, sharing of electrons still completely fills every outermost electronic shell.

The four bonds in Si are directed along the $\langle 111 \rangle$ directions. The $\langle 111 \rangle$ direction is special in that the x, y, and z directions are symmetric with respect to it. Then in the Si bonds, there are equal probabilities finding p_x, p_y, or p_z orbitals. But as we mentioned, the outer electronic shell configuration for a Si atom is $3s^2 3p^2$. For the three p orbitals, only two are occupied. To choose either two out of the three p orbitals to participate in the bond formation will destroy the symmetry of the bond. This problem is solved by promoting one

s electron to the p orbital, such that the outer shell contains s, p_x, p_y, and p_z orbitals. This is called orbital hybridization (Fig. 4.1). The two $3s$ electrons and two $3p$ electrons hybridize into four sp^3 orbitals, which are called sp^3

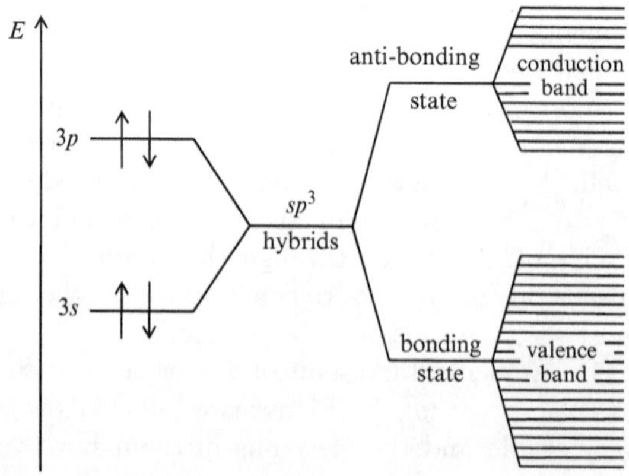

Fig. 4.1. From atomic orbitals to semiconductor band structures. In tetrahedral semiconductors, s and p atomic orbitals first hybridize, then evolve into bonding and antibonding states, which comprise the valence and conduction bands, respectively

hybrids. The single s, p orbitals and sp^3 hybrids are shown in Fig. 4.2a and b. This process needs extra energy compared to the $3s^2 3p^2$ configuration for a

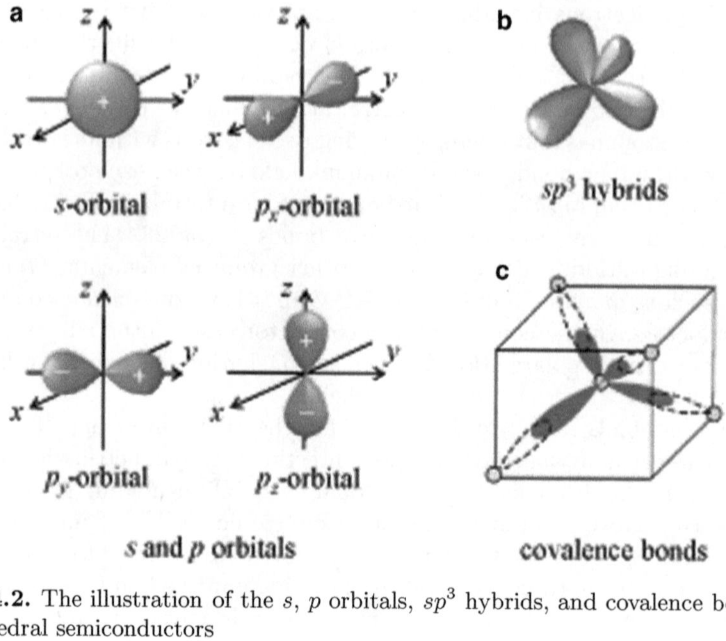

Fig. 4.2. The illustration of the s, p orbitals, sp^3 hybrids, and covalence bonds in tetrahedral semiconductors

single atom, but it is more than compensated in the subsequent formation of the covalent bonds. The four sp^3 hybrids, each contain one quarter of the four s, p_x, p_y, and p_z components as follows:

$$
\begin{aligned}
|h_1\rangle &= [|s\rangle + |p_x\rangle + |p_y\rangle + |p_z\rangle]/2, \\
|h_2\rangle &= [|s\rangle + |p_x\rangle - |p_y\rangle - |p_z\rangle]/2, \\
|h_3\rangle &= [|s\rangle - |p_x\rangle + |p_y\rangle - |p_z\rangle]/2, \\
|h_4\rangle &= [|s\rangle - |p_x\rangle - |p_y\rangle + |p_z\rangle]/2,
\end{aligned}
\tag{4.1}
$$

and orientated along [111], $[1\bar{1}\bar{1}]$, $[\bar{1}1\bar{1}]$, and $[\bar{1}\bar{1}1]$, respectively, where we use the Dirac ket to signify the atomic orbitals. The hybrid orientation is realized by addition and cancelation between the p orbitals due to their polar nature. After hybridization, the electron density is highest only along the orientation of the hybrids.

For covalent bonding, each hybrid couples with the other hybrid on the adjacent atomic site, which points toward the former as shown in Fig. 4.2c. Like that shown in Fig. 3.6a, the coupling between the two hybrids results in a bonding state and an antibonding state. The bonding state has lower energy and forms the covalent bond as shown in Fig. 4.2c. The covalent bonds are electron-pair bonds, similar to the hydrogen molecule bonds. The four bonds one atom has contains eight electrons. Then through the sharing of electrons in the bonds, the outermost electronic shell of the atoms is fully filled. Before bonding, each atomic energy level is a sharp level. Upon bonding, each atom in the crystal is connected to the other atoms all over the entire crystal by bonding to the neighboring atoms. The network-like coupling of the electronic wave functions broadens one single level in a free atom into a band in crystals. This process is schematically shown in Fig. 4.3a, whereas Fig. 4.3b schematically shows the band the single level spans in the crystal, where E_{a} is the energy for an arbitrary atomic energy level. For a semiconductor crystal

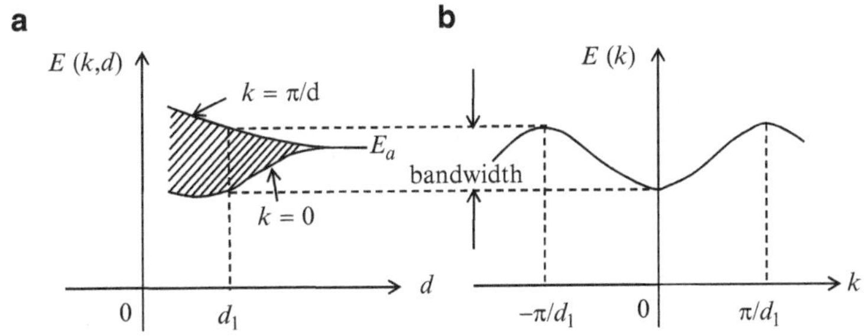

Fig. 4.3. (**a**) Atomic orbitals broaden into bands when their spacing distance shortens. d in the figure is the interatomic spacing in the semiconductor lattice. (**b**) Schematic band diagram evolved from an atomic orbital shown in (**a**)

consisting of $2N$ atoms (N primitive cells) each of which has four sp^3 hybrids, after bonding, the crystal has $4N$ bonding states and $4N$ antibonding states, as shown in Fig. 4.1. At zero temperature, the lower bonding states are fully filled by $8N$ valence electrons and comprise the valence bands, and the upper antibonding states are completely empty and comprise the conduction bands. The splitting between the bonding and antibonding states is the bandgap. This band formation procedure is generally valid for all semiconductors. However, the realistic formation of bands in a specific semiconductor is much more complicated involving its own particular characteristics. For example in Si, the valence electrons have finite probability to be excited to the $3d$ orbitals and $4s$ orbitals. Only considering $3s$ and $3p$ orbitals is not adequate to give an accurate conduction band dispersion, although it can reproduce the valence bands satisfactorily. The current discussion that only involves one s and three p orbitals is referred to as the sp^3 tight-binding model. Nevertheless, the sp^3 model is sufficient to give a qualitative physical picture.

4.2.2 Overlap Integrals

The atomic orbital overlap can be through two patterns, the σ-bonding and π-bonding, as shown in Fig. 4.4. The σ-bonding is by the head-on overlap of orbitals, and the π-bonding is by the side-on overlap of orbitals. Usually

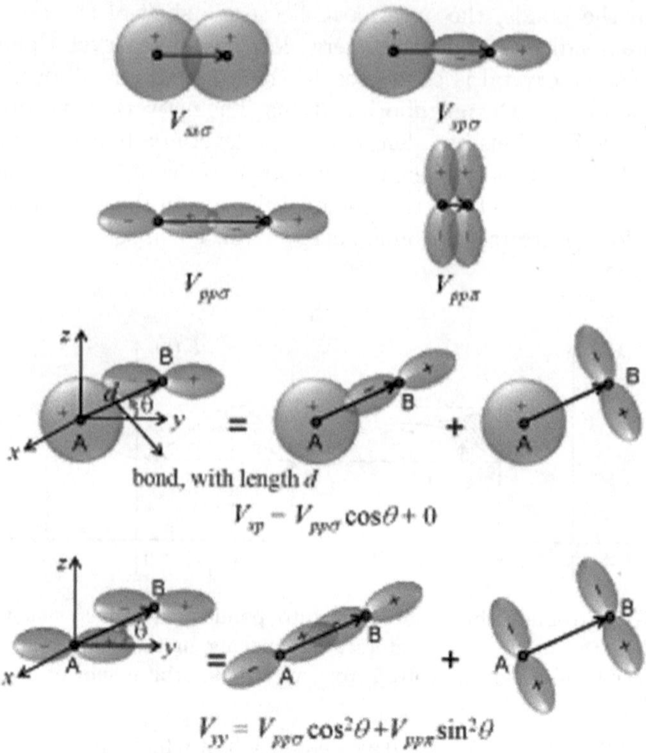

Fig. 4.4. Orbital overlap types and decomposition of the overlap integrals

the σ-bonding is much stronger than the π-bonding. The coupling between nearest neighbors can be decomposed into four basic overlap parameters, the σ-bonding between two s orbitals, $V_{ss\sigma}$, the σ-bonding between one s orbital and one p orbital, $V_{sp\sigma}$, the σ-bonding between two p orbitals, $V_{pp\sigma}$, and the π-bonding between two p orbitals, $V_{pp\pi}$. The bonding between s orbitals and the bonding between an s orbital and a p orbital can only be through the σ type due to the homopolar nature of the s orbital. The coupling strength between two sp^3 hybrids can be directly expressed using the four overlap parameters.

The significance of the orbital overlapping determines both the bonding–antibonding splitting and the bandwidth of a band. The bandwidth is the energy range a band spans in the Brillouin zone, as shown in Fig. 4.3b. Stronger overlapping will result in both larger bonding–antibonding splitting and bandwidth relevant to an orbital. The band with a larger bandwidth generally has larger band curvature at the Γ point, and thus it has a smaller electron effective mass. For a band originated from a single atomic level, the effective mass is inversely proportional to the bandwidth. In all semiconductors, interaction between the next-nearest neighbors for the valence electrons is much smaller than between the nearest neighbors, because of their localized nature of the electronic wave functions. So in this book, we only concentrate on the interaction between nearest neighbors.

Here we want to mention two observations. First, the curve shapes of the energy dispersion, or the bandwidth, are determined partly by the real-space crystal structure. If the atoms in a certain direction are far apart, then the bandwidth will be narrow along that direction. Second, among the Group IV elements C, Si, Ge, and Sn, C has the largest band gap and smallest lattice constant. Sn is a metal and has the largest lattice constant. The band gap decreases from C to Sn, while the lattice constant increases from C to Sn. As we know the band gap is related to the bonding–antibonding splitting, which is determined by the interatomic overlap integral. So overall, the overlap integral increases with reduction of interatomic distance. Harrison (Froyen and Harrison, 1979; Harrison, 1989) found that the overlap integrals in covalent solids and the bond length have an approximately universal relation, $V = \eta\hbar^2/(m_0 d^{-2})$, where η is a structural factor that only depends on crystal structures, m_0 is the electron mass, and d is the bond length. This relation is truly important and useful, because only one set of parameters is adequate to obtain the band structures of various materials. When focusing on one specific covalent crystal, and when the bond length changes due to distortion, the overlap integral can also be considered to be scaled following the d^{-2} principle. We may look at bonding state as a potential well with width d when an electron is bonded in a covalent bond. Then the ground state in the potential well has a d^{-2} relation with the well width. This d^{-2} relation makes much sense when it comes to the electron transport through the effective mass dependence on the interatomic overlap integral. When the interatomic distance is reduced, the hopping of electrons between atomic sites becomes easier, which may be ascribed to a reduced effective mass. A reduced effective mass means an increased bandwidth, which in turn is determined by the increased overlap integral. This is precisely what the reduced interatomic distance should bring about. On the other hand, if the interatomic distance increases, electron hopping between atoms becomes difficult. This may be ascribed to a larger effective mass and thus a reduced overlap integral. Hydrostatic strain reduces the atomic distance uniformly in the entire crystal, and thus the bandwidth is increased in all directions. Shear strain can increase the interatomic distance in some direction and reduce the interatomic distance in some other directions. In such cases, the bandwidth shifts are different among different directions. Also, the electron effective mass variations are different for different directions. We will discuss this point later when coming to the $\langle 110 \rangle$ uniaxial stress case. Sometimes the hybrids are misaligned by some shear distortions. In such a case, the sp^3 hybrid orientation does not change, since it is constructed on separate atoms. The neighboring hybrids that form a bond in the undistorted condition now have an angle relative to the bond. Following the decomposition rules of the overlap integral, extra energy added in the bond because of the angle distortion can be obtained.

4.2.3 Properties of Electronic Wave Functions

Even though we are seeking the E–k relations in band structure calculations, the electronic wave function plays a more fundamental role than the band energy itself in the process of band formation, because after all, energy bands are created by atomic orbital overlap. In the earlier subsections, four hybrids are constructed on each atom, and the coupling of these hybrids results in the bonding and antibonding states, which comprise the valence and conduction bands. Because there are two atoms in one primitive cell in Si, so there are eight different hybrids in a Si primitive cell, with four of them constructed on one atomic site, which we may label as h_i^1, where $i = 1, 2, 3, 4$, and the other four hybrids on the other atom may be labeled as h_j^2, where $j = 1, 2, 3, 4$. Before the implementation of the tight-binding formalism, we need first to construct a set of Bloch basis upon which the final electronic states are spanned. A Bloch basis function $\psi_{\mathbf{k}}^{\alpha,l}$ based on an atomic orbital ϕ_l is

$$|\psi_{\mathbf{k}}^{\alpha,l}\rangle = \psi_{\mathbf{k}}^{\alpha,l}(\mathbf{r}) = \frac{1}{\sqrt{N}} \sum_m e^{i\mathbf{k}\cdot\mathbf{R_m}} \phi_l(\mathbf{r} - \mathbf{R_m} - \mathbf{r}_\alpha), \tag{4.2}$$

where $\mathbf{R_m}$ is the coordinate of the mth primitive cell, and m runs over the whole crystal. Function $\phi_l(\mathbf{r} - \mathbf{R_m} - \mathbf{r}_\alpha)$ is the orbital of the αth atom in the mth primitive cell. In tetrahedral semiconductors that have two atoms in the primitive cell, $\alpha = 1, 2$. Based on the eight hybrids, eight Bloch basis are constructed, and the diagonalization of the tight-binding Hamiltonian gives eight energy bands. The electronic states of the bands are linear superpositions of the eight Bloch basis. The selection of the eight hybrids for basis is natural, because these hybrids are illustrative, and they conform the usual knowledge of the chemical bonds. However, this does not bring any convenience into band structure calculations. Usually it is more convenient to discuss the wave function properties using the original s, p_x, p_y, and p_z orbitals due to their distinctive symmetry properties. Quantum mechanically, choosing the atomic orbitals or hybrids as the basis brings no difference to the band energies. When using the atomic orbitals as the basis set, the function ϕ's in (4.2) are replaced by the s and p orbitals. Since at each atom in the primitive cell, there are four atomic orbitals, and there are two atoms in one primitive cell, so there are also eight Bloch basis based on the atomic orbitals.

When using the hybrids as the basis, the valence bands are formed by the bonding states of the hybrids. At or around the highest symmetry point, the Γ point, the electronic wave functions are superpositions of these hybrids. But these superpositions are decoupled along the $\langle 100 \rangle$ directions when using the atomic orbitals as the basis, and the valence band states are the bonding states of these atomic orbitals. Therefore, there are four valence bands according to the s, p_x, p_y, and p_z orbitals. The band structure of Si calculated using the sp^3 tight-binding model is shown in Fig. 4.5, where only the nearest-neighbor interaction is considered. The conduction band edge is still located at the Γ point rather than at the Δ point where it should be in the current sp^3 tight-binding framework. Higher energy atomic orbitals have to be included to correctly reproduce the Si conduction band. This work was done by Vogl (Vogl et al, 1983), who used one more orbital, s^*, to include the effects of the $4s$ orbital and partially $3d$ orbital.

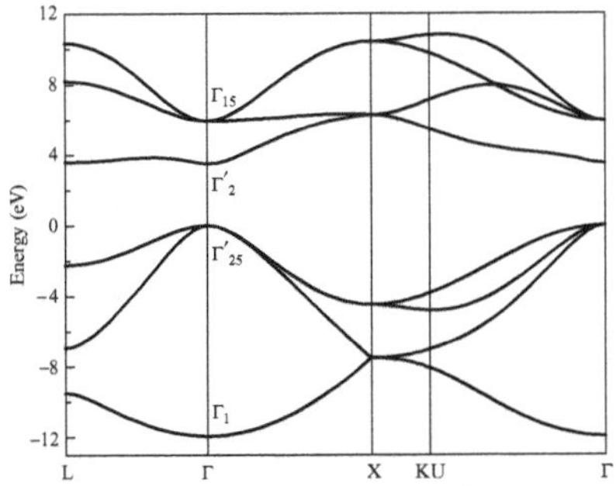

Fig. 4.5. sp^3 band structure of Si

The bands reflect the characters of the atomic levels that make them up and are traditionally marked by the same atomic orbital symmetry. At the bottom is the s bonding state. For cubic semiconductors, the s bonding state almost always has much lower energy than the rest of the bands and is fully occupied. The valence band edge states are composed of the three p bonding states. The properties of the hole state polarization along the [001] direction are shown in Fig. 4.6b. It is easily understandable that along any ⟨100⟩

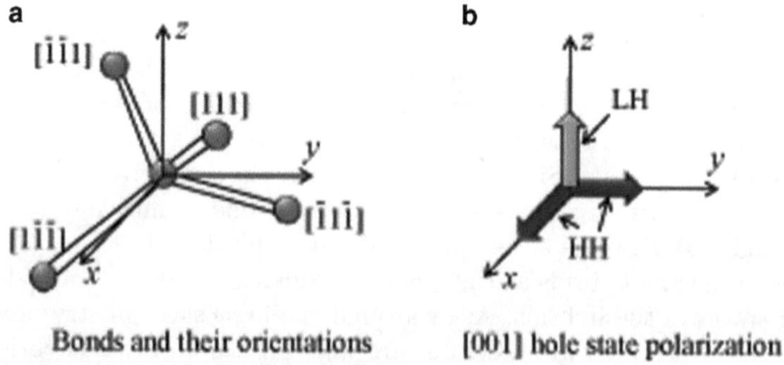

Fig. 4.6. (**a**) Bonds and their orientations and (**b**) hole state polarization

direction, two bands with states composed of two perpendicular p orbitals are degenerate. These two bands have smaller curvature and are the HH bands. The third band with states composed of the parallel p orbital is the LH band with larger curvature. There is no coupling between the HH and the LH bands. When away from the Γ point, the HH wave functions are still composed of pure p-orbitals, but the LH wave function is quickly mixed with the s-component. The admixture of the s-orbital in the LH band results in a larger bandwidth and subsequent larger band curvature at the Γ point. Along other high symmetry directions than ⟨100⟩, the valence bands are similar. The HH states are composed of the orbitals perpendicular to the direction under study, and LH state is parallel to this direction. Wave function properties for the conduction band vary for different materials. For direct gap semiconductors such as GaAs, its conduction band is primarily composed of the s antibonding states. The Si conduction band edge is composed of antibonding states consisting of both s and p components. Because the Si Δ valleys are directionally located, the mixing of the electronic wave function for a Δ valley also depends on the Δ direction. For example, the electronic wave function for Δ valley along z is composed of s and p_z antibonding states.

The wave functions of the valence bands may be visualized by the bond diagram shown in Fig. 4.6a. Since the bonds are composed of the bonding states, and each bond consists of equal probability of the three p components for unstrained semiconductors, we may project the bonds into one direction and obtain the orbital composition in the bonds along this direction. Then according to the wave function properties of the HH and LH states, it is evident that the p-component of the bond projection along one direction composes the LH state along this specific direction, and that of the projection at the perpendicular plane composes the HH states. When strain is applied, it alters the bonds by changing their length or angle. When this occurs, it also changes the composition of the three p orbitals in the bonds and consequently the interatomic wave function overlap. Energy band shift by strain can be straightforwardly seen by studying bond variation.

4.2.4 Strain Effects on Tight-Binding Band Structures

Let us inspect how bond changes under some specific types of strain. Under hydrostatic strain, only bond lengths are reduced. Reduced interatomic distance enhances the interatomic electronic wave function overlap, and the overlap interaction is strengthened. This results in increased splitting between the bonding and antibonding states, i.e., the bandgap is widened. This is the primary effect of hydrostatic strain. Bandwidth, which critically depends on the overlap integrals, also increases with reduced bond length. But the band curvature change is a small effect for the usual bond length variation under hydrostatic strain.

Shear strain has different effect on bands with different symmetries. For s bands such as the GaAs conduction band, only hydrostatic strain has pronounced effect to shift it. Shear strain has no effect shifting it and very small effect warping it. The warping comes from the strain-changed interband coupling from the other bands. This can be visualized by that even when shear strain rotates one s orbital with respect the other, if it does not change the interatomic distance, it does not affect the overlap between the two s orbitals. However, by inspecting Fig. 4.4, we can see that when shear strain alters the angles, the overlap parameters such as V_{sp} and V_{yy} are modified. Thus, shear strain has strong effects on p bands. Fortunately, the bond projection is a good illustrative tool for analyzing the p composition change in these bands.

Next, we will use this bond projection picture to study two technologically important stress cases, the biaxial stress case and $\langle 110 \rangle$ stress case, using Si as a prototype material. Under biaxial stress, the four bonds all rotate toward or away from the x-y plane depending on whether it is compressive or tensile. For biaxial tension, all the four bonds are equivalent and rotate toward the x-y plane, as shown in the upper panel of Fig. 4.7a. With such a rotation, the weight of the p_x and p_y orbitals in the bonds increases and that of the p_z-orbital decreases. Along the [001] direction, this results in increased overlap integrals between in-plane orbitals and lowered HH bands and decreased overlap integrals between the p_z orbitals and ascended LH bands. Along the [100] and [010] direction, the LH band is lowered because it is composed mainly of the p_x and p_y orbitals, respectively, and the topmost bands are HH bands. Under such a band alteration, there is no pure HH or LH band. The top valence band under biaxial tension is LH-like out-of-plane and HH-like in-plane along z. The valence band structure under biaxial tensile stress is schematically shown in the lower panel of Fig. 4.7a.

The wave function of the Si conduction band along the Δ-axis is primarily composed of antibonding p states. For the two Δ valleys along z, due to the

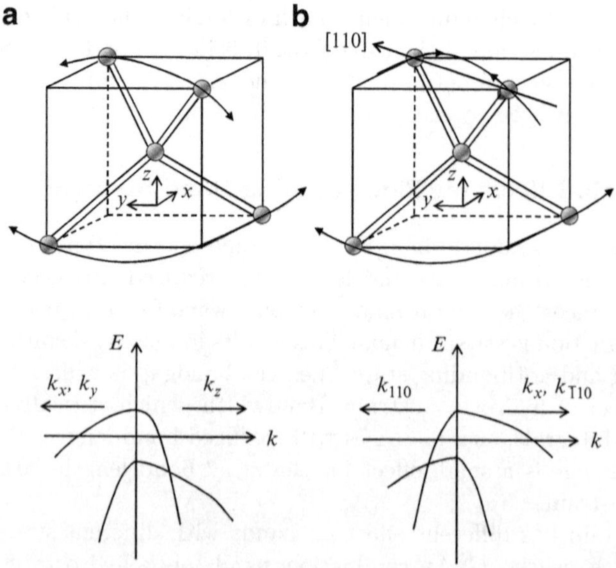

Fig. 4.7. Bond rotation under (**a**) biaxial, and (**b**) uniaxial strain, and their corresponding valence band shifts and warping

decrease of overlap integrals between the p_z orbitals under biaxial tension, they lower in energy. On the contrary, the other four valleys ascend in energy.

For strain tensor induced by [110] uniaxial stress, the most evident difference is the nonvanishing e_{xy} tensor element. Unlike the biaxial strain, this shear term changes the symmetry in the x-y plane by changing both the bond angles and lengths. For the four nearest neighbors connected by the bonds, two of them lie along the [110] direction and the other two lie along the $[\bar{1}10]$ direction. Under compressive stress, the bonds that project along the [110] direction are shortened and pushed together and the bonds that project along the $[\bar{1}10]$ direction are elongated and pulled apart, which is illustrated in the upper panel of Fig. 4.7b. Despite the shortened bond length which increases the interaction between the origin atom with the atoms along [110], the rotation of the two bonds increases the weight of the perpendicular orbitals and reduces that of the parallel orbitals with respect to the [110] direction. Thus, lowered HH states and a raised LH state along [110] result. Along $[\bar{1}10]$, the situation is reversed because compression along [110] is equivalent to tension along $[\bar{1}10]$. The fact that the two bonds are pulled apart increases the parallel projection and reduces the perpendicular projection. The final band splitting can be considered as a combined result of both types of shear strain, and thus after band splitting, we have a HH band along [001] and $[\bar{1}10]$, and a LH band along [110] for the top valence band, just as shown in the lower panel of Fig. 4.7b. Quantitative results indicate the bands along [110] have anticrossing effect, not exactly like that as shown in Fig. 4.7b. We will discuss this later in the $\mathbf{k} \cdot \mathbf{p}$ section.

For a uniaxial stress along [110], its effects on Si conduction valleys are determined by e_{xx}, e_{yy}, and e_{zz}, which can be obtained from (2.36). These tensor elements have the sign determined just by "tensile" or "compressive," not related to "biaxial" or "uniaxial." So like in the biaxial tension case, under uniaxial tension, the two Δ valley along the z axis descend and the four in-plane (x-y plane) conduction valleys ascend.

Therefore, based on the analysis above, compression shifts the LH band up and the HH band down along the compressive stress axis (biaxial tensile stress is equivalent to compressive stress along z, as far as only shear stress is concerned), for both biaxial and uniaxial stress. This occurs because compression rotates the bonds toward the plane perpendicular to the stress axis and makes the interatomic interaction in the perpendicular plane stronger than that parallel to the stress axis. The analysis of other cubic semiconductors may follow the same procedure discussed above.

4.2.5 Determining Deformation Potentials Using Tight-Binding Method

Deformation potentials can be fairly easily understood through the tight-binding framework by means of the band shift caused by strain. Deformation potential theory has been first proposed by Bardeen and Shockley within the effective mass approximation to study the interaction of electrons and acoustic phonons. Acoustic phonon creates local strain, which induces energy band shift, and this shift couples electrons and acoustic phonons. Band shift caused by homogeneous strain can be described using the same deformation potential theory. Basically, band energy shift and strain are related by

$$\Delta E = \sum_{ij} \Xi_{ij} \varepsilon_{ij}, \tag{4.3}$$

where Ξ are deformation potentials. For a general k point in the Brillouin zone, three deformation potentials are needed to describe the band energy shift. In high symmetry point, this number lowers. For example, for the s symmetry conduction band edge such as the conduction band of GaAs, only one deformation potential, a_c or Ξ_d^{Γ}, which is the hydrostatic deformation potential for the conduction band, is needed. Using the band parameters in the tight-binding model, the deformation potentials can be expressed in terms of the overlap integrals. For example, for s type bands, the band edge energies are given by

$$E_{\pm s} = E_s \pm 4|V_{ss}|, \tag{4.4}$$

where E_s is the on-site energy for s orbitals, and $V_{ss} = V_{ss\sigma}$ is the next-neighbor overlap integral between two s orbitals.[1] The "±" signs signify the bonding and antibonding state energies. Obviously, the conduction band edge energy is $E_{-s} = E_s + |V_{ss}|$. When a hydrostatic strain $\varepsilon_{xx} = \varepsilon_{yy} = \varepsilon_{zz} = \varepsilon$ is applied, the interatomic distance is changed from d_0 to $d_0(1+\varepsilon)$. Assuming that the overlap integrals obey the d^{-2} principle, then the conduction band edge shift is $\Delta E = (\varepsilon_{xx} + \varepsilon_{yy} + \varepsilon_{zz}) \times (-8V_{ss})$. Thus, the conduction band deformation potential is given by $\Delta E/3\varepsilon = -8V_{ss}/3$. Normally, the deformation potentials are experimentally measured. Then, the experimentally measured deformation potentials in return can help determine the overlap integrals in the tight-binding method.

4.2.6 Summary for the Qualitative Overview

The simple sp^3 tight-binding formalism can give a clear physical picture of how semiconductor energy band forms and the origin of the symmetry of a band. The strain effects on band structures are vividly depicted by bond stretching and rotation. Simply, a compressive stress will result in raised LH and lowered HH bands along the stress axis, due to the altered overlap between atomic orbitals. In the following, we introduce the complete procedure of how to calculate the band structure using tight-binding method. Strain effects are discussed numerically by means of the change of overlap integrals caused by strain.

4.3 BRIEF INTRODUCTION TO PLANE WAVE EXPANSION METHOD

Because their overwhelming success of the plane-wave expansion based methods, such as the pseudopotential method, to calculate the crystal band structures, we first give a brief introduction to the general theory on which these methods are based upon. Readers can find the references cited in this section for further reading. Pseudopotential method has also been extensively applied to investigate the strain effects on electronic structures of semiconductors (Uchida et al, 2005; Fischetti and Laux, 1996). Sometimes, it is the only way to theoretically study the warping of the band, e.g., the Si conduction band.

[1] In Slater and Koster's work (Slater and Koster, 1954) and Harrison's book (Harrison, 1989), symbols E were used for overlap integrals. While in Chadi and Cohen's work (Chadi and Cohen, 1975), they used V for overlap integrals. The relation between E and V is $E = 4V$. In this book, we adopted Harrison's definition, but write the on-site energies as E, and overlap interals as V.

The core task of a band theory is to solve the single-electron Schrödinger equation with a periodic crystal potential:

$$
\begin{aligned}
H\psi_{\mathbf{k}}(\mathbf{r}) &= E(\mathbf{k})\psi_{\mathbf{k}}(\mathbf{r}), \\
H &= -\frac{\hbar^2}{2m}\nabla^2 + V(\mathbf{r}),
\end{aligned} \tag{4.5}
$$

where $V(\mathbf{r}) = V(\mathbf{r}+\mathbf{R})$, with $\mathbf{R}$ representing the crystal lattice point, and $\psi_{\mathbf{k}}(\mathbf{r})$ the Bloch function is written as

$$
\begin{aligned}
\psi_{\mathbf{k}}(\mathbf{r}) &= u_{\mathbf{k}}(\mathbf{r})e^{i\mathbf{k}\cdot\mathbf{r}}, \\
u_{\mathbf{k}}(\mathbf{r}+\mathbf{R}) &= u_{\mathbf{k}}(\mathbf{r}).
\end{aligned} \tag{4.6}
$$

The requirement of the electronic wave function inside crystals to be Bloch waves is significant since these Bloch functions are basically periodically modulated plane waves that extend over the entire crystal, i.e., the electronic states are extended states instead of localized states around the atomic cores. The key of all band theories is to find a reasonable framework of approximations to represent $u_{\mathbf{k}}(\mathbf{r})$, then the eigenenergies are computed within this framework.

Obviously it is natural to expand $u_{\mathbf{k}}(\mathbf{r})$ using a series of plane waves by Fourier transformation as in (3.17), based on the fact that $u_{\mathbf{k}}(\mathbf{r})$ is a periodical function of the crystal lattice. Based on plane wave expansion, we obtain from (4.6),

$$
\psi_{\mathbf{k}}(\mathbf{r}) = u_{\mathbf{k}}(\mathbf{r})e^{i\mathbf{k}\cdot\mathbf{r}} = \sum_{\mathbf{G}} C_{\mathbf{k}-\mathbf{G}}e^{i(\mathbf{k}-\mathbf{G})\cdot\mathbf{r}}, \tag{4.7}
$$

where the Bloch wave functions are expanded using plane waves $e^{i(\mathbf{k}-\mathbf{G})\cdot\mathbf{r}}$ and summation is over the reciprocal lattice. Equation (4.7) is the starting point of a set of band structure calculating methods including the augmented plane wave method (APW), the orthogonal plane wave method (OPW), and also the famous pseudopotential method.

For simplicity, we label the plane wave states using the Dirac bra–ket as

$$
\langle\mathbf{r}|\mathbf{k}-\mathbf{G}\rangle \equiv \frac{1}{\sqrt{\Omega}}e^{i(\mathbf{k}-\mathbf{G})\cdot\mathbf{r}}, \tag{4.8}
$$

where Ω is the volume of the crystal. Equation (4.7) then is written as

$$
|\psi_{\mathbf{k}}\rangle = \sum_{\mathbf{G}} C_{\mathbf{k}-\mathbf{G}}|\mathbf{k}-\mathbf{G}\rangle. \tag{4.9}
$$

Substituting into the Schrödinger equation (4.5), we obtain

$$
\sum_{\mathbf{G}} C_{\mathbf{k}-\mathbf{G}}(H-E)|\mathbf{k}-\mathbf{G}\rangle = 0. \tag{4.10}
$$

Multiplying by $\langle \mathbf{k}-\mathbf{G}'|$, and applying the orthogonality of plane waves,

$$\langle \mathbf{k}-\mathbf{G}'|\mathbf{k}-\mathbf{G}\rangle = \delta_{\mathbf{G},\mathbf{G}'} \tag{4.11}$$

we obtain a set of coupled equations with the unknown variables $C_{\mathbf{k}-\mathbf{G}}$,

$$\sum_{\mathbf{G}} \left[\left(\frac{\hbar^2}{2m}(\mathbf{k}-\mathbf{G})^2 - E \right) \delta_{\mathbf{G},\mathbf{G}'} + \langle \mathbf{k}-\mathbf{G}'|V(\mathbf{r})|\mathbf{k}-\mathbf{G}\rangle \right] C_{\mathbf{k}-\mathbf{G}'} = 0, \tag{4.12}$$

with the Fourier components of the crystal potential

$$\langle \mathbf{k}-\mathbf{G}'|V(\mathbf{r})|\mathbf{k}-\mathbf{G}\rangle = \frac{1}{\Omega}\int e^{-i(\mathbf{k}-\mathbf{G}')\cdot\mathbf{r}} V(\mathbf{r}) e^{i(\mathbf{k}-\mathbf{G})\cdot\mathbf{r}} d\mathbf{r}, \tag{4.13}$$

being the function of $(\mathbf{k}-\mathbf{G}')$.

The condition for (4.7) to have nontrivial solutions is to satisfy the following secular equation

$$\det \left| \left[\left(\frac{\hbar^2}{2m}(\mathbf{k}-\mathbf{G})^2 - E \right) \delta_{\mathbf{G},\mathbf{G}'} + \langle \mathbf{k}-\mathbf{G}'|V(\mathbf{r})|\mathbf{k}-\mathbf{G}\rangle \right] \right| = 0. \tag{4.14}$$

Through this equation, both the wave functions and eigenenergies can be obtained. However, plane wave expansion converges slowly due to the sharp oscillation of the crystal potential around the localized ion site. Since the main properties of the solids result from the electronic dynamics of the electrons around the Fermi surface, it is justifiable to divide the electrons into tightly-bound core electrons and loosely-bound valence electrons, which form the core electronic bands and the valence and conduction bands, respectively. This classification of electrons prompted the emergence of the augmented plane wave (APW) (Callaway, 1976) and orthogonal plane wave (OPW) (Callaway, 1976) methods, which see the core electrons as part of the localized ions and focus on the behavior of the valence electrons that experience much smoother potential, thus resulting in quicker convergence. Based on the OPW method, the pseudopotential method (Phillips and Kleinman, 1959; Harrison, 1976; Callaway, 1976) has been developed and applied to various types of crystals and has gained huge success.

However, due to the extending property of plane waves, and as the crystal potential is expanded in a series of plane waves, the role of interatomic interaction in the plane-wave-expansion-based methods is not straightforwardly seen. These band methods need to treat the strained crystals as new structures and thus the effect of the interatomic interaction alteration due to strain is not obvious. One other framework of band theory, namely, the tight-binding method, directly treats the interatomic interactions and thus provides a real space picture of the interatomic interaction between atoms, which gives rise to particular features of the energy bands, density of states, etc. This is extremely useful in studies of how these features change when the electronic configuration is altered by strain.

4.4 TIGHT-BINDING METHOD

4.4.1 A General Introduction

We first review the general procedure of the tight-binding method. Let us assume that the solutions to the Schrödinger equation of the free atoms that form the crystal

$$\left[-\frac{\hbar^2}{2m}\nabla^2 + U(\mathbf{r})\right]\chi_i(\mathbf{r}) = E_i\chi_i(\mathbf{r}) \tag{4.15}$$

are known, where $U(\mathbf{r})$ is the atomic potential, and $\chi_i(\mathbf{r})$ is the eigenstate for an electron in the atomic energy level E_i. Now assume that the atoms are brought together to form the crystal for which $V(\mathbf{r})$ is the periodic potential, and $\psi(\mathbf{r})$ and $E(\mathbf{k})$ are, respectively, the wave functions and energy eigenvalues for the electron in the crystal:

$$H\psi(\mathbf{r}) = \left[-\frac{\hbar^2}{2m}\nabla^2 + V(\mathbf{r})\right]\psi(\mathbf{r}) = E(\mathbf{k})\psi(\mathbf{r}), \tag{4.16}$$

where in the tight-binding approximation we write $V(\mathbf{r})$ as a sum of atomic potentials:

$$V(\mathbf{r}) \simeq \sum_n U(\mathbf{r} - \mathbf{R_n}). \tag{4.17}$$

For one electron located at $\mathbf{R_n}$, the potential it feels is still the atomic potential at the same site, but slightly altered by the potential from its neighbors. The perturbative potential to this electron is

$$H^{'}(\mathbf{r} - \mathbf{R_n}) = \sum_{m\neq n} U(\mathbf{r} - \mathbf{R_m}), \tag{4.18}$$

which is just the sum over the potential of all atoms apart from the one at $\mathbf{R_n}$. Without this perturbation, each atomic state has a degeneracy of N, which is the number of the atoms in the crystal. However, the perturbation, i.e., the interaction between the atoms lifts this degeneracy. The energy eigenvalue can be found through the Ritz variational procedure. Multiplying (4.16) by $\psi^*(\mathbf{r})$ and integrating over the crystal volume, we obtain

$$E = \frac{\langle\psi|H|\psi\rangle}{\langle\psi|\psi\rangle}. \tag{4.19}$$

If we insert a trial wave function instead of the true wave function, the energy we obtain will always be larger than the true energy. However, we can minimize E by better approximating the wave function. For example, if we choose a basis to expand the wave function, the condition for minimizing E in (4.19) ($\partial E/\partial c^* = 0$, where c being one expansion coefficient) readily gives a set of coupled equations, which can be solved for optimized expansion coefficients.

We now construct the wave functions for the unperturbed problem as a linear combination of each atomic functions $\chi(\mathbf{r}-\mathbf{R_n})$, labeled by quantum number j,

$$\psi^j_{\mathbf{k}}(\mathbf{r}) = \frac{1}{\sqrt{N}}\sum_{n=1}^{N} e^{i\mathbf{k}\cdot\mathbf{R_n}}\chi_j(\mathbf{r}-\mathbf{R_n}), \tag{4.20}$$

where N is the number of primitive cells in the crystal. The coefficient $e^{i\mathbf{k}\cdot\mathbf{R_n}}$ is needed to make the wave function a Bloch wave, where k is the wave vector in the first Brillouin zone whose values are determined by the periodic boundary conditions. This equation is just the sum of the electronic orbital $\chi_j(\mathbf{r}-\mathbf{R_n})$ of atom located at $\mathbf{R_n}$, and the plane wave factor is required by the translational symmetry of the crystal.

In a nondegenerate problem, we can substitute (4.20) alone into (4.19), and define $\mathbf{R_{nm}} = \mathbf{R_n} - \mathbf{R_m}$, then for the denominator we have

$$\begin{aligned}
\langle\psi^j_{\mathbf{k}}|\psi^j_{\mathbf{k}}\rangle &= \frac{1}{N}\sum_{n,m} e^{i\mathbf{k}\cdot(\mathbf{R_n}-\mathbf{R_m})}\int \chi^*_j(\mathbf{r}-\mathbf{R_m})\chi_j(\mathbf{r}-\mathbf{R_n})\mathrm{d}\mathbf{r} \\
&= \frac{1}{N}\sum_{n,m} e^{i\mathbf{k}\cdot\mathbf{R_{nm}}}\int \chi^*_j(\mathbf{r}-\mathbf{R_m})\chi_j(\mathbf{r}-\mathbf{R_m}-\mathbf{R_{nm}})\mathrm{d}\mathbf{r} \\
&= \frac{1}{N}\sum_{n,m} e^{i\mathbf{k}\cdot\mathbf{R_{nm}}}\int \chi^*_j(\mathbf{r}')\chi_j(\mathbf{r}'-\mathbf{R_{nm}})\mathrm{d}\mathbf{r}'.
\end{aligned} \tag{4.21}$$

By the approximation of nearly-localized electrons, the integral only has significant values when $\mathbf{R_{nm}} = 0$. Thus, to a first-order approximation, we may just retain terms with $n = m$, and thus we obtain

$$\langle\psi^j_{\mathbf{k}}|\psi^j_{\mathbf{k}}\rangle \simeq \frac{1}{N}\sum_{n=1}^{N}\int \chi^*_j(\mathbf{r})\chi_j(\mathbf{r})\mathrm{d}\mathbf{r} = 1, \tag{4.22}$$

Following the same procedure, and making use of the fact that the solutions to (4.15) are known, the nominator of (4.19) gives

$$\begin{aligned}
\langle\psi|H|\psi\rangle &\simeq \frac{1}{N}\sum_{n,m} e^{i\mathbf{k}\cdot(\mathbf{R_n}-\mathbf{R_m})}\int \chi^*_j(\mathbf{r}-\mathbf{R_m})(E_j + H'(\mathbf{r}-\mathbf{R_n}))\chi_j(\mathbf{r}-\mathbf{R_n})d\mathbf{r} \\
&= \frac{1}{N}\sum_{n,m} e^{i\mathbf{k}\cdot\mathbf{R_{nm}}}\int \chi^*_j(\mathbf{r}')(E_j + H'(\mathbf{r}'-\mathbf{R_{nm}}))\chi_j(\mathbf{r}'-\mathbf{R_{nm}})d\mathbf{r}'.
\end{aligned} \tag{4.23}$$

In the above equation, when neglecting the overlap integral of atomic orbitals on different sites, sum of E_j over all atomic sites gives NE_j. Integration of H' depends on the form of the interatomic interaction and the overlap of atomic orbitals on different atomic sites. Sometimes the overlap integral up to the nearest neighbors is good enough to give satisfactory results; sometimes one needs to include also the next-nearest neighbors. If we retain only the overlap integral from the nearest neighbors, the energy dispersion is

$$E(\mathbf{k}) = E_j - J_0 - J_1 \sum_{NN} e^{i\mathbf{k}\cdot\mathbf{R_{NN}}}, \tag{4.24}$$

where "NN" stands for "nearest-neighbor" and

$$J_0 = -\int \chi_j^*(\mathbf{r})H'(\mathbf{r})\chi_j(\mathbf{r})d\mathbf{r} \tag{4.25}$$

and

$$J_1 = -\int \chi_j^*(\mathbf{r})H'(\mathbf{r}-\mathbf{R})\chi_j(\mathbf{r}-\mathbf{R})d\mathbf{r}, \tag{4.26}$$

where R is the spacing between nearest neighbors.

From the earlier discussion, it can be seen that the atomic orbitals, $\chi_j(\mathbf{r})$, are not ideal for the purposes of analysis, as the orbitals on different atomic sites are not orthogonal to one another. Although we made approximations to assume that the overlap is small, but it is hard to quantify the effect on the band of neglecting the orbital overlap. Löwdin (Löwdin, 1951) provided a scheme for creating an orthogonal set from a nonorthogonal one, in such a way as to preserve the symmetry properties of the original set. Löwdin functions, which are defined as

$$\phi_j(\mathbf{r}-\mathbf{R_n}) = \sum_{i,\mathbf{R_m}} S^{-1/2}_{j\mathbf{R_n},i\mathbf{R_m}} \chi_{i,\mathbf{R_m}}, \tag{4.27}$$

where S is the overlap matrix, have a greater extent in space than atomic orbitals, implying that the Hamiltonian matrix will have elements significantly different from zero between atoms that are second- or third-nearest neighbors. The Bloch sums can then be formed from these Löwdin functions, which have the same symmetry of the corresponding atomic orbitals. Accordingly, the matrix elements of the atomic Hamiltonian between the Löwdin functions are not the same as the atomic energies. However, this is not important, since after all, the tight-biding Hamiltonian is formed in a parametrization scheme. Using the Löwdin functions to replace the atomic orbitals, the result in (4.22) is exact.

The above procedure gives one energy band whose wave function is solely characterized by the atomic orbital χ_j (or the Löwdin function ϕ_j. In the following we use only the Löwdin functions ϕ_j, but we still call them the atomic orbitals). This method may be appropriate for the calculation of band structures of alkaline metals, whose valence electrons are mostly s-electrons. But for the band structure of semiconductors, a single atomic orbital is not enough to describe both the conduction and valence bands. We need multiple atomic orbitals and construct the zeroth order of crystal electronic basis using them following (4.20), and then employ the linear superposition of these zeroth order basis as trial functions in (4.19) to find both the wave function and energy dispersions.

In the following we discuss two simple examples.

Band formed by s-orbital in the simple cubic lattice. The s-orbital is spherically symmetrical, then the overlap integral in every direction is the same, thus in (4.24) J_1 has the same value and it is

$$J_1 = -V_{ss\sigma}. \tag{4.28}$$

The s-orbital has the even parity, i.e., $\phi_s(\mathbf{r}) = \phi_s(-\mathbf{r})$. In (4.26), the overlap integral is positive if we assume the potential of an electron is zero when it is infinitely far away from the atomic core, since the perturbative potential H' is negative under this assumption.

The coordinates of the nearest neighbors for a simple cubic lattice are as follows:

$$\begin{array}{ccc} (\mathrm{a},0,0) & (0,\mathrm{a},0) & (0,0,\mathrm{a}) \\ (-\mathrm{a},0,0) & (0,-\mathrm{a},0) & (0,0,-\mathrm{a}). \end{array}$$

Substituting these coordinates of the nearest neighbors into 4.24, we obtain

$$E(k) = E_s - J_0 + 2V_{ss\sigma}(\cos k_x a + \cos k_y a + \cos k_z a). \tag{4.29}$$

The energy at the Γ, X, and R point is as follows:

$$\begin{aligned} \Gamma: \quad & k = (0,0,0), \\ & E^{\Gamma} = E_s - J_0 + 6V_{ss\sigma}. \end{aligned} \tag{4.30}$$

$$\begin{aligned} X: \quad & k = (\frac{\pi}{a},0,0), \\ & E^{X} = E_s - J_0 + 2V_{ss\sigma}. \end{aligned} \tag{4.31}$$

$$\begin{aligned} R: \quad & k = (\frac{\pi}{a},\frac{\pi}{a},\frac{\pi}{a}), \\ & E^{X} = E_s - J_0 - 6V_{ss\sigma}. \end{aligned} \tag{4.32}$$

Because $J_1 > 0$, thus the Γ and R points correspond to the band bottom and top, respectively. The band width is determined by J_1, which is in turn determined by the overlap integral between the nearest neighbors.

Let us assume a changes by δ, i.e., $a' = a(1+\delta)$. Following the d^{-2} principle, $V_{ss\sigma}^{'} = V_{ss\sigma}(1+\delta)^{-2}$. The bandwidth is then reduced by $(1+\delta)^{-2}$, and the inverse of the effective mass along one direction, for instance, [100], is

$$\frac{1}{m'^*} = \frac{1}{\hbar^2}\frac{\partial^2 E}{\partial k_x^2} = 2\frac{a'^2 V'_{ss\sigma}}{\hbar^2}\cos k_x a'. \tag{4.33}$$

Two points can be seen from this equation. First, the effective mass at $k_x = 0$ is inversely proportional to the bandwidth (multiples of $V_{ss\sigma}$). Second, when a changes (equivalent to hydrostatic strain), the band curvature does not change, since $a'^2 V'_{ss\sigma} = a^2 V_{ss\sigma}$.

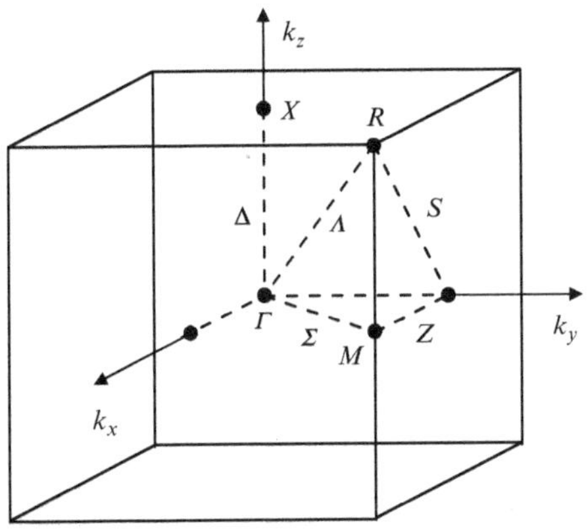

Fig. 4.8. Brillouin zone of the simple cubic lattice

Bands formed by p-orbitals in the simple cubic lattice. The atomic p-states is 3-fold degenerate. We may write the three p-orbitals as:

$$\phi_{p_x} = xf(r), \ \phi_{p_y} = yf(r), \ \phi_{p_z} = zf(r), \tag{4.34}$$

where $f(r)$ is a function of the radial distance only. It can be proven that each p-orbital forms a band, whose wave function is in form of (4.20), i.e.,

$$\psi_{\mathbf{k}}^{p_x}(\mathbf{r}) = \frac{1}{\sqrt{N}} \sum_n e^{i\mathbf{k}\cdot\mathbf{R_n}} \phi_{p_x}(\mathbf{r} - \mathbf{R_n}),$$
$$\psi_{\mathbf{k}}^{p_y}(\mathbf{r}) = \frac{1}{\sqrt{N}} \sum_n e^{i\mathbf{k}\cdot\mathbf{R_n}} \phi_{p_y}(\mathbf{r} - \mathbf{R_n}),$$
$$\psi_{\mathbf{k}}^{p_z}(\mathbf{r}) = \frac{1}{\sqrt{N}} \sum_n e^{i\mathbf{k}\cdot\mathbf{R_n}} \phi_{p_z}(\mathbf{r} - \mathbf{R_n}).$$

The band dispersion can still be expressed using (4.24), but we shall note that the nearest-neighbor overlap integral is not the same for different orbitals. Take the ϕ_{p_x} orbital for instance. The probability distribution of electron mainly concentrates along the x-axis, so among the overlap integral to its six nearest neighbors, the integrals with $(a, 0, 0)$ and $(-a, 0, 0)$ are larger (we may call this the σ-bonding), which we define as J_1

$$J_1 = -V_{pp\sigma}. \tag{4.35}$$

The other four near-neighbor overlap integrals (we may call this the π-bonding) are smaller and equal to each other, which we define as J_2

$$J_2 = -V_{pp\pi}. \tag{4.36}$$

The overlap integrals $V_{ss\sigma}$, $V_{sp\sigma}$, $V_{pp\sigma}$, and $V_{pp\pi}$ are illustrated in Fig. 4.4. Thus,

$$E_k^{p_x} = E_p - J_0 + 2V_{pp\sigma} \cos k_x a + 2V_{pp\pi}(\cos k_y a + \cos k_z a). \tag{4.37}$$

Identically,

$$E_k^{p_y} = E_p - J_0 + 2V_{pp\sigma} \cos k_y a + 2V_{pp\pi}(\cos k_z a + \cos k_x a), \tag{4.38}$$

and

$$E_k^{p_z} = E_p - J_0 + 2V_{pp\sigma} \cos k_z a + 2V_{pp\pi}(\cos k_x a + \cos k_y a). \tag{4.39}$$

The p-orbitals have the odd parity, thus the σ-type integral $J_1 < 0$ and the π-type integral $J_2 > 0$. According to (4.29), (4.37)–(4.39), the bands formed by s- and p-orbitals are plotted in Fig. 4.9 along the ΓX direction (Δ-axis). The bottom curve is the band formed by the s-orbital, and the curve in the middle is the band formed by p_x-orbital, and the top curve is the bands formed by p_y- and p_z-orbitals, which along the [100] direction are degenerate. The labels Γ_1 and Γ_{15} denote the symmetry group the bands belong to.

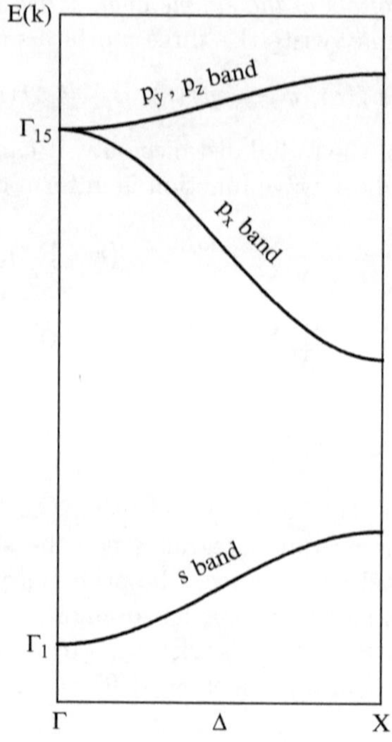

Fig. 4.9. The s- and p-bands in the simple cubic lattice

4.4.2 The sp^3 Tight-Binding Model

In the above we have discussed the simplest situation in the tight-binding method, i.e., one atomic energy level corresponds to one energy band in the solids. For core electrons, due to their localized character, the wave function overlap is small, and thus the band width is very narrow, so that the atomic energy levels and the bands in solids may have simple correspondence. However, sometimes there does not exist the simple correspondence. The atomic orbitals may mix when forming crystals, thus there usually exists interaction between different orbitals. The band formed by a single orbital as calculated above is only justifiable when the interorbital interaction is much smaller than the energy difference between atomic levels. Therefore, for the valence electrons whose wave function is more delocalized, the single orbital, nondegenerate procedure is not adequate to calculate the band structure. However, we may consider that the bands of interest are formed by several energetically close atomic orbitals, neglecting the other remote orbitals. For example, the valence and conduction bands of C, Si, Ge, etc., group IV, and InAs, GaAs, etc., III–V semiconductors may be considered formed by s and p orbitals. As discussed in Chap. 3, group IV and III–V semiconductors are generally tetrahedral crystals, which have complex lattice structure, i.e., there are two atoms in one primitive cell. The two atoms are not equivalent even in group IV semiconductors, so the atomic orbitals contributed by each atom need to be taken into account. Generally, when there are l atoms in the primitive cell, and their coordinates are

$$\mathbf{R}_\mathbf{m} + \mathbf{r}_\alpha, \alpha = 1, 2, 3, \ldots l, \tag{4.40}$$

where $\mathbf{r}$ represents the position of one atom inside the primitive cell, the Bloch sum for atomic orbital ϕ_j is constructed as

$$\psi_\mathbf{k}^{\alpha,j} = \frac{1}{\sqrt{N}} \sum_m e^{i\mathbf{k}\cdot\mathbf{R}_\mathbf{m}} \phi_j(\mathbf{r} - \mathbf{R}_\mathbf{m} - \mathbf{r}_\alpha). \tag{4.41}$$

In a multiple-orbital tight-binding method, the electronic wave functions can be considered as a linear superposition of these Bloch sums:

$$\Psi_\mathbf{k} = \sum_{m,\alpha} c_{m,\alpha} \psi_\mathbf{k}^{\alpha,j}. \tag{4.42}$$

Substituting this equation to (4.16), and multiplying $\Psi_\mathbf{k}^*$ from the left and integrate over space, it yields the coupled equations for the expansion coefficients $c_{m,\alpha}$:

$$\sum_{m,\alpha} \left(H_{ml,m'\alpha'} - E(\mathbf{k})\delta_{mm'}\delta_{\alpha\alpha'}\right) c_{m'\alpha'}(\mathbf{k}) = 0, \tag{4.43}$$

where the δ-functions enter by neglecting the orbital overlap on different atomic sites as has been done in (4.22). The energy dispersion and electronic wave function can be obtained through diagonalizing the Hamiltonian $H_{ml,m'\alpha'}$.

In tetrahedral semiconductors, we label the two types of atoms as type "a" (anion) and "c" (cation), and in the primitive cell, they are displaced by

$\mathbf{d} = \frac{1}{4}(a, a, a)$, where a is the lattice constant. If we choose one a atom as the origin, then its four nearest neighbors are all type c, located at

$$\begin{aligned}
\mathbf{d}_1 &= [111]a/4, \\
\mathbf{d}_2 &= [1\bar{1}\bar{1}]a/4, \\
\mathbf{d}_3 &= [\bar{1}1\bar{1}]a/4, \\
\mathbf{d}_4 &= [\bar{1}\bar{1}1]a/4.
\end{aligned} \tag{4.44}$$

So if we consider one s-orbital and three p-orbitals from each atom in the primitive cell, then there are eight Bloch sums, namely,

$$\begin{aligned}
\psi_{\mathbf{k}}^{s^a} &= \frac{1}{\sqrt{N}} \sum_m e^{i\mathbf{k}\cdot\mathbf{R_m}} \phi_{s^a}(\mathbf{r} - \mathbf{R_m}), \\
\psi_{\mathbf{k}}^{p_x^a} &= \frac{1}{\sqrt{N}} \sum_m e^{i\mathbf{k}\cdot\mathbf{R_m}} \phi_{p_x^a}(\mathbf{r} - \mathbf{R_m}), \\
\psi_{\mathbf{k}}^{p_y^a} &= \frac{1}{\sqrt{N}} \sum_m e^{i\mathbf{k}\cdot\mathbf{R_m}} \phi_{p_y^a}(\mathbf{r} - \mathbf{R_m}), \\
\psi_{\mathbf{k}}^{p_z^a} &= \frac{1}{\sqrt{N}} \sum_m e^{i\mathbf{k}\cdot\mathbf{R_m}} \phi_{p_z^a}(\mathbf{r} - \mathbf{R_m}), \\
\psi_{\mathbf{k}}^{s^c} &= \frac{1}{\sqrt{N}} \sum_m e^{i\mathbf{k}\cdot\mathbf{R_m}} \phi_{s^c}(\mathbf{r} - \mathbf{R_m} - \mathbf{d}), \\
\psi_{\mathbf{k}}^{p_x^c} &= \frac{1}{\sqrt{N}} \sum_m e^{i\mathbf{k}\cdot\mathbf{R_m}} \phi_{p_x^c}(\mathbf{r} - \mathbf{R_m} - \mathbf{d}), \\
\psi_{\mathbf{k}}^{p_y^a} &= \frac{1}{\sqrt{N}} \sum_m e^{i\mathbf{k}\cdot\mathbf{R_m}} \phi_{p_y^c}(\mathbf{r} - \mathbf{R_m} - \mathbf{d}), \\
\psi_{\mathbf{k}}^{p_z^c} &= \frac{1}{\sqrt{N}} \sum_m e^{i\mathbf{k}\cdot\mathbf{R_m}} \phi_{p_z^c}(\mathbf{r} - \mathbf{R_m} - \mathbf{d}).
\end{aligned} \tag{4.45}$$

We now use these eight Bloch sums as the basis to construct the Hamiltonian matrix. Consider first $H_{\alpha\alpha}$, the matrix elements on the same site for the same Bloch sums. Equation (4.23) gives $E_j - J_0$. Note this is not just the energy of one atomic level, but also includes the contribution of the crystal potential. For anion s-orbital, we write it as E_s^a. Similarly, there are also E_s^c, E_p^a, and E_p^c, representing the on-site atomic energy for cation s-orbital, anion p-orbital and cation p-orbital. Since we only consider the overlap integral up to the nearest neighbors, there are no off-diagonal elements between orbitals, or Bloch sums on same type of atoms, because different orbitals on the same atomic site are orthogonal. Next, we consider $H_{\alpha\beta}$, the off-diagonal matrix elements for which in (4.23), the bra and ket Bloch functions are not the same and are on different types of atoms. For $H_{s^a s^c}$, because the s-orbitals are spherically symmetrical, the overlap integral is defined as

$$V_{ss\sigma} = \langle \phi_{s^a} | H | \phi_{s^c} \rangle. \tag{4.46}$$

Thus, each of the NNs enter the matrix element $V_{ss\sigma}$ with a phase factor $e^{i\mathbf{k}\cdot\mathbf{d}}$, where $\mathbf{d}$ is given in (4.44). For $H_{s^a p_x^c}$, due to the spherical symmetry of the s-orbital, first we define

$$V_{sp\sigma} = \langle \phi_{s^a} | H | \phi_{p_x^c} \rangle \tag{4.47}$$

for the case that the p_x-orbital symmetry axis pointing to s. However, because the p_x-orbital on the nearest neighbors is oriented along the x-axis, and its orientation is not perpendicular to the $\langle 111 \rangle$ direction, we need to resolve the p_x-orbital into components along and perpendicular to the $\langle 111 \rangle$ direction. The angular decomposition of the overlap integrals is shown in Fig. 4.4. Thus, we have the σ-type overlap but the π-type overlap vanishes. Therefore, in tetrahedral semiconductors, one s-orbital interacts with the p-orbitals by interaction constant $\pm V_{sp\sigma}/\sqrt{3}$. The factor $1/\sqrt{3}$ is the magnitude of the directional cosine of the NNs with respect to the origin. The sign of the interaction constants depends on whether the positive or negative lobe of the p-orbital points to the central s-orbital. For matrix elements between p-orbitals, the general situation for one atom located at the origin and the other atom located at $d(\cos\theta_x, \cos\theta_y, \cos\theta_z)$, where d is the distance between the two p-orbitals, and $\cos\theta_x$, $\cos\theta_y$, and $\cos\theta_z$ are the directional cosines of the second p-orbital, the interaction constants are:

$$\begin{aligned} V_{xx} &= V_{pp\sigma} \cos^2\theta_x + V_{pp\pi} \cos^2\theta_x, \\ V_{xy} &= (V_{pp\sigma} - V_{pp\pi}) \cos\theta_x \cos\theta_y, \end{aligned} \tag{4.48}$$

where $V_{pp\sigma}$ and $V_{pp\pi}$ are defined similarly as $V_{ss\sigma}$ and $V_{sp\sigma}$. Thus, we have the following interaction constants defined as (Harrison, 1989):

$$\begin{aligned} V_{ss} &= V_{ss\sigma}, \\ V_{sp} &= -V_{sp\sigma}/\sqrt{3}, \\ V_{xx} &= 1/3V_{pp\sigma} + 2/3V_{pp\pi}, \\ V_{xy} &= 1/3V_{pp\sigma} - 1/3V_{pp\pi}. \end{aligned} \tag{4.49}$$

The four sums of phase factors are given by (Harrison, 1989)

$$\begin{aligned} g_0(\mathbf{k}) &= e^{i\mathbf{k}\cdot\mathbf{d}_1} + e^{i\mathbf{k}\cdot\mathbf{d}_2} + e^{i\mathbf{k}\cdot\mathbf{d}_3} + e^{i\mathbf{k}\cdot\mathbf{d}_4}, \\ g_1(\mathbf{k}) &= e^{i\mathbf{k}\cdot\mathbf{d}_1} + e^{i\mathbf{k}\cdot\mathbf{d}_2} - e^{i\mathbf{k}\cdot\mathbf{d}_3} - e^{i\mathbf{k}\cdot\mathbf{d}_4}, \\ g_2(\mathbf{k}) &= e^{i\mathbf{k}\cdot\mathbf{d}_1} - e^{i\mathbf{k}\cdot\mathbf{d}_2} + e^{i\mathbf{k}\cdot\mathbf{d}_3} - e^{i\mathbf{k}\cdot\mathbf{d}_4}, \\ g_3(\mathbf{k}) &= e^{i\mathbf{k}\cdot\mathbf{d}_1} - e^{i\mathbf{k}\cdot\mathbf{d}_2} - e^{i\mathbf{k}\cdot\mathbf{d}_3} + e^{i\mathbf{k}\cdot\mathbf{d}_4}. \end{aligned} \tag{4.50}$$

Finally the sp^3 Hamiltonian is obtained in Table 4.1. For group IV semiconductors, apparently $E_{s^c} = E_{s^a}$, $E_{p^c} = E_{p^a}$, and $V_{s^c p} = V_{s^a p}$. The on-site energy E_{s^c}, E_{s^a}, E_{p^c}, and E_{p^a}, and interaction constants $V_{ss\sigma}$, $V_{sp\sigma}$, $V_{pp\sigma}$, and $V_{pp\pi}$ are generally fitting parameters obtained by comparing the tight-binding band structures with those obtained by the other methods such as pseudopotential method. These parameters are given in Table 4.2 (Harrison, 1989).

Table 4.1. The sp^3 tight-binding Hamiltonian with two-center integral and nearest-neighbor approximations for tetrahedral semiconductors

	s^c	s^a	p_z^c	p_z^a	p_x^c	p_x^a	p_y^c	p_y^a
s^c	E_{s^c}	$V_{ss}g_0$	0	$V_{s^cp}g_3$	0	$V_{s^cp}g_1$	0	$V_{s^cp}g_2$
s^a	$V_{ss}g_0^*$	E_{s^a}	$-V_{s^ap}g_3^*$	0	$-V_{s^ap}g_1^*$	0	$-V_{s^ap}g_2^*$	0
p_z^c	0	$-V_{s^ap}g_3$	E_{p^c}	$V_{xx}g_0$	0	$V_{xy}g_2$	0	$V_{xy}g_1$
p_z^a	$V_{s^cp}g_3^*$	0	$V_{xx}g_0^*$	E_{p^a}	$V_{xy}g_2^*$	0	$V_{xy}g_1^*$	0
p_x^c	0	$-V_{s^ap}g_1$	0	$V_{xy}g_2$	E_{p^c}	$V_{xx}g_0$	0	$V_{xy}g_3$
p_x^a	$V_{s^cp}g_1^*$	0	$V_{xy}g_2^*$	0	$V_{xx}g_0^*$	E_{p^a}	$V_{xy}g_3^*$	0
p_y^c	0	$-V_{s^ap}g_2$	0	$V_{xy}g_1$	0	$V_{xy}g_3$	E_{p^c}	$V_{xx}g_0$
p_y^a	$V_{s^cp}g_2^*$	0	$V_{xy}g_1^*$	0	$V_{xy}g_3^*$	0	$V_{xx}g_0^*$	E_{p^a}

Table 4.2. The sp^3 tight-binding band parameters for some tetrahedral semiconductors. All values are in eV. Reproduced from Harrison (Harrison, 1989)

Band parameters	C	Si	Ge	GaAs	ZnSe
$E_p - E_s$	6.8	7.20	4.80	(6.44, 9.64)	(5.88, 12.4)
$E_{p^c} - E_{p^a}$	0	0	0	0.96	3.29
$-V_{ss\sigma}$	5.55	2.03	1.70	1.70	1.54
$V_{sp\sigma}$	5.91	2.55	2.30	(2.4, 1.9)	(2.6, 1.4)
$V_{pp\sigma}$	7.78	4.55	4.07	3.44	3.20
$-V_{pp\pi}$	2.50	1.09	1.05	0.89	0.92

4.4.3 Tight-Binding Band Structure

Electronic structure can be obtained by diagonalizing the sp^3 Hamiltonian as in Table 4.1. For a general k point in the Brillouin zone, the energy can be obtained only through numerical calculation. At high symmetry points or along high symmetry lines, energy dispersion may be obtained analytically. The Γ point is of special importance, since it is the point that contains the band degeneracy information, and the wave functions at the Γ point also shed light on the wave functions at k points away from the zone center.

The Γ point. For the Γ point, $k = 0$. From (4.50), $g_0 = 4$, and $g_1 = g_2 = g_3 = 0$. Then all the off-diagonal matrix elements in Hamiltonian Table 4.1 are zero except those coupling s^c with s^a, those coupling p_x^c with p_x^a, and so on. Thus, the Hamiltonian decouples into four 2×2 blocks each of which contains only s-, p_x-, p_y, and p_z-orbitals. The upper one block gives

$$E = \frac{E_{s^c} + E_{s^a}}{2} \pm \sqrt{\left(\frac{E_{s^c} - E_{s^a}}{2}\right)^2 + (4V_{ss})^2}, \tag{4.51}$$

which becomes $E = E_s \pm 4V_{xx}$ for group IV elementary semiconductors, for which the wave functions are

$$\psi_\Gamma = \frac{1}{\sqrt{2}}(\psi_\Gamma^{s^a} \pm \psi_\Gamma^{s^c}), \tag{4.52}$$

where $\psi_\Gamma^{s^a}$ and $\psi_\Gamma^{s^c}$ are the Bloch sums defined in (4.45). For zinc-blende semiconductors, only the coefficients of the two s Bloch sums are different, due to different on-site energies. These two singly degenerate bands are the s bonding ("−" sign in energy and "+" sign in wave function) and antibonding ("+" sign in energy and "−" sign in wave function) states. The s bonding state lies in the very bottom of the energy band belonging to the Γ_1 symmetry and the antibonding state generally gives the conduction band in zinc-blende semiconductors belonging to the Γ_2' symmetry. The lower three blocks give the same results,

$$E = \frac{E_{p^c} + E_{p^a}}{2} \pm \sqrt{\left(\frac{E_{p^c} - E_{p^a}}{2}\right)^2 + (4V_{xx})^2}. \tag{4.53}$$

For either the upper sign or the lower sign, this equation gives triply degenerate p bands, i.e., three bands formed by the p_x, p_y, and p_z orbital respectively. The antibonding p states correspond to the "+" sign and have higher energies, and the bonding p states correspond to the "−" sign and have lower energies. Because of the odd parity of the p orbitals, the wave function for p bonding states is in the form $\frac{1}{\sqrt{2}}(\psi_\Gamma^{p^c} - \psi_\Gamma^{p^a})$, i.e., the bonding states are symmetrical with respect to the middle point of bond, which connects one atom to its NNs. The eigenenergies and wave functions showed above illustrate that the Hamiltonian diagonalizes at the Γ point if we choose the bonding and antibonding states as basis. In other words, we may think that the atomic orbitals first mix to compose the bonding and antibonding states, then starting from this, they form the electronic band structures. In both group IV and zinc-blende semiconductors, the s and p bonding states form the valence bands, and the valence band edge is the p-states. To simplify the discussion, we only concentrate on the group IV semiconductors, which gives the same physics with less complications.

The $\langle 001\rangle$ *direction.* Along the [001] direction, $k_x = k_y = 0$, $g_0 = 4\cos(k_z a/4)$, $g_3 = i4\sin(k_z a/4)$, and $g_1 = g_2 = 0$. The Hamiltonian decouples into two block matrices. Along the [001] direction, the s-orbitals couple with z-orbitals, and the x-orbitals couple with y-orbitals. The x- and y-orbitals do not couple with z- or s-orbitals. The $\langle 001\rangle$ is a principal crystal axis with C_{4v} symmetry, and the p_x, p_y, and p_z are the three symmetrized orbitals with respect to the $\langle 001\rangle$ axis, we then can transform the single orbital basis into s and p bonding and antibonding states as follows:

$$|s_+\rangle = \frac{1}{\sqrt{2}}(\psi_{\mathbf{k}}^{s^c} + \psi_{\mathbf{k}}^{s^a}),\ |s_-\rangle = \frac{1}{\sqrt{2}}(\psi_{\mathbf{k}}^{s^c} - \psi_{\mathbf{k}}^{s^a}), \tag{4.54}$$

and

$$|p_-\rangle = \frac{1}{\sqrt{2}}(\psi_{\mathbf{k}}^{p^c} - \psi_{\mathbf{k}}^{p^a}),\ |p_+\rangle = \frac{1}{\sqrt{2}}(\psi_{\mathbf{k}}^{p^c} + \psi_{\mathbf{k}}^{p^a}),\ p = p_x, p_y, p_z, \tag{4.55}$$

where $|s_+\rangle$ and $|p_-\rangle$ correspond to the bonding states and $|s_-\rangle$ and $|p_+\rangle$ correspond to the antibonding states. Note that the bonding and antibonding states are also symmetrical and antisymmetrical states with respect to interexchange of the two atoms in the primitive cell. Under this transformation, defining

$$E_s = (E_{s^c} + E_{s^a})/2, \tag{4.56a}$$
$$E_p = (E_{p^c} + E_{p^a})/2, \tag{4.56b}$$
$$V_{sp} = (V_{s^cp} + V_{s^ap})/2, \tag{4.56c}$$

and

$$\Delta E_s = (E_{s^c} - E_{s^a})/2, \tag{4.57a}$$
$$\Delta E_p = (E_{p^c} - E_{p^a})/2, \tag{4.57b}$$
$$\Delta V_{sp} = (V_{s^cp} - V_{s^ap})/2, \tag{4.57c}$$

the Hamiltonian becomes as shown in Table 4.3. For group IV semiconductors such as Si and Ge, $\Delta E_s = \Delta E_p = \Delta V_{sp} = 0$, and the Hamiltonian is further block-diagonalized into four blocks. The energy dispersion can be analytically obtained, and here we give the results by Chadi and Cohen (Chadi and Cohen, 1975) while neglecting the next-nearest-neighbor interaction:

- The singly degenerate bands:

$$\begin{aligned} E(k) &= \frac{1}{2}(E_s + E_p) \pm \frac{1}{2}(V_{ss} + V_{xx}) \cos \frac{ak}{4} \\ &\pm \frac{1}{2}\sqrt{\left(E_p - E_s \pm (V_{ss} - V_{xx}) \cos \frac{ak}{4}\right)^2 + \left(2V_{sp} \sin \frac{ak}{4}\right)^2}, \end{aligned} \tag{4.58}$$

  which gives the energy dispersions for the two s bands (Γ_1 and Γ_2') and two z bands, one of which is the LH band.
- The doubly degenerate bands:

$$E(k) = E_p \pm \sqrt{\left(V_{xx} \cos \frac{ak}{4}\right)^2 + \left(V_{xy} \sin \frac{ak}{4}\right)^2}, \tag{4.59}$$

  which include the HH valence bands and the doubly degenerate p-type conduction bands.

The three p states along the three principal axes do not mix at the Γ point. The HH bands are composed of the p_x- and p_y-orbitals. The LH bands are composed of the p_z-orbital, which couples with the s-states at finite k. Thus along [001], the HH states are oriented along the x and y directions, and the LH state is oriented along the z direction. The trend of the HH state oriented perpendicular to and the LH state being parallel to the diagonalization axis is due to the fact that the symmetry couples only the z-orbital with the s-orbital along [001], and the σ-type bonding of s-p_z and s-s is very strong.

In Si, the s bonding and antibonding splitting are very large compared to both s-p splitting and the p bonding and antibonding splitting, so the Si

Table 4.3. The sp^3 tight-binding Hamiltonian for tetrahedral semiconductors block-diagonalized along [001]

	s_+	p_{z+}	s_-	p_{z-}	p_{x+}	p_{y-}	p_{y+}	p_{x-}
s_+	$E_s + V_{ss}g_0$	$V_{sp}g_3$	ΔE_s	$\Delta V_{sp}g_3$	0	0	0	0
p_{z+}	$-V_{sp}g_3$	$E_p + V_{xx}g_0$	$\Delta V_{sp}g_3$	ΔE_p	0	0	0	0
s_-	ΔE_s	$-\Delta V_{sp}g_3$	$E_s - V_{ss}g_0$	$V_{sp}g_3$	0	0	0	0
p_{z-}	$-\Delta V_{sp}g_3$	ΔE_p	$-V_{sp}g_3$	$E_p - V_{xx}g_0$	0	0	0	0
p_{x+}	0	0	0	0	$E_p + V_{xx}g_0$	$-V_{xy}g_3$	0	ΔE_p
p_{y-}	0	0	0	0	$V_{xy}g_3$	$E_p - V_{xx}g_0$	ΔE_p	0
p_{y+}	0	0	0	0	0	ΔE_p	$E_p + V_{xx}g_0$	$-V_{xy}g_3$
p_{x-}	0	0	0	0	ΔE_p	0	$V_{xy}g_3$	$E_p - V_{xx}g_0$

conduction band is primarily p antibonding states, located at the $0.85\frac{2\pi}{a}$ along the Δ axes (which is not successfully reproduced by the sp^3 tight-binding model). While in GaAs, the s bonding and antibonding splitting are not as large, so the conduction band is composed of the s antibonding states, located at the Γ point. This situation is shown in Fig. 4.10.

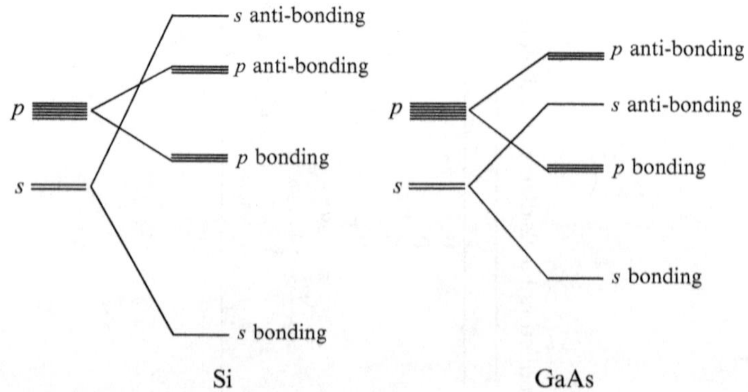

Fig. 4.10. Band characteristic determined by interplay between bonding and antibonding splitting and s-p splitting. In Si, the conduction and valence bands are both composed of p states; in GaAs, the valence bands are composed of the p states while the conduction band is composed of s state

The Hamiltonian along the other symmetry axes can also be greatly simplified by constructing the bonding and antibonding states, but with respect to those specific axes, e.g., the $\langle 110\rangle$ and $\langle 111\rangle$ axes.

The $\langle 111\rangle$ *direction.* For example, along the $\langle 111\rangle$ direction, due to the C_{3v} rotation symmetry that transforms the three p orbitals into each other, first we construct three new symmetrized orbitals according to the trigonal symmetry for the p^c orbitals as following:

$$\begin{aligned}|X_1\rangle &= [|p_x^c\rangle - |p_y^c\rangle]/\sqrt{2},\\ |Y_1\rangle &= [|p_x^c\rangle + |p_y^c\rangle - 2|p_z^c\rangle]/\sqrt{6},\\ |Z_1\rangle &= [|p_x^c\rangle + |p_y^c\rangle + |p_z^c\rangle]/\sqrt{3}.\end{aligned} \tag{4.60}$$

Similar symmetrized orbitals, $|X_2\rangle$, $|Y_2\rangle$, and $|Z_2\rangle$ are also defined for the p^a orbitals. The symmetrized orbitals along $\langle 001\rangle$, $\langle 110\rangle$, and $\langle 111\rangle$ are shown in Fig. 4.11, where we can see that the symmetrized orbitals are orthogonal to each other and one is along the quantization axis.

Next we construct the bonding and antibonding orbitals based on the symmetrized orbitals following exactly (4.55) while replacing the p orbitals using corresponding X, Y, and Z orbitals defined above, and we defined these bonding and antibonding orbitals as

$$\begin{aligned}&X_+=(X_1+X_2)/\sqrt{2},\quad Y_+=(Y_1+Y_2)/\sqrt{2},\quad Z_+=(Z_1-Z_2)/\sqrt{2},\\ &X_-=(X_1-X_2)/\sqrt{2},\quad Y_-=(Y_1-Y_2)/\sqrt{2},\quad Z_-=(Z_1-Z_2)/\sqrt{2}.\end{aligned} \tag{4.61}$$

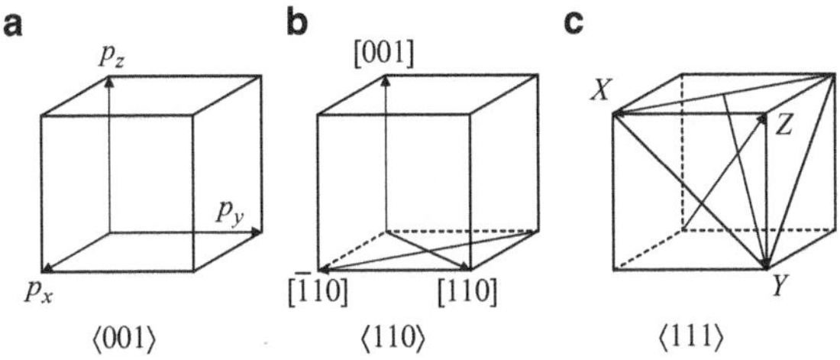

Fig. 4.11. Symmetrized orbitals for diagonalization direction along (**a**) ⟨100⟩, (**b**) ⟨110⟩, and (**c**) ⟨111⟩

Using these six orbitals as basis, the 6×6 valence band Hamiltonian is block-diagonalized into three 2×2 blocks. The block between X_+ and X_- is the same as the block between Y_+ and Y_-, given by

	X_+	X_-
X_+	$E_p + V_{xx}\Re(g_0) - V_{xy}\Re(g_1)$	$-\mathrm{i}V_{xx}\Im(g_0) + \mathrm{i}V_{xy}\Im(g_1)$
X_-	$\mathrm{i}V_{xx}\Im(g_0) - \mathrm{i}V_{xy}\Im(g_1)$	$E_p - V_{xx}\Re(g_0) + V_{xy}\Re(g_1)$

(4.62)

where along ⟨111⟩ direction $\mathbf{k} = 1/\sqrt{3}(k,k,k))$, $g_0 = 4\cos^3(\frac{ak}{4\sqrt{3}}) - 4\mathrm{i}\sin^3(\frac{ak}{4\sqrt{3}})$, and $g_1 = g_2 = g_3 = -4\cos(\frac{ak}{4\sqrt{3}})\sin^2(\frac{ak}{4\sqrt{3}}) - 4\mathrm{i}\sin(\frac{ak}{4\sqrt{3}})\cos^2(\frac{ak}{4\sqrt{3}})$, and $\Re(g)$ and $\Im(g)$ represent the real and imaginary parts of the g phase factors. The block matrix between Z_+ and Z_- orbitals is

	Z_+	Z_-
Z_+	$E_p + V_{xx}\Re(g_0) + 2V_{xy}\Re(g_1)$	$-\mathrm{i}V_{xx}\Im(g_0) - 2\mathrm{i}V_{xy}\Im(g_1)$
Z_-	$\mathrm{i}V_{xx}\Im(g_0) + 2\mathrm{i}V_{xy}\Im(g_1)$	$E_p - V_{xx}\Re(g_0) - 2V_{xy}\Re(g_1)$

(4.63)

At the Γ point, the bonding and antibonding states are also decoupled. Similar to the ⟨001⟩ direction, doubly degenerate bands composed of X_- and Y_- orbitals are the two HH bands, and the singly degenerate band composed of Z_- orbital is the LH band.

The sp^3 band structure of Si is illustrated in Fig. 4.12, compared to Si band structure calculated by the empirical nonlocal pseudopotential method (Chelikowsky and Cohen, 1976). The overall valence band structure coincides very well. This is not surprising, since the valence electronic waves are fairly localized. The sp^3 model cannot successfully reproduce the Si conduction band, since this band is well known also consisting of the 4s and 3d features. For successful modeling of the Si conduction band, more orbitals need to be included in the basis set. Vogl *et al.* (Vogl et al, 1983) proposed an sp^3s^* tight-binding model, which includes a model s^* orbital that has the s symmetry but takes into account the interaction from both the 4s and 3d orbitals. This model successfully reproduces the Si conduction band, giving rise to six conduction band valleys at the Δ-axes. Meanwhile, the valence bands are not significantly affected.

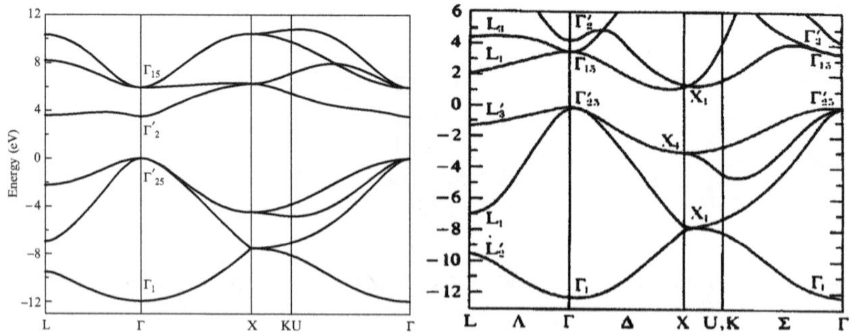

Fig. 4.12. Band structures for Si calculated by sp^3 tight-binding method (left) and pseudo-potential method (right, after Chelikowsky and Cohen (1976))

4.4.4 The sp^3 Hybridization and Bond Orbital Approximation

If we compute the valence electron distribution probability in Si using the wave functions we obtained in the earlier band structure calculations, we will find that the highest electron density is at the mid point of the lines connecting the neighboring Si atoms, i.e., valence electrons are mainly located along the bonds. The band structure of the tetrahedral crystals may be understood in terms of the bonds. For this, we first form the sp^3 hybrids. The hybrid is the real state of existence of the s and p orbitals in the tetrahedral crystal, with one s electron promoted to the p state. Actually, for each type of atoms, the sp^3 hybrids are just a unitary transformation of the one s and three p orbitals. We rewrite the definition of the sp^3 hybrids here again:

$$\begin{aligned}|h_1\rangle &= [|s\rangle + |p_x\rangle + |p_y\rangle + |p_z\rangle]/2,\\ |h_2\rangle &= [|s\rangle + |p_x\rangle - |p_y\rangle - |p_z\rangle]/2,\\ |h_3\rangle &= [|s\rangle - |p_x\rangle + |p_y\rangle - |p_z\rangle]/2,\\ |h_4\rangle &= [|s\rangle - |p_x\rangle - |p_y\rangle + |p_z\rangle]/2. \end{aligned} \tag{4.64}$$

They are oriented along [111], $[1\bar{1}\bar{1}]$, $[\bar{1}1\bar{1}]$, and $[\bar{1}\bar{1}1]$, respectively. The superposition of any two p orbitals is still a p orbital, but directed along a different direction. Thus, all these four sp^3 hybrids may also be written as:

$$|h\rangle = [|s\rangle + \sqrt{3}|p\rangle]/2, \tag{4.65}$$

but directed along four different $\langle 111\rangle$ directions. Because the p orbitals are polarized with one lobe being positive and one lobe being negative and the s orbital is homopolar, they will add up at the positive side of the p orbital, and cancel out at the negative side, thus the sp^3 hybrids lean toward the positive side. It is easy to verify that the four hybrids are orthogonal to each other. The sp^3 hybrids are formed on both atom types. Expanding the electronic states for the crystal in terms of these hybrids on the atoms is entirely equivalent to expanding them in terms of the atomic states that constitute the hybrids.

The hybrids on the neighboring atoms (say, $|h_1\rangle$ and $|h_1'\rangle$) leaning toward the same bond have a very strong coupling. Using (4.65) it is readily given by

$$V_2 = -(V_{ss\sigma} - 2\sqrt{3}V_{sp\sigma} - 3V_{pp\sigma})/4. \tag{4.66}$$

It is negative and called the covalent energy. The is the reason why the covalent bonding of the tetrahedral crystals is so strong.

Therefore, we may construct the bond orbitals as

$$|\psi\rangle = [|h\rangle + |h'\rangle]/\sqrt{2} \tag{4.67}$$

for homopolar semiconductors such as Si and Ge. For zinc-blende semiconductors, the coefficients before each hybrid are not the same, $1/\sqrt{2}$, but with a larger number before the cation hybrid and a smaller number before the anion hybrid. Similarly, the antibond orbitals can also be constructed with both hybrids pointing outward of the bond. Thus, out of the eight s and p atomic orbitals, we can construct four bond orbitals and four antibond orbitals. The bond energy E_b is evaluated for homopolar semiconductors as

$$E_b = (E_s + 3E_p)/4 - V_2, \tag{4.68}$$

and the antibond energy is

$$E_a = (E_s + 3E_p)/4 + V_2. \tag{4.69}$$

The coupling between the hybrid $|h_1'\rangle$ one the second atom and the hybrids $|h_2\rangle$, $|h_3\rangle$, and $|h_4\rangle$ on the first atom is given by

$$V_1^x = -(V_{ss\sigma} - 2\sqrt{3}/3V_{sp\sigma} + V_{pp\sigma})/4, \tag{4.70}$$

which can be verified to be much smaller than the covalent energy V_2. The other important coupling is that between the two hybrids on the same atom, which is called the metallic energy and given by

$$V_1 = (E_p - E_s)/4. \tag{4.71}$$

V_1 is much larger than V_1^x and is the primary coupling between the bonds. It can also be readily seen that the primary coupling between a bond orbital and each neighboring antibond orbital is also given by V_1.

Using the bonds and antibonds as basis, the Hamiltonian can be illustratively written as (Harrison, 1999)

$$\begin{matrix} E_b & V & V & V & 0 & V' & V' & V' \\ V & E_b & V & V & V' & 0 & V' & V' \\ V & V & E_b & V & V' & V' & 0 & V' \\ V & V & V & E_b & V' & V' & V' & 0 \\ 0 & V' & V' & V' & E_a & V & V & V \\ V' & 0 & V' & V' & V & E_a & V & V \\ V' & V' & 0 & V' & V & V & E_a & V \\ V' & V' & V' & 0 & V & V & V & E_a. \end{matrix} \tag{4.72}$$

We used a V to represent the matrix element between any pair of bonds or between the antibonds (they may be different and also include the phase factors) and used a V' to represent the matrix element between a pair of bond orbital and neighboring antibond orbital. The bond orbital approximation totally neglects the coupling between the bonds and antibonds, since the splitting between the bonds and antibonds is much larger than the bond-antibond coupling. For Si, the total probability density on the bond is 96%, and that on the neighboring antibonds is only about 4% (Harrison, 1989). The matrix then decouples into two 4×4 blocks, being the Hamiltonian matrices only between bond orbitals and antibond orbitals. The crystal band structure can then be considered evolved from these bonds and antibonds as shown in Fig. 4.1.

The bond orbitals form the valence bands, including the Γ_1 (s band) and Γ_{25}' bands (p band). The bond–antibond coupling neglected in the band orbital approximation may be treated in perturbation theory.

The bond orbitals are not the eigenstates of the crystal systems. However, we may connect the bonds to the eigenstates by projecting the bonds onto a plane and along the plane normal. Along the normal direction, the p components in the plane compose the HH bands, and the p component along the plane normal composes the LH band. The increase or decrease of the projection components implies the enhancement or reduction of the overlap integrals, which determine the valence band edge energy. Thus, the bonds in the tetrahedral crystals may serve as an intuitive means to understand the change of band structures with the change of the atomic configuration with strain.

4.5 STRAIN EFFECTS IN TIGHT-BINDING FRAMEWORK

Strain has two effects on the bonds. One is to change the bond length, the other is to change the bond angle. The change of bond length alters the interatomic interaction constants between the s and p orbitals, e.g., $V_{ss\sigma}$, $V_{sp\sigma}$, $V_{pp\sigma}$, and $V_{pp\pi}$, which only depends on the interatomic spacing. The change of bond angle alters the decomposition of the bond into the three cartesian axes, and also changes the interaction constants following (4.48) and (4.49). Band structure is then altered by both the bond length and bond angle changes.

4.5.1 Hydrostatic Strain: d^{-2} Principle

Hydrostatic strain changes the bond lengths by the same ratio while keeps the bond angles unchanged. The interaction constants in the tight-binding Hamiltonian then are totally determined by the interatomic σ and π overlap integrals. We can demonstrate that these coupling parameters have a very simple relation with the nearest-neighbor spacing in one semiconductor, and in different semiconductors, they can also be expressed as a simple function of the nearest-neighbor distance multiplied by a geometric factor. This result not only provides a very straightforward way to investigate the strain effects in tight-binding framework, but also makes the tight-binding method very powerful for predicting the electronic, elastic, and dielectric properties of various materials.

To obtain the relation between the overlap integrals and the interatomic distance, we may first inspect the energy dispersion relations obtained by the tight-binding method and by the nearly-free electron model. The valence bands obtained by these two models are actually very similar. For a simple illustration, let us first inspect the s band in (4.29) in a simple cubic crystal. Along the [100] direction, the band width is $E_X - E_\Gamma = 4V_{ss\sigma}$. On the other hand, the nearly free electron model gives a band width of $\hbar^2\pi^2/2md^2$. Equating the band widths obtained by these two different methods we get

$$4V_{ss\sigma} = \frac{\hbar^2\pi^2}{2md^2}. \tag{4.73}$$

For p bands, the similarity between the tight-binding bands and the nearly-free electron bands gives the same relation with a different coefficient. This result suggests that the overlap constants depend on the bond length d as d^{-2}. Generally, all four overlap parameters for the s and p orbitals can be expressed in the form

$$V_{ll'm} = \eta_{ll'm}\frac{\hbar^2}{md^2}, \tag{4.74}$$

where $\eta_{ll'm}$ is a factor that depends on the crystal symmetry. From (4.29), we see that $\eta_{ss\sigma} = \pi^2/8$. From (4.37), or (4.39), we obtain $\eta_{pp\sigma} = 3\pi^2/8$, and $\eta_{pp\pi} = -\pi^2/8$. In Table 4.4, we list the values $\eta_{ll'm}$ for the simple cubic and tetrahedral crystal systems. For the diamond and zinc-blende crystals, Harrison has treated the factors $\eta_{ll'm}$ as adjustable parameters in fitting the

Table 4.4. The geometric factor $\eta_{ll'm}$, determining the overlap integrals for the s and p bands. The last column represents the adjusted values obtained by fitting the energy bands of Si and Ge

	Simple cubic	Tetrahedral	Adjusted values
$\eta_{ss\sigma}$	$-\pi^2/8 = -1.23$	$-9\pi^2/64 = -1.39$	-1.40
$\eta_{sp\sigma}$	$(\pi/2)\sqrt{(\pi^2/4) - 1} = 1.90$	$(9\pi^2/32)\sqrt{1 - (16/32\pi^2)} = 1.88$	1.84
$\eta_{pp\sigma}$	$3\pi^2/8 = 3.70$	$21\pi^2/64 = 3.24$	3.24
$\eta_{pp\pi}$	$-\pi^2/8 = -1.23$	$-3\pi^2/32 = -0.93$	-0.81

energy bands of Si and Ge. He found excellent agreement between the calculated values and the adjusted values for three of the parameters. The only exception is $\eta_{pp\pi}$; here the fitted value of -0.81 is somewhat lower than the calculated one.

Table 4.4 together with (4.74) and the lattice constants are all that is needed to calculate the overlap parameters for computing the valence bands and the lowest conduction bands in many tetrahedral semiconductors. Similarly, for the same semiconductor under hydrostatic strain, the overlap parameters follow the same trend of dependence on the bond length.

First we inspect the energy dispersion relation of the s band and p bands in (4.29), (4.37), and (4.39). Assuming under hydrostatic strain, the lattice constant a dilates by δ, i.e., the new lattice constant is now $a(1+\delta)$. Following the d^{-2} relation, the overlap integrals all change to

$$V' = V(1+\delta)^{-2}. \tag{4.75}$$

Then the band widths are changed at first and the energies at the Γ point are shifted. Noticing that the energy dispersion is some constants plus the triangular functions of ak multiplied by the overlap integrals, the second derivative of E to k of any band is given by an equation like

$$\frac{\partial^2 E}{\partial k^2} \sim Va^2 \cos(ka), \tag{4.76}$$

no matter the system is strained or not. However with dilated lattice constant,

$$Va^2 \to V'a'^2 = V(1+\delta)^{-2}[a(1+\delta)]^2 = Va^2. \tag{4.77}$$

Therefore, the strained band curvature will equal the unstrained band curvature at

$$k' = k/(1+\delta). \tag{4.78}$$

That is to say, the band is scaled according to the shift of the Brillouin zone boundary, which is inversely proportional to the lattice constant.

In the sp^3 model, the band shift can be used to determine the hydrostatic deformation potential. For example, the energy at the valence band edge, $E_p - 4V_{xx}$, will change to $E_p - 4V_{xx}(1+\delta)^{-2}$ with hydrostatic dilation δ. The valence band edge shifts by $8V_{xx}\delta$ at the small strain limit. Thus, the hydrostatic deformation potential for the valence band at the Γ point is $8V_{xx}/3$. Normally, experiments can only measure the relative hydrostatic deformation potential, $a_c - a_v$, where a_c is the conduction band, and a_v is the valence band deformation potential, since it is easy to measure the band gap change under strain, but difficult to determine the absolute shifts of the band edges. However, the Si conduction and valence band edges are not located at the same k point. For GaAs, the relative deformation potential is readily evaluated following the same reasoning as $(-8/3)(V_{xx} - V_{ss})$.

The entire sp^3 band structure is not precisely scaled as the single band. This is because in the single band, the shift of energy does not affect the band curvature; however, in multiple bands, the relative shift changes the coupling between different bands, especially for degenerate bands, where a small energy shift can have great effects. Nevertheless, compared to shear strain, band structure change due to hydrostatic strain is not significant. At least, degeneracy in the band structure is not affected.

4.5.2 Shear Strain: Bond Rotation

As we discussed in Chap. 2, there are two types of shear strain in cubic crystals. One is related to the change of lengths along the axes, and the other is related to the rotation of the axes of the crystal. We find that the shear strain induced by the biaxial stress corresponds precisely to the first type of shear, and the shear strain induced by the uniaxial stress along [110] contains both types of shear, which we may investigate separately. Qualitative discussion of these two types of strain has already been given in Sect. 4.2.4. Here, we want to investigate these strain effects more in a quantitative manner.

First we study the biaxial stress-induced strain effects. The strain tensor (2.34) can be decomposed into hydrostatic and shear strain following (2.37) as:

$$\begin{pmatrix} e_{xx} & 0 & 0 \\ 0 & e_{xx} & 0 \\ 0 & 0 & e_{zz} \end{pmatrix} = \begin{pmatrix} \delta & 0 & 0 \\ 0 & \delta & 0 \\ 0 & 0 & \delta \end{pmatrix} + \begin{pmatrix} \alpha & 0 & 0 \\ 0 & \alpha & 0 \\ 0 & 0 & -2\alpha \end{pmatrix}, \tag{4.79}$$

where $\delta = (2e_{xx} + e_{zz})/3$ is the hydrostatic strain element, and $\alpha = (e_{xx} - e_{zz})/3$ is the shear strain element. Now that we have already discussed the effects of the hydrostatic strain effects on band structures, we only focus on the shear strain in this section. The effect of this shear strain is to lower the cubic symmetry of the crystal to tetragonal symmetry by making the z direction inequivalent to the x and y directions. As a result, the originally triply degenerate valence band wave functions $|p_x\rangle$, $|p_y\rangle$, and $|p_z\rangle$ will split into a doublet and a singlet. To calculate this splitting it is necessary to evaluate the overlap integrals $V_{xx}(= V_{yy})$ and V_{zz} in the presence of the shear strain.

Under this shear strain, the original locations of the nearest neighbors in (4.44) with respect to an atom in the origin change to:

$$\begin{aligned} \mathbf{d}'_1 &= [1+\alpha, 1+\alpha, 1-2\alpha]a/4, \\ \mathbf{d}'_2 &= [1+\alpha, -1-\alpha, -1+2\alpha]a/4, \\ \mathbf{d}'_3 &= [-1-\alpha, 1+\alpha, -1+2\alpha]a/4, \\ \mathbf{d}'_4 &= [-1-\alpha, -1-\alpha, 1-2\alpha]a/4. \end{aligned} \tag{4.80}$$

It is easy to check that the bond lengths are not changed to the first order of α, but the bond angles are altered. Indeed, the three new coordinates of the neighboring atoms are just the numerators of the directional cosines with a denominator $\sqrt{3}$. Following (4.48), the new sets of interaction constants are reevaluated as

$$\begin{aligned} V'_{xx} &= V'_{yy} = \frac{1}{3}V_{pp\sigma}(1+2\alpha) + \frac{2}{3}V_{pp\pi}(1-\alpha), \\ V'_{zz} &= \frac{1}{3}V_{pp\sigma}(1-4\alpha) + \frac{2}{3}V_{pp\pi}(1+2\alpha). \end{aligned} \tag{4.81}$$

The splitting between the doublets and the singlet at the Γ point is then given by $-4(V_{xx} - V_{zz}) = -8\alpha(V_{pp\sigma} - V_{pp\pi})$. Whether the doublets or the singlet is on the valence band edge is determined by the sign of α. For biaxial tension ($\alpha > 0$), the splitting is negative and the singlet $|p_z\rangle$ band is on the band edge. Remembering that the p_z band is actually the LH band, and the p_x and p_y bands are the doubly degenerate HH band, biaxial tension raises the LH band and lowers the HH band in the [001] direction.

The band splitting and shifts can be understood even without going to detailed calculation as we did earlier, by just the inspecting the bond rotation. Under biaxial strain, the four bonds all rotate toward or away from the x-y plane depending on the sign of α. For biaxial tension, all the four bonds are equivalent and rotate toward the x-y plane, as shown in Fig. 4.7a. With such a rotation, the weight of the p_x- and p_y-orbitals in the bonds increases and that of the p_z-orbital decreases. Along the [001] direction, this results in increased overlap integrals between in-plane orbitals and lowered HH bands and decreased overlap integrals between the p_z-orbitals and a weakened LH band that ascends in energy (becomes the top valence band). Along the [100] and [010] direction, the LH band is lowered because it is composed mainly of the p_x- and p_y-orbitals, respectively, and the topmost bands are HH bands. Therefore, the top valence band under biaxial tension is LH-like out-of-plane and HH-like in-plane.

The wave function of the Si conduction band along the Δ-axis is composed of antibonding s- and antibonding p-states directed along this axis, e.g., the conduction band state along the [001] direction is composed of $|s_-\rangle$ and $|p_{z+}\rangle$ states. We may write it arbitrarily as

$$|\psi_z^{\mathrm{c}}\rangle = A|s_-\rangle + B|p_{z+}\rangle, \quad A^2 + B^2 = 1, \tag{4.82}$$

where ψ_z^{c} represents the conduction band wave function in the z axis. The matrix element between ψ_z^{c} gives

$$\langle\psi_z^{\mathrm{c}}|H|\psi_z^{\mathrm{c}}\rangle = A^2(E_s - V_{ss}g_0) + B^2(E_p + V_{zz}g_0). \tag{4.83}$$

Without bond length change, V_{ss} does not shift. Because of the decrease of overlap integrals between the p_z orbitals under biaxial tension, the conduction band valleys along the z axis descend.

Strain tensor for [110] uniaxial stress contains two shear terms that are

$$\begin{pmatrix} \alpha & 0 & 0 \\ 0 & \alpha & 0 \\ 0 & 0 & -2\alpha \end{pmatrix} + \begin{pmatrix} 0 & \beta & 0 \\ \beta & 0 & 0 \\ 0 & 0 & 0 \end{pmatrix}, \tag{4.84}$$

where $\beta = e_{xy}$. Note that the first shear term is identical in form to that in the biaxial strain case with possible value difference. Thus the effect of this term is the same as discussed earlier for the biaxial strain. The conspicuous difference of the uniaxial strain from the biaxial strain is the second shear term which only has the off-diagonal elements that the biaxial strain lacks.

This shear strain not only changes the bond length and angles as the biaxial does, but also creates anisotropy in the x-y plane. It distinguishes between the [110] and [$\bar{1}$10] directions. The qualitative picture is already discussed in subsection 4.2.4. Readers can follow the procedures for biaxial strain to work out the quantitative analysis for the uniaxial strain to see how overlap integrals change with strain and how the change shifts the bands.

The shift of Si conduction valleys is primarily determined by the first type of shear strain. For a uniaxial stress along [110], $\alpha < 0$, therefore, the two conduction valleys along the z axis ascend and the four in-plane (x-y plane) conduction valleys descend.

The effects on band structures from other types of strain may also be studied following the same procedure. In summary, a compressive strain raises the LH band and lowers the HH band along the strain axis. This is due to the altered orbital overlap between atoms.

4.6 SUMMARY FOR THE TIGHT-BINDING METHOD

The sp^3 tight-binding model is a very coarse band structure model. For the valence bands, the results are satisfactory. The higher energy bands cannot be successfully reproduced since they also consist of more delocalized orbitals. As we mentioned earlier, an s^* orbital added to the sp^3 model would result in a descending band out of the Γ_{15} conduction band of Si and give the conduction valley along the Δ axes near the X point. This s^* takes into account partly of the 4s and partly of the 3d orbital effects. Since 3d states are not spherically symmetrical, their interaction with the 3s and 3p states cannot be entirely accounted, thus the position of the conduction band valleys deviates a little from the commonly believed one. A more sophisticated model, $sp^3s^*d^5$ model, considers the five d orbitals explicitly and can successfully reproduce the valence bands and the low-energy conduction bands. However, so far we have not considered spin. Spin interacts with the electronic orbital motion and has important effects on the degenerate bands, such as the valence bands. Without spin, the valence bands are triply degenerate. If there is no spin–orbit interaction, the degeneracy order is actually six if spin degeneracy is accounted. Spin–orbit interaction removes the sixfold degeneracy at the valence band edge, leaving fourfold degenerate HH and LH bands at the edge, and one twofold degenerate split-off hole band with a lower energy. The splitting of the split-off band from the valence band edge plays critical role no matter in optical transitions or carrier transport. Including spin into the tight-binding model doubles the basis states. For an $sp^3s^*d^5$ model, the number of basis is 40. Using this model, a quantitative description of the bands can be obtained. However, a 40×40 Hamiltonian is cumbersome to deal with and difficult to analyze. For most cases, an entire band picture is not necessary because in applications to semiconductors or MOSFETs, electrons or holes around the Fermi surface are generally only located at the conduction or valence band edges. Thus, it may not be necessary to obtain the detailed knowledge of the entire band to study the carrier transport. For an accurate yet simple description of the band structure around some special k point, say, the Γ point, the $\mathbf{k} \cdot \mathbf{p}$ method is an excellent theoretical means, which usually treats the band structure around a special point in k-space.

4.7 THE K · P METHOD

The $\mathbf{k} \cdot \mathbf{p}$ method was introduced by Bardeen (Bardeen, 1938) and Seitz (Seitz, 1940), and developed later by Luttinger (Luttinger and Kohn, 1955) and Kane (Kane, 1957). It is a perturbation theory based method, often called effective mass theory in the literature. It employs various experimental parameters, such as the band gap, E_g, split-off energy, Δ, the conduction-valence band coupling element, E_p, and the electron and hole effective masses, etc. These parameters can be accurately determined by optical or magneto-optical experiments, thus the $\mathbf{k} \cdot \mathbf{p}$ treats the band structure with high precision. One other big advantage of the $\mathbf{k} \cdot \mathbf{p}$ method is that it usually uses a very small basis set, and the Hamiltonian is easy to diagonalize and analyze. Therefore, the $\mathbf{k} \cdot \mathbf{p}$ method finds important applications in semiconductor optics, magnetism, and transport. We will introduce the $\mathbf{k} \cdot \mathbf{p}$ theory in this book also because of two other important considerations. One is that the strain effects in the $\mathbf{k} \cdot \mathbf{p}$ framework are very straightforwardly treated, without considering the symmetry reduction-induced geometrical and phase factor variation. The other is that in investigating the low-dimensional semiconductors, such as the quantum wells and wires, the $\mathbf{k} \cdot \mathbf{p}$ method is easy to implement. Tight-binding and pseudopotential method need to treat the strained semiconductors as a new crystal system. However, strain is just considered as an extra coupling term, which is directly added to the unperturbed Hamiltonian in the $\mathbf{k} \cdot \mathbf{p}$ framework. The $\mathbf{k} \cdot \mathbf{p}$ method combined with the Poisson equation is extensively used to investigate the electric field and charge distribution in MOSFETs and other device structures.

4.7.1 Effective Mass

We can always expand one band around one of its extrema k_0 as follows,

$$E(k) = E(k_0) + \left.\frac{dE}{dk}\right|_{k=k_0} (k - k_0) + \frac{1}{2}\left.\frac{d^2E}{dk^2}\right|_{k=k_0} (k - k_0)^2 + \dots, \qquad (4.85)$$

where we neglected the expansion terms with order greater than $(k - k_0)^2$ for sufficiently small $(k - k_0)$. The first-order derivative of E to k vanishes automatically, leaving us

$$E(k)|_{k\to k_0} \simeq E(k_0) + \frac{1}{2}\frac{d^2E}{dk^2}\bigg|_{k=k_0}(k-k_0)^2 = E_0 + \frac{\hbar^2(k-k_0)^2}{2m^*}, \quad (4.86)$$

where m^* is the (density-of-states) effective mass defined as

$$\frac{1}{m^*} = \frac{1}{\hbar^2}\frac{d^2E}{dk^2}\bigg|_{k=k_0}. \quad (4.87)$$

The effective mass can be measured by experiments such as cyclotron resonance where the resonance absorption is observed for a semiconductor sample put under a light radiation by sweeping the magnetic field, as shown in Fig. 4.13. Suppose the resonance field is B, the light frequency is ω, the carrier effective mass is given by

$$m^* = \frac{eB}{\omega}. \quad (4.88)$$

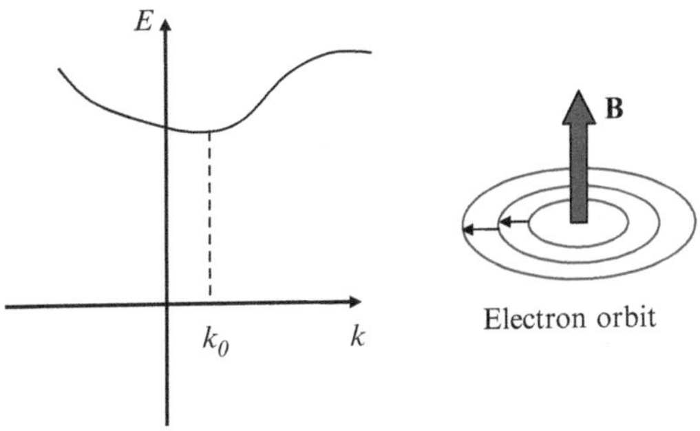

Fig. 4.13. Curvature determined effective mass, and cyclotron resonance method to experimentally obtain it

This is what is normally done for the effective mass of a single band such as the Si or GaAs conduction band. However, this simple effective mass is only valid for parabolic bands where in Eq. (4.85) the higher expansion order can be neglected. Also for degenerate bands, carriers belonging to different band have different effective masses, and furthermore, for one band, the effective mass may be anisotropic. One simple effective mass cannot give any complex band structure and thus is not adequate. In studying semiconductor optical properties, and especially in studying the strain effects, which is our focus, the physical mechanisms that determine the effective mass need to be explored. The **k** · **p** theory is ideal for this purpose.

4.7.2 k · p Hamiltonian

The core idea of the $\mathbf{k} \cdot \mathbf{p}$ method is to solve the Schrödinger equation using perturbation theory. We now write down again the Hamiltonian for an electron in a semiconductor

$$H = \frac{p^2}{2m_0} + V(\mathbf{r}), \tag{4.89}$$

here $\mathbf{p} = -i\hbar\nabla$ is the momentum operator, m_0 refers to the free electron mass, and $V(\mathbf{r})$ is the potential including the effective lattice periodic potential caused by the ions and core electrons or the potential due to the exchange interaction, impurities, etc. If we consider $V(\mathbf{r})$ to be periodic, the solution of the Schrödinger equation is Bloch waves, i.e.,

$$\psi_{\mathbf{k}}(\mathbf{r}) = e^{i\mathbf{k}\cdot\mathbf{r}} u_{\mathbf{k}}(\mathbf{r}), \tag{4.90}$$

where

$$u_{\mathbf{k}}(\mathbf{r} + \mathbf{R}) = u_{\mathbf{k}}(\mathbf{r}), \tag{4.91}$$

and $\mathbf{R}$ is a lattice vector, and $\mathbf{k}$ is the wave vector.

The eigenvalues of the Schrödinger equation split into a series of bands. Consider the Schrödinger equation in the nth band with a wave vector $\mathbf{k}$,

$$\left[\frac{p^2}{2m_0} + V(\mathbf{r})\right] \psi_{n\mathbf{k}}(\mathbf{r}) = E_n(\mathbf{k})\psi_{n\mathbf{k}}(\mathbf{r}). \tag{4.92}$$

Inserting the Bloch function (4.90) into (4.92), we obtain

$$\left[\frac{p^2}{2m_0} + V(\mathbf{r}) + \frac{\hbar^2 k^2}{2m_0} + \frac{\hbar}{m_0}\mathbf{k}\cdot\mathbf{p}\right] u_{n\mathbf{k}}(\mathbf{r}) = E_n(\mathbf{k}) u_{n\mathbf{k}}(\mathbf{r}). \tag{4.93}$$

The above equation can be expanded near a particular point k_0 of interest in the band structure. When $k_0 = 0$ (the Γ point), the above equation becomes

$$(H_0 + H_1) u_{n\mathbf{k}}(\mathbf{r}) = \left[E_n(\mathbf{k}) - \frac{\hbar^2 k^2}{2m_0}\right] u_{n\mathbf{k}}(\mathbf{r}), \tag{4.94}$$

where

$$H_0 = \frac{p^2}{2m_0} + V(\mathbf{r}), \tag{4.95}$$

$$H_1 = \frac{\hbar}{m_0}\mathbf{k}\cdot\mathbf{p}, \tag{4.96}$$

and we define

$$W_{n\mathbf{k}} = E_n(\mathbf{k}) - \frac{\hbar^2 k^2}{2m_0}. \tag{4.97}$$

When k is small, the $\mathbf{k} \cdot \mathbf{p}$ term H_1 can be treated as perturbation. This is the reason why this formalism is called the "$\mathbf{k} \cdot \mathbf{p}$" method.

If the Hamiltonian H_0 has a complete set of orthonormal eigenfunctions at $k = 0$, u_{n0}, i.e.,

$$H_0 u_{n0}(\mathbf{r}) = E_{n0} u_{n0}(\mathbf{r}), \tag{4.98}$$

then theoretically any lattice periodic function can be expanded using eigenfunctions u_{n0}, i.e., away from the Γ point (or k_0, if it is of interest), we can always have

$$u_{n\mathbf{k}} = \sum_m c_m^n(\mathbf{k}) u_{m0}. \tag{4.99}$$

As a matter of fact, expanding the wave function using the known complete set is the central spirit of the **k · p** method, as shown in Fig. 4.14. Then the band energy away from the Γ point can be perturbatively obtained.

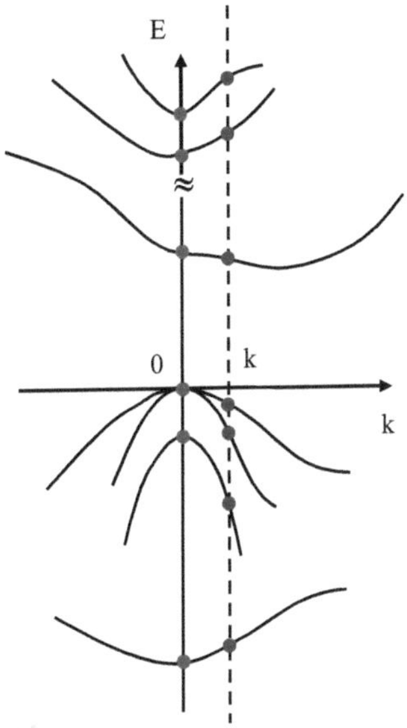

Fig. 4.14. The schematic of how **k · p** method works to obtain the band structure. The bands at $k = 0$ are known, then the bands at the adjacent k points are obtained by linear expansion of the wave functions at the $k = 0$ point

4.7.3 Single Band Perturbation Expansion

If the band of interest is a single band, such as the Si and GaAs conduction band, we can use single band perturbation theory to obtain the band structure near the band edge if the energies and wave functions at the band edge are known. To proceed, we first rewrite (4.94) (then we are looking into the band structure around the Γ point) as

$$(H_0 + \lambda H_1) u_{n\mathbf{k}} = W_{n\mathbf{k}} u_{n\mathbf{k}}, \tag{4.100}$$

where for convenience, a perturbation parameter number λ is introduced. We can always recover the original problem later by setting $\lambda = 1$. Then we look for the solutions of the form

$$u_{n\mathbf{k}} = u_{n\mathbf{k}}^{(0)} + \lambda u_{n\mathbf{k}}^{(1)} + \lambda^2 u_{n\mathbf{k}}^{(2)} + \ldots, \tag{4.101}$$

$$W_{n\mathbf{k}} = W_{n\mathbf{k}}^{(0)} + \lambda W_{n\mathbf{k}}^{(1)} + \lambda^2 W_{n\mathbf{k}}^{(2)} + \ldots, \tag{4.102}$$

where the energy and wave function corrections are written in orders. Substituting the above expressions $u_{n\mathbf{k}}$ and $W_{n\mathbf{k}}$ to (4.100), and collecting the terms with equal power of λ, we obtain

$$\text{zeroth order} \quad (H_0 - W_{n\mathbf{k}}^{(0)})u_{n\mathbf{k}}^{(0)} = 0, \tag{4.103}$$

$$\text{first order} \quad (H_0 - W_{n\mathbf{k}}^{(0)})u_{n\mathbf{k}}^{(1)} = (W_{n\mathbf{k}}^{(1)} - H_1)u_{n\mathbf{k}}^{(0)}, \tag{4.104}$$

$$\text{second order} \quad (H_0 - W_{n\mathbf{k}}^{(0)})u_{n\mathbf{k}}^{(2)} = (W_{n\mathbf{k}}^{(1)} - H_1)u_{n\mathbf{k}}^{(1)} + W_{n\mathbf{k}}^{(2)}u_{n\mathbf{k}}^{(0)}, \tag{4.105}$$

and so on.

Zeroth Order Solutions. Equation (4.103) gives

$$W_{n\mathbf{k}}^{(0)} = E_{n0}, \tag{4.106}$$

and

$$u_{n\mathbf{k}}^{(0)} = u_{n0}, \tag{4.107}$$

which are exactly the solution of the Schrödinger equation at the Γ point.

First Order Solutions. Taking scalar product of (4.104) with u_{n0} gives the first energy correction, which is

$$W_{n\mathbf{k}}^{(1)} = \langle u_{n0}|H_1|u_{n0}\rangle = \frac{\hbar\mathbf{k}}{m_0} \cdot \langle u_{n0}|\mathbf{p}|u_{n0}\rangle. \tag{4.108}$$

In cubic crystals, the wave function of any single band or degenerate bands at the Γ point has definite parity, while the operator $\mathbf{p}$ changes parity under inversion, so the matrix element $\mathbf{p}_{n0,n0} = \langle u_{n0}|\mathbf{p}|u_{n0}\rangle$ vanishes. Thus, the first order of energy correction is zero. For energy correction, we have to go to second-order perturbation.

Taking scalar product of (4.104) with u_{m0} where $m \neq n$, we obtain the first-order wave function correction,

$$\langle u_{m0}|u_{n\mathbf{k}}\rangle = \frac{\langle u_{m0}|H_1|u_{n0}\rangle}{E_{n0} - E_{m0}}. \tag{4.109}$$

That is to say,

$$u_{n\mathbf{k}}^{(1)} = \sum_m u_{m0}\langle u_{m0}|H_1|u_{n0}\rangle = \sum_{m\neq n} \frac{\langle u_{m0}|H_1|u_{n0}\rangle}{E_{n0} - E_{m0}} u_{m0}, \tag{4.110}$$

i.e., although the first energy correction is zero, the wave function is admixed from the other bands. However, note that the admixture cannot be from the bands whose wave functions have the same parity. For single s band, the wave mixture mainly results from the p bands, with energetically close bands contributing the most.

Second Order Solutions. Taking scalar product of (4.105) with u_{n0}, and using the first-order wave function correction, we obtain the second-order energy correction,

$$W_{n\mathbf{k}}^{(2)} = \sum_{m \neq n} \frac{|\langle u_{m0}|H_1|u_{n0}\rangle|^2}{E_{n0} - E_{m0}}. \tag{4.111}$$

So the band energy at the point $\mathbf{k}$ to the second order of k is given by

$$E_n(\mathbf{k}) = E_{n0} + \frac{\hbar^2 k^2}{2m_0} + \frac{\hbar^2}{m_0^2} \sum_{m \neq n} \frac{|\langle u_{m0}|\mathbf{k}\cdot\mathbf{p}|u_{n0}\rangle|^2}{E_{n0} - E_{m0}}. \tag{4.112}$$

Since the eigenstates of H_0 have definite parity for cubic semiconductors, it is easy to prove that

$$\langle u_{n0}|p_\alpha|u_{m0}\rangle\langle u_{m0}|p_\beta|u_{n0}\rangle = 0, \quad \text{for } \alpha \neq \beta, \tag{4.113}$$

and

$$|\langle u_{n0}|p_x|u_{m0}\rangle|^2 = |\langle u_{n0}|p_y|u_{m0}\rangle|^2 = |\langle u_{n0}|p_z|u_{m0}\rangle|^2. \tag{4.114}$$

Thus, the band energy in (4.112) can be rewritten as

$$E_n(\mathbf{k}) = E_{n0} + \frac{\hbar^2 k^2}{2m^*}, \tag{4.115}$$

where

$$\frac{1}{m^*} = \frac{1}{m_0}\left[1 + \frac{2}{m_0} \sum_{m \neq n} \frac{|\langle u_{m0}|p_x|u_{n0}\rangle|^2}{E_{n0} - E_{m0}}\right]. \tag{4.116}$$

The effective mass here is a constant, due to the symmetry at the Γ point. At the other point other than the zone center, the effective mass is generally a symmetric tensor called the effective mass tensor. We will illustrate this point later when we discuss the Si conduction band. In (4.116), the effective mass is determined by the coupling from the other bands. Among the various remote bands, generally only one band or one set of (degenerate) bands has the strongest coupling. For normal direct gap semiconductors, the coupling from the valence bands dominates. If we neglect the coupling from the other bands, then the sum in (4.112) only has one term with the denominator being the band gap E_g. If the band gap is small, then the effective mass will also be small. This is the reason why narrow gap semiconductors always have small electron effective masses. In (4.116), the value of $(m_0/m^* - 1)E_g$ shall be nearly constant if we only consider the valence band coupling. In Table 4.5, we list this value for several semiconductors, and we can see that they are indeed very close.

Table 4.5. Bandgap, electron effective mass, and their numerical relation for some direct-gap semiconductors

Material	$E_g(0K)$ (eV)	$m^*(m_0)$	$(m_0/m^* - 1)E_g$
GaAs	1.52	0.067	21
InP	1.42	0.076	17
GaSb	0.81	0.041	19
InAs	0.42	0.024	17
InSb	0.24	0.014	17

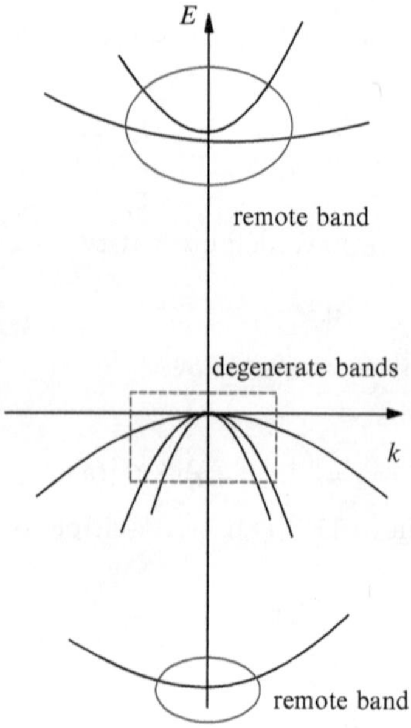

Fig. 4.15. Bands are classified to a degenerate set and remote bands. The remote band effect can be considered later after the degenerate sets are solved

4.7.4 Degenerate Band Perturbation Expansion

When dealing with the degenerate bands, the earlier discussions for a single band cannot be applied, since the coupling between the degenerate bands is actually infinite if we apply (4.110). We need to employ the degenerate perturbation theory.

Assume N degenerate states $|1\rangle$, $|2\rangle$, $|3\rangle$, ..., $|N\rangle$ with energy $E_1 = E_2 = E_3 = \cdots = E_N$. Because of the degeneracy, we cannot assign one state to one particular band. The zeroth order wave function is the linear combination of the N degenerate basis states

$$|n\rangle' = \sum_{m=1}^{N} |m\rangle\langle m|n\rangle'. \tag{4.117}$$

The degenerate perturbation procedure is to find N new orthonormal eigenstates $|n\rangle'$ that evolve continuously into nondegenerate eigenstates as we leave the zone center.

We now divide the bands at the Γ point into two classes, one class consists of the N degenerate bands, the other is the remote bands as shown in Fig. 4.15. Taking scalar product of (4.104) with $|j\rangle$, we have

$$\sum_{m=1}^{N} \langle j|H_1|m\rangle\langle m|n\rangle' - W_{n\mathbf{k}}^{(1)}\langle j|n\rangle' = 0, \quad \text{for} \quad j \in N, \tag{4.118}$$

$$\langle j|n\rangle' = \sum_{m=1}^{N} \frac{\langle j|H_1|m\rangle\langle m|n\rangle'}{E_1 - E_j}, \quad \text{for} \quad j \notin N. \tag{4.119}$$

The Equation (4.118) is an N-coupled equations which can be normally solved for N new eigenstates, but with one condition, i.e., H_1 couples the N degenerate bands. However, as we stated earlier, $\langle j|H_1|m\rangle = 0$ for all j, $m \in N$. This results in $W_{n\mathbf{k}}^{(1)} = 0$. Second-order degenerate perturbation is necessary for energy correction.

Repeating the above procedure with the second-order (4.105), we obtain

$$\sum_{l=1}^{N} \sum_{\alpha \notin N} \frac{\langle j|H_1|\alpha\rangle\langle \alpha|H_1|l\rangle}{E_1 - E_\alpha}\langle l|n\rangle' - W_{n\mathbf{k}}^{(2)}\langle j|n\rangle' = 0, \quad \text{for} \quad j \in N. \tag{4.120}$$

This equation is actually a set of N-coupled equations that can be solved for $W_{n\mathbf{k}}^{(2)}$ and $|n\rangle'$. Using the above equation, we can rewrite the original Schrödinger equation in a matrix form as

$$H(\mathbf{k})A = E_n(\mathbf{k})A, \tag{4.121}$$

where

$$H_{ij} = \left[E_1 + \frac{\hbar^2 k^2}{2m_0}\right]\delta_{ij} + \frac{\hbar^2}{m_0^2}\sum_{\alpha \notin N} \frac{\mathbf{k}\cdot\mathbf{p}_{i\alpha}\mathbf{k}\cdot\mathbf{p}_{\alpha j}}{E_1 - E_\alpha}, \tag{4.122}$$

and

$$A_j = \langle j|n\rangle' \tag{4.123}$$

is the coefficient of the expansion of wave function $|n\rangle'$ in state $|j\rangle$. The wave function expansion involves only the degenerate set, and the summation in the Hamiltonian only involves the coupling from the remote bands. Band structure away from the Γ point evolved from the degenerate bands is obtained by diagonalizing the Hamiltonian (4.122).

4.8 LUTTINGER HAMILTONIAN

4.8.1 Luttinger Hamiltonian Without Spin–orbit Coupling

First we investigate the valence band structure without considering the spin–orbit coupling, which splits the valence bands into HH, LH, and split-off hole bands. Thus, we can have a direct comparison with the results we obtained using the tight-binding method. From the discussion in the tight-binding theory, the valence band edge is composed of the p-states. The three degenerate states can be written as $|X\rangle$, $|Y\rangle$, and $|Z\rangle$. From the tight-binding discussion, we know they actually correspond to the bonding states of $|p_x\rangle$, $|p_y\rangle$ and $|p_z\rangle$. However, the precise expressions of these states are not important. We only need to know the transformation properties of them under space symmetry operations, or its parity. The three states transform exactly like the atomic p orbitals, and they have odd parities. Using the second-order degenerate perturbation theory, the energy of the nth band is

$$E_{n\mathbf{k}} = E_v + \frac{\hbar^2 k^2}{2m_0} + W_{n\mathbf{k}}^{(2)}, \quad n = 1, 2, 3, \tag{4.124}$$

where E_v is the valence band edge energy. With the three valence band basis functions, the Hamiltonian is 3×3 in the following

$$H_{ij} = \left(E_v + \frac{\hbar^2 k^2}{2m_0} \right) \delta_{ij} + \frac{\hbar^2}{m_0^2} \sum_{\alpha > 3} \frac{\mathbf{k} \cdot \mathbf{p}_{i\alpha} \mathbf{k} \cdot \mathbf{p}_{\alpha j}}{E_v - E_\alpha}. \tag{4.125}$$

The matrix elements can be simplified by symmetry. For example for H_{11},

$$|\langle X|H_1|\alpha\rangle|^2 = \frac{\hbar^2}{m_0^2} \left(|\langle X|p_x|\alpha\rangle|^2 k_x^2 + |\langle X|p_y|\alpha\rangle|^2 k_y^2 + |\langle X|p_z|\alpha\rangle|^2 k_z^2 \right), \tag{4.126}$$

So we have

$$H_{11} = \langle X|H|X\rangle = E_v + \sum_{j=x,y,z} \left[\frac{\hbar^2}{2m_0} + \frac{\hbar^2}{m_0^2} \sum_{\alpha>3} \frac{|\langle X|p_j|\alpha\rangle|^2}{E_v - E_\alpha} \right] k_j^2. \tag{4.127}$$

Since due to symmetry,

$$|\langle X|p_y|\alpha\rangle|^2 = |\langle X|p_z|\alpha\rangle|^2, \tag{4.128}$$

we can rewrite H_{11} as

$$H_{11} = E_1 + Lk_x^2 + M(k_y^2 + k_z^2) \tag{4.129}$$

with

$$L = \frac{\hbar^2}{2m_0} + \frac{\hbar^2}{m_0^2}\sum_{\alpha>3}\frac{|\langle X|p_x|\alpha\rangle|^2}{E_v - E_\alpha} \tag{4.130}$$

and

$$M = \frac{\hbar^2}{2m_0} + \frac{\hbar^2}{m_0^2}\sum_{\alpha>3}\frac{|\langle X|p_y|\alpha\rangle|^2}{E_v - E_\alpha}. \tag{4.131}$$

Similarly, the matrix element H_{12} can be written as

$$H_{12} = \langle X|H|Y\rangle = Nk_xk_y \tag{4.132}$$

with

$$N = \frac{\hbar^2}{m_0^2}\sum_{\alpha>3}\frac{\langle X|p_x|\alpha\rangle\langle\alpha|p_y|Y\rangle + \langle X|p_y|\alpha\rangle\langle\alpha|p_x|Y\rangle}{E_v - E_\alpha}. \tag{4.133}$$

Thus, the 3×3 Luttinger Hamiltonian is

$$H = \begin{bmatrix} E_v + Lk_x^2 + M(k_y^2 + k_z^2) & Nk_xk_y & Nk_xk_z \\ Nk_yk_x & E_v + Lk_y^2 + M(k_z^2 + k_x^2) & Nk_yk_z \\ Nk_zk_x & Nk_zk_y & E_v + Lk_z^2 + M(k_x^2 + k_y^2) \end{bmatrix}. \tag{4.134}$$

The energy E_v can be chosen as the energy zero as a convention. Therefore this Hamiltonian for cubic semiconductor valence bands can be obtained solely by symmetry considerations. The cubic symmetry requires three independent parameters (L, M and N) to describe the valence band structure.

The valence band dispersions can be analytically obtained. Along $\langle 100\rangle$, the two HH bands are $E_{\mathrm{HH}} = Mk_z^2$, and the LH band is $E_{\mathrm{LH}} = Lk_z^2$. Along $\langle 110\rangle$, the three valence bands are not degenerate with dispersions given by $E_k = (L+M \pm N)k^2/2$, and $E(k) = Mk^2$. Along $\langle 111\rangle$, the two HH bands are $E_{\mathrm{HH}} = (L + 2M - N)k^2/3$, and the LH band is $E_{\mathrm{LH}} = (L + 2M + 2N)k^2/3$. The band dispersion and degeneracy resemble exactly those obtained from the tight-binding method. One common misunderstanding is that the HH bands are composed of $|X\rangle$ and $|Y\rangle$, and the LH band is composed of $|Z\rangle$. This is true, but only along the [001] direction, which we choose as the quantization axis above. According to the cubic symmetry, along [100], the HH bands are composed of $|Y\rangle$ and $|Z\rangle$, and along [010], they are composed of $|Z\rangle$ and $|X\rangle$. The LH band is always composed of the basis state parallel to the quantization axis.

4.8.2 Luttinger Hamiltonian with Spin–Orbit Coupling

When we consider spin, which is represented by two independent states

$$\left|\frac{1}{2},\frac{1}{2}\right\rangle = |\uparrow\rangle, \tag{4.135a}$$

$$\left|\frac{1}{2},-\frac{1}{2}\right\rangle = |\downarrow\rangle, \tag{4.135b}$$

the number of basis vectors for describing the valence bands becomes six, namely,

$$|X\uparrow\rangle,\quad |Y\uparrow\rangle,\quad |Z\uparrow\rangle, \tag{4.136a}$$

$$|X\downarrow\rangle,\quad |Y\downarrow\rangle,\quad |Z\downarrow\rangle. \tag{4.136b}$$

Spin–orbit coupling can be understood by the classic picture that spin feels the magnetic field induced by the orbital motions of the electrons, thus spin couples with the orbital motion by so called $\mathbf{L}\cdot\mathbf{S}$ coupling, where $\mathbf{L}$ is the orbital angular momentum, and $\mathbf{S}$ is the electron spin. The spin–orbit coupling can be naturally derived by the Dirac equation and is a relativistic term. However, detailed discussion of this is beyond the scope of this book. The spin–orbit coupling term is given by:

$$H_{\mathrm{SO}} = \lambda \mathbf{L}\cdot\mathbf{S}, \tag{4.137}$$

and the total angular momentum is

$$\mathbf{J}^2 = (\mathbf{L}+\mathbf{S})^2 = \mathbf{L}^2 + \mathbf{S}^2 + 2\mathbf{L}\cdot\mathbf{S}. \tag{4.138}$$

Thus,

$$\mathbf{L}\cdot\mathbf{S} = \frac{1}{2}(\mathbf{J}^2 - \mathbf{L}^2 - \mathbf{S}^2) = \frac{\hbar}{2}[j(j+1) - l(l+1) - s(s+1)]. \tag{4.139}$$

Therefore, the eigenstates of the spin–orbit coupling are the eigenstates of angular momentums. The valence band edge are composed of p states, which have orbital angular momentum $l = 1$. The eigenstates of both the orbital angular momentum and its component l_z, which is the projection of $\mathbf{L}$ on the z axis, are given by

$$|1,\pm 1\rangle = |X \pm iY\rangle/\sqrt{2}, \tag{4.140a}$$

$$|1,0\rangle = |Z\rangle. \tag{4.140b}$$

When they couple with the electron spin ($s = 1/2$), the resulted states decoupled into two groups with $j = 3/2$ and $j = 1/2$ and are given by

$j = 3/2$ set:

$$\left| j = \frac{3}{2}, m = \frac{3}{2} \right\rangle = -\frac{1}{\sqrt{2}} |(X + iY) \uparrow\rangle, \tag{4.141a}$$

$$\left| j = \frac{3}{2}, m = \frac{1}{2} \right\rangle = -\frac{1}{\sqrt{6}} (|(X + iY) \downarrow\rangle - 2|Z \uparrow\rangle), \tag{4.141b}$$

$$\left| j = \frac{3}{2}, m = -\frac{1}{2} \right\rangle = \frac{1}{\sqrt{6}} (|(X - iY) \uparrow\rangle + 2|Z \downarrow\rangle), \tag{4.141c}$$

$$\left| j = \frac{3}{2}, m = -\frac{3}{2} \right\rangle = \frac{1}{\sqrt{2}} |(X - iY) \downarrow\rangle, \tag{4.141d}$$

$j = 1/2$ set:

$$\left| j = \frac{1}{2}, m = \frac{1}{2} \right\rangle = \frac{1}{\sqrt{3}} (|(X + iY) \downarrow\rangle + |Z \uparrow\rangle), \tag{4.141e}$$

$$\left| j = \frac{1}{2}, m = -\frac{1}{2} \right\rangle = \frac{1}{\sqrt{3}} (|(X - iY) \uparrow\rangle - |Z \downarrow\rangle), \tag{4.141f}$$

where j is the quantum number of the total angular momentum, and m is the quantum number of the projection of **J** on the z axis. Note that these six states are the unitary transformation of the simple superposition of the orbital state and spin states in (4.136). The simple p states can also be written as the linear combination of the eigenstates of the total angular momentum **J**. for example,

$$|X \uparrow\rangle = \frac{1}{\sqrt{2}} \left[-\left| \frac{3}{2}, \frac{3}{2} \right\rangle + \frac{1}{\sqrt{3}} \left| \frac{3}{2}, -\frac{1}{2} \right\rangle - \sqrt{\frac{2}{3}} \left| \frac{1}{2}, -\frac{1}{2} \right\rangle \right], \tag{4.142a}$$

$$|X \downarrow\rangle = \frac{1}{\sqrt{2}} \left[\left| \frac{3}{2}, -\frac{3}{2} \right\rangle - \frac{1}{\sqrt{3}} \left| \frac{3}{2}, \frac{1}{2} \right\rangle - \sqrt{\frac{2}{3}} \left| \frac{1}{2}, \frac{1}{2} \right\rangle \right], \tag{4.142b}$$

and so on. Finally, we write the spin–orbit Hamiltonian as

$$H_{\rm SO} = \frac{\lambda\hbar}{2} [j(j + 1) - l(l + 1) - s(s + 1)], \tag{4.143}$$

where for p states, $l = 1$, and $s = 1$, while j is given by the first index of the $\mathbf{L} \cdot \mathbf{S}$ coupled states in (4.141). Using (4.142), it can be shown that the only nonvanishing matrix elements between the simple p states are

$$\langle X \uparrow |H_{SO}| Y \uparrow\rangle = -i\frac{\Delta}{3}, \tag{4.144a}$$

$$\langle X \uparrow |H_{SO}| Z \downarrow\rangle = \frac{\Delta}{3}, \tag{4.144b}$$

$$\langle Y \uparrow |H_{SO}| Z \downarrow\rangle = -i\frac{\Delta}{3}, \tag{4.144c}$$

$$\langle X \downarrow |H_{SO}| Y \downarrow\rangle = i\frac{\Delta}{3}, \tag{4.144d}$$

$$\langle X\downarrow|H_{SO}|Z\uparrow\rangle=-\frac{\Delta}{3}, \tag{4.144e}$$

$$\langle Y\downarrow|H_{SO}|Z\uparrow\rangle=-i\frac{\Delta}{3}, \tag{4.144f}$$

where $\Delta = 3\lambda\hbar^2/2$ is the spin–orbit splitting. The spin–orbit splitting Δ is comparable to Δ of its constituent atoms. In diamond structure semiconductors, there is only one type of atom, so there is no confusion. In zinc-blende semiconductors, there are two types of atoms, which contribute to the spin–orbit splitting with different weight according to the electron probability distribution (and thus Δ is more close to that of the anions). However, in $\mathbf{k}\cdot\mathbf{p}$ theory, Δ is considered as an input parameter, which takes into account contributions from both types of atoms.

With the spin–orbit coupling, the $\mathbf{k}\cdot\mathbf{p}$ Hamiltonian in (4.93) becomes

$$H=\frac{p^2}{2m_0}+V(\mathbf{r})+\frac{\hbar^2k^2}{2m_0}+\frac{\hbar}{m_0}\mathbf{k}\cdot\mathbf{p}+\lambda\mathbf{L}\cdot\mathbf{S}. \tag{4.145}$$

All terms apart from the spin–orbit coupling term operates only on the orbital degree of freedom, thus the matrix elements of them between different spin states are zero. Using the six eigenstates of angular momentum as basis, the Luttinger Hamiltonian with spin–orbit coupling can be obtained as

$$\begin{bmatrix} -P-Q & S & -R & 0 & \frac{1}{\sqrt{2}}S & -\sqrt{2}R \\ S^+ & -P+Q & 0 & -R & \sqrt{2}Q & -\sqrt{\frac{3}{2}}S \\ -R^+ & 0 & -P+Q & -S & -\sqrt{\frac{3}{2}}S^+ & -\sqrt{2}Q \\ 0 & -R^+ & -S^+ & -P-Q & \sqrt{2}R^+ & \frac{1}{\sqrt{2}}S^+ \\ \frac{1}{\sqrt{2}}S^+ & \sqrt{2}Q^+ & -\sqrt{\frac{3}{2}}S & \sqrt{2}R & -P-\Delta & 0 \\ -\sqrt{2}R^+ & -\sqrt{\frac{3}{2}}S^+ & -\sqrt{2}Q^+ & \frac{1}{\sqrt{2}}S & 0 & -P-\Delta \end{bmatrix} \begin{matrix} |HH\uparrow\rangle \\ |LH\uparrow\rangle \\ |LH\downarrow\rangle \\ |HH\downarrow\rangle \\ |SO\uparrow\rangle \\ |SO\downarrow\rangle \end{matrix} \tag{4.146}$$

where superscript + means Hermitian conjugate, and

$$P=\frac{\hbar^2}{2m_0}\gamma_1(k_x^2+k_y^2+k_z^2), \tag{4.147a}$$

$$Q=\frac{\hbar^2}{2m_0}\gamma_2(k_x^2+k_y^2-2k_z^2), \tag{4.147b}$$

$$S=\frac{\hbar^2}{m_0}\sqrt{3}\gamma_3(k_x-ik_y)k_z, \tag{4.147c}$$

$$R=-\frac{\hbar^2}{2m_0}\sqrt{3}\left(\gamma_2(k_x^2-k_y^2)-2i\gamma_3k_xk_y\right), \tag{4.147d}$$

where γ_1, γ_2, and γ_3 are the Luttinger parameters, which relate to the other three parameters L, M, and N by

$$-\frac{\hbar^2}{2m_0}\gamma_1 = \frac{1}{3}(L+2M), \tag{4.148a}$$

$$-\frac{\hbar^2}{2m_0}\gamma_2 = \frac{1}{6}(L-M), \tag{4.148b}$$

$$-\frac{\hbar^2}{2m_0}\gamma_3 = \frac{N}{6}. \tag{4.148c}$$

The six basis are the eigenfunctions of Hamiltonian (4.146) at $k = 0$ with eigenenergies $0, 0, 0, 0, -\Delta, -\Delta$. They correspond to the state of HH spin-up state ($|HH\uparrow\rangle$), LH spin-up state ($|LH\uparrow\rangle$), LH spin-down state ($|LH\downarrow\rangle$), HH spin-down state ($|HH\downarrow\rangle$), split-off hole spin-up state ($|SO\uparrow\rangle$), and split-off hole spin-down state ($|SO\downarrow\rangle$).

4.8.3 4 × 4 Analytical Energy Dispersion

When the energy range of interest is sufficiently smaller than the split-off energy, it is usual to assume that the coupling from the split-off bands can be ignored. In this case, the electronic structures of the HH and LH bands can approximately be described by a 4×4 Hamiltonian [the upper-left 4×4 matrix block in the Hamiltonian (4.146)]. The eigenenergies of this 4×4 Hamiltonian can be analytically obtained, and are given by

$$\begin{aligned} E(k) &= -P \pm \sqrt{|Q|^2+|S|^2+|R|^2} \\ &= -\frac{\hbar^2}{2m_0}\left[\gamma_1 k^2 \pm \sqrt{4\gamma_2^2k^4 + 12(\gamma_3^2-\gamma_2^2)(k_x^2k_y^2+k_y^2k_z^2+k_z^2k_x^2)}\right] \\ &= Ak^2 \pm \sqrt{B^2k^4 + C^2(k_x^2k_y^2+k_y^2k_z^2+k_z^2k_x^2)}, \end{aligned} \tag{4.149}$$

By diagonalizing the 4×4 Hamiltonian, the analytic expressions for the effective masses for both the HH and LH bands along $\langle 001\rangle$ and $\langle 110\rangle$ are given by

$\langle 001\rangle$:

$$\frac{m_{hh}^*}{m_0} = \frac{1}{\gamma_1 - 2\gamma_2}, \tag{4.150a}$$

$$\frac{m_{lh}^*}{m_0} = \frac{1}{\gamma_1 + 2\gamma_2}. \tag{4.150b}$$

$\langle 110\rangle$:

$$\frac{m_{hh}^*}{m_0} = \frac{1}{\gamma_1 - \sqrt{\gamma_2^2+3\gamma_3^2}}, \tag{4.151a}$$

$$\frac{m_{lh}^*}{m_0} = \frac{1}{\gamma_1 + \sqrt{\gamma_2^2+3\gamma_3^2}}. \tag{4.151b}$$

The Luttinger parameters γ_2 and γ_3 determine the interband coupling. In this Hamiltonian, the HH couples with LH band through parameter γ_2 along $\langle 100\rangle$, while they couple each other through γ_3 along $\langle 110\rangle$. Generally $\gamma_2 < \gamma_3$, and thus the band repulsion along $\langle 110\rangle$ is stronger, and the HH effective mass is larger and LH effective mass is smaller along $\langle 110\rangle$ than along $\langle 100\rangle$. The relative difference between γ_2 and γ_3 indicates the anisotropy of the valence bands. If $\gamma_2 = \gamma_3$, the valence bands are isotropic. This is evident also by inspecting the energy dispersion (4.149), where $C = 0$ when $\gamma_2 = \gamma_3$.

4.8.4 Coordinate Transformation

For bulk systems, the energy dispersion along an arbitrary direction, which may be described by a pair of angles, θ and ϕ, can be obtained fairly easily by resolving the vector $\mathbf{k} = (k_x, k_y, k_z)$ along x, y, and z directions, $\mathbf{k} = k(\sin\theta\cos\phi, \sin\theta\sin\phi, \cos\theta)$, and substituting the vector components into the $\mathbf{k}\cdot\mathbf{p}$ Hamiltonian, where θ and ϕ are the angles in the corresponding spherical coordinate system. However, it is more convenient to make a coordinate transformation in some cases, e.g., in a confined low-dimensional system where the confining direction is not along the $\langle 100\rangle$ direction, or, to make an integration over a certain portion of solid angle. Suppose to transfer $\mathbf{k}_z$ to $\mathbf{k}'_z$, which is the wave vector along z of the new coordinate system, pointing into $(\theta,\ \phi)$ direction in the original coordinate system. This transformation can be achieved by a two-step rotation. First rotate about z axis by an angle ϕ, then rotate about the new y axis by an angle θ. The new z axis then points to the (θ, ϕ) direction in the old coordinate system. However, a directionally pointed z axis does not necessarily correspond to only one coordinate system. An arbitrary rotation about the new z can make a different coordinate system. Thus, an arbitrary rotation, or, the relation between two arbitrary coordinate systems in 3D space needs to be described by three angles. The third angle, ψ, describes the rotation about the new z. These three angles are called Euler angles, corresponding to rotation zyz. Using s_i and c_i to represent sine and cosine functions, where $i = 1, 2, 3$ to represent ϕ, θ, ψ, respectively, the transformation between k vectors is

$$\begin{pmatrix} k_{x'} \\ k_{y'} \\ k_{z'} \end{pmatrix} = \begin{pmatrix} c_1c_2c_3 - s_1s_3 & c_2c_3s_1 + c_1s_3 & -c_3s_2 \\ -c_3s_1 - c_1c_2s_3 & c_1c_3 - c_2s_1s_3 & s_2s_3 \\ c_1s_2 & s_1s_2 & c_2 \end{pmatrix} \begin{pmatrix} k_x \\ k_y \\ k_z \end{pmatrix}. \tag{4.152}$$

This transformation is useful to calculate the subband structures of semiconductors with surface orientation other than (100), such as, the (110) surface-orientated MOSFETs. However, in often cases, the in-plane axis directions are not important to us so that we can usually neglect the last step of rotation. Under such a case, $\psi = 0$, and the complexity of the transformation matrix is greatly reduced. For example, for (110) oriented wafer surface, if we choose the (110) direction as the new z direction, then $\theta = \pi/2$ and $\phi = \pi/4$. The new coordinate axes are $x' = -z$, $y' = (y - x)/\sqrt{2}$, and $z' = (x + y)/\sqrt{2}$. Under this two-step rotation, the basis functions are transformed as

$$\begin{pmatrix} X' \\ Y' \\ Z' \end{pmatrix} = \begin{pmatrix} \cos\theta\cos\phi & \cos\theta\sin\phi & -\sin\theta \\ -\sin\phi & \cos\phi & 0 \\ \sin\theta\cos\phi & \sin\theta\sin\phi & \cos\theta \end{pmatrix} \begin{pmatrix} X \\ Y \\ Z \end{pmatrix}. \tag{4.153}$$

Spin is invariant under a rotation of 4π, and the transformation is more complex than Cartesian coordinates. But with a total angular momentum of $\hbar/2$, its components along any arbitrary direction are $\hbar/2$. In the transformed coordinate system, the new spin states may be just simply represented by $|\uparrow'\rangle$ and $|\downarrow'\rangle$, following the same commutation rule: $\langle\uparrow' \mid \downarrow'\rangle = 0$, and $\langle\uparrow' \mid \uparrow'\rangle = \langle\downarrow' \mid \downarrow'\rangle = 1$.

4.9 KANE'S MODEL WITH REMOTE BAND COUPLING

When the band gap is large and the valence bands are far away from the other bands, Luttinger model is appropriate by taking into account the remote band coupling perturbatively. While when the band gap is small, which is the case for narrow gap III–V semiconductors, the coupling between the conduction and valence bands is strong. Taking the coupling from the conduction band into account by simple perturbation is not adequate. Under this situation, we may include the conduction band explicitly into the degenerate set. The conduction band of direction gap semiconductors is composed of s states, thus we add two more basis states

$$|iS\uparrow\rangle, \quad |iS\downarrow\rangle \tag{4.154}$$

to the six basis states in the Luttinger model, where a constant i is added before the state $|S\rangle$ for convenience to construct the Hamiltonian matrix later. The Hamiltonian matrix elements between the s states are

$$\langle S\uparrow|H|S\uparrow\rangle = E_g + \frac{\hbar^2k^2}{2m_0}, \tag{4.155a}$$

$$\langle S\uparrow|H|S\downarrow\rangle = 0, \tag{4.155b}$$

since the spin–orbit coupling does not affect the s states (the orbital angular momentum for the s state is 0). By symmetry, the only nonvanishing matrix elements between the s and p states are in the form of

$$P = \frac{-i\hbar}{m_0}\langle S|p_z|Z\rangle, \tag{4.156}$$

which is the Kane's parameter, representing the coupling strength between the conduction and valence bands. Kane's model only considers the coupling between the conduction and valence bands, while neglecting the remote bands' interaction. The explicit Hamiltonian matrix of Kane's model is omitted here, but it is easy to obtain.

From Kane's model, the conduction band energy dispersion for the direct gap semiconductors can be obtained, and the electron effective mass is found related to the band gap and split-off energy by

$$\frac{1}{m^*} = \frac{1}{m_0} + \frac{P^2}{3m_0}\left(\frac{2}{E_g} + \frac{1}{E_g + \Delta}\right). \tag{4.157}$$

However, due to the neglecting of the remote band coupling, the valence band structure is not correct. Nevertheless, taking into account the remote band coupling is easy, by following the discussion for the Luttinger model. There is one more parameter, though, for the coupling between the s states and the remote bands, which is defined as

$$F = \frac{1}{m_0}\sum_{\alpha} \frac{|\langle S|p_x|\alpha\rangle|^2}{E_c - E_\alpha}. \tag{4.158}$$

With $|iS\uparrow\rangle$ and $|iS\downarrow\rangle$ taken as the first and second basis states, the 8×8 Kane's Hamiltonian with the remote band coupling taken into account is then given by

$$\begin{bmatrix}
E_g+\frac{\hbar^2k^2}{2m_0}\gamma_4 & 0 & \frac{-1}{\sqrt{2}}Vk_+ & -\sqrt{\frac{2}{3}}Vk_z & \frac{1}{\sqrt{6}}Vk_- & 0 & \frac{1}{\sqrt{3}}Vk_z & \frac{-1}{\sqrt{3}}Vk_- \\
0 & E_g+\frac{\hbar^2k^2}{2m_0}\gamma_4 & 0 & \frac{1}{\sqrt{6}}Vk_+ & \sqrt{\frac{2}{3}}Vk_z & \frac{-1}{\sqrt{2}}Vk_- & \frac{1}{\sqrt{3}}Vk_+ & \frac{1}{\sqrt{3}}Vk_z \\
\frac{-1}{\sqrt{2}}Vk_- & 0 & -P-Q & S & -R & 0 & \frac{1}{\sqrt{2}}S & -\sqrt{2}R \\
-\sqrt{\frac{2}{3}}Vk_z & \frac{1}{\sqrt{6}}Vk_- & S^+ & -P+Q & 0 & -R & \sqrt{2}Q & -\sqrt{\frac{3}{2}}S \\
\frac{1}{\sqrt{6}}Vk_+ & \sqrt{\frac{2}{3}}Vk_z & -R^+ & 0 & -P+Q & -S & -\sqrt{\frac{3}{2}}S^+ & -\sqrt{2}Q \\
0 & \frac{-1}{\sqrt{2}}Vk_+ & 0 & -R^+ & -S^+ & -P-Q & \sqrt{2}R^+ & \frac{1}{\sqrt{2}}S^+ \\
\frac{1}{\sqrt{3}}Vk_z & \frac{1}{\sqrt{3}}Vk_- & \frac{1}{\sqrt{2}}S^+ & \sqrt{2}Q^+ & -\sqrt{\frac{3}{2}}S & \sqrt{2}R & -P-\Delta & 0 \\
\frac{-1}{\sqrt{3}}Vk_+ & \frac{1}{\sqrt{3}}Vk_z & -\sqrt{2}R^+ & -\sqrt{\frac{3}{2}}S^+ & -\sqrt{2}Q^+ & \frac{1}{\sqrt{2}}S & 0 & -P-\Delta
\end{bmatrix} \tag{4.159}$$

where

$$k_+ = k_x + ik_y, \tag{4.160a}$$

$$k_- = k_x - ik_y, \tag{4.160b}$$

$$\gamma_4 = 1 + 2F, \tag{4.160c}$$

$$V = \sqrt{\frac{\hbar^2}{m_0}\frac{E_p}{2}}, \tag{4.160d}$$

and the Luttinger parameters in P, Q, M, and L are to be replaced by

$$\gamma_1' = \gamma_1 - \frac{E_p}{3E_g}, \tag{4.161a}$$

$$\gamma_2' = \gamma_2 - \frac{E_p}{6E_g}, \tag{4.161b}$$

$$\gamma_3' = \gamma_3 - \frac{E_p}{3E_g}, \tag{4.161c}$$

to remove from the original Luttinger parameters the coupling to the conduction band, which is explicitly accounted for in the above Hamiltonian. The parameter E_p is defined as

$$E_p = \frac{2m_0}{\hbar^2} P^2 \tag{4.162}$$

related to the Kane's parameter P defined in Eq. (4.156). The relation between Kane's model, Luttinger model, and Kane's model with remote band coupling is illustrated in Fig. 4.16.

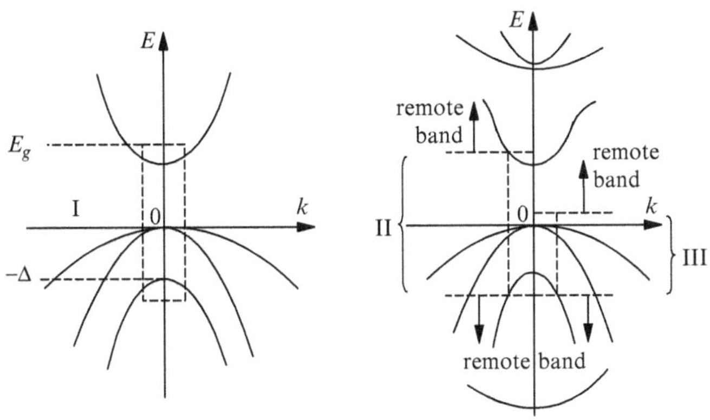

Fig. 4.16. The bands considered in Kane's model (*left*) and in Luttinger's model (*right*). In Kane's model, both conduction and valence bands are considered, while the remote band effects are neglected. In Luttinger's model, the valence bands are considered with the remote band's coupling

4.10 BAND STRUCTURE OF SELECTED SEMICONDUCTORS

In this section, we demonstrate the band structure for some typical semiconductors. For Si, Ge, GaAs, etc. semiconductors whose band gaps are remarkably larger than the split-off energy, the six-band Luttinger Hamiltonian is suitable and we only give the valence band structures, while for InAs, and

Table 4.6. Band parameters for selected semiconductors

	$E_{g,dir}$ (eV)	$E_{g,ind}$ (eV)	Δ (eV)	m_n/m_0	γ_1	γ_2	γ_3	E_P (eV)
Si	4.15	1.12 (Δ)	0.044	0.92/0.19	4.22	0.39	1.44	–
Ge	0.81	0.664 (L)	0.29	1.59/0.082	13.4	4.24	5.69	–
GaAS	1.42	1.71 (L)	0.34	0.063	6.98	2.06	2.93	28.8
InAs	0.354	1.53 (L)	0.38	0.022	20.0	8.5	9.2	21.5
InSb	0.172	1.03 (L)	0.85	0.014	34.8	15.5	16.5	23.3

Electron effective mass is for the lowest conduction band, and E_P is only for the direct gap semiconductors.

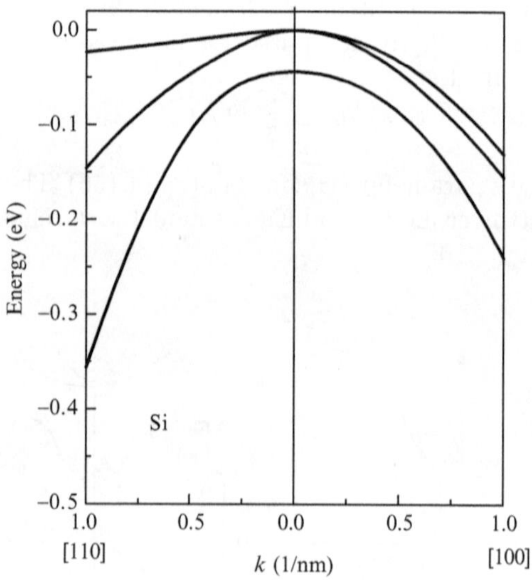

Fig. 4.17. Si valence bands calculated using Luttinger model

InSb etc. narrow gap semiconductors we use the eight-band Kane's model including the remote band coupling to compute the conduction and valence band structures altogether. The band parameters for various semiconductors are listed in Table 4.6.

Band structure of Si. The valence band structure of Si is shown in Fig. 4.17. Si has a small split-off energy with only 44 meV, thus the interaction between the top valence bands, including both the HH and LH bands, and the split-off bands is strong. This makes the HH band of Si pronouncedly anisotropic. From the parameter values listed in the table, we can see that smaller split-off energy leads to larger difference between γ_2 and γ_3, thus more warped valence bands. The HH effective mass for Si is $0.61m_0$ along $\langle 110\rangle$ and $0.29m_0$ along $\langle 100\rangle$. The LH is more isotropic than the HH band in any semiconductors. The LH effective mass for Si is $0.20m_0$ along $\langle 110\rangle$ and $0.15m_0$ along $\langle 100\rangle$.

The conduction band edge of Si is not located at the Γ point. However, we can use the single band perturbation theory discussed earlier around the Γ point. Suppose the extremum is located at $\mathbf{k_0} \neq 0$ ($\mathbf{k_0}$ is always along symmetry lines, thus the effective mass can be written in the coordinate system coinciding with the principle crystal axes), similarly we obtain the effective mass as

$$\frac{1}{m_i^*} = \frac{1}{m_0} + \frac{2}{m_0^2} \sum_{\alpha} \frac{\langle n\mathbf{k_0}|p_i|\alpha\mathbf{k_0}\rangle\langle\alpha\mathbf{k_0}|p_i|n\mathbf{k_0}\rangle}{E_n(\mathbf{k_0}) - E_\alpha(\mathbf{k_0})}, \tag{4.163}$$

where $i = x, y, z$. This effective mass is generally anisotropic. Along the symmetry axis, e.g., the Δ axis in the case for Si, the effective mass is called the longitudinal mass m_l, and the effective mass in the orthogonal direction is called the transverse mass m_t. The difference between m_l and m_t can be understood from the symmetry considerations. The band $|n\mathbf{k_0}\rangle$ itself and the remote bands in the longitudinal and transverse directions are different. Even though the denominator in (4.163) is the same for one remote band, the numerator is different along different axes. For instance, if we inspect the momentum matrix element between the S component of the $|n\mathbf{k_0}\rangle$ band along [100] and the HH band, which is composed of the $|X\rangle$ state, it is clear that $\langle S|p_x|X\rangle \neq 0$, while $\langle S|p_y|X\rangle = \langle S|p_z|X\rangle = 0$. Si conduction band consists of six Δ valleys, each of them is an ellipsoid with the longitudinal mass of 0.92 m_0, and the transverse mass of 0.19 m_0.

Band structure of Ge. The valence band structure of Ge is shown in Fig. 4.18. The split-off energy of Ge is 0.29 eV, much larger than Si. Correspondingly, the HH band of Ge is much less warped than Si. The HH effective mass for Ge is 0.38 m_0 along $\langle 110\rangle$ and 0.20 m_0 along $\langle 100\rangle$. The LH effective mass for Si is 0.042 m_0 along $\langle 110\rangle$ and 0.046 m_0 along $\langle 100\rangle$. The LH mass for Ge is the lightest among semiconductors.

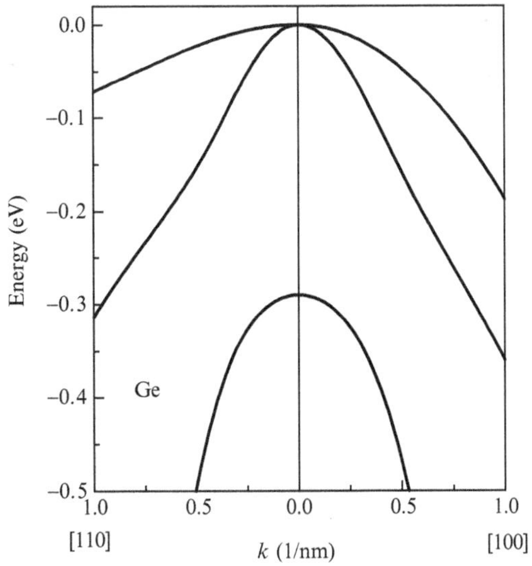

Fig. 4.18. Ge valence bands calculated using Luttinger model

The conduction band edge of Ge is located at the L point. The two L points at the two facing hexagon surfaces differ by a vector of $\frac{2\pi}{a}(1,1,1)$, which corresponds to one reciprocal lattice vector, and thus are equivalent. Therefore, the Ge conduction band has four valleys. Each conduction valley is an ellipsoid with the longitudinal direction along one $\langle 111 \rangle$ direction. The longitudinal mass of 1.66 m_0, and the transverse mass is 0.082 m_0.

Band structure of GaAs. The valence band structure of GaAs is shown in Fig. 4.19. The split-off energy of Ge is 0.34 eV, similar to that of Ge. Also, the HH band of GaAs is not warped very much. The HH effective mass for GaAs is very large, being 0.67 m_0 along $\langle 110 \rangle$ and 0.39 m_0 along $\langle 100 \rangle$. The LH is close to isotropic with effective mass being 0.081 m_0 along $\langle 110 \rangle$ and 0.090 m_0 along $\langle 100 \rangle$.

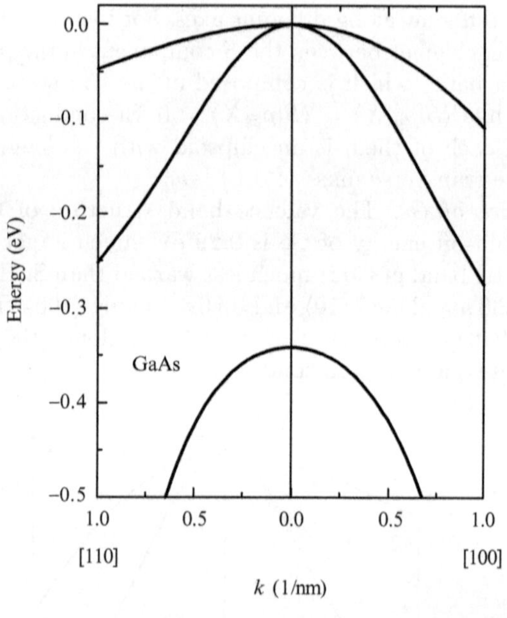

Fig. 4.19. GaAs valence bands calculated using Luttinger model

GaAs is a direct gap semiconductor with conduction band edge located at the Γ point. As we mentioned earlier, a single band at the Γ point is isotropic around the band extremum, so the energy surface for GaAs conduction band is a sphere at low energy limit. The electron effective mass is about 0.063 m_0 at room temperature. The second conduction band is located at the L point. The energy difference between the Γ valley and the L valley is only 0.29 eV. The longitudinal mass of the L valley is 1.9m_0 and the transverse mass is 0.075m_0. When the electron energy or density in question is high, there is a very large chance that the L valley is also occupied, and thus the L valley occupation shall also be considered.

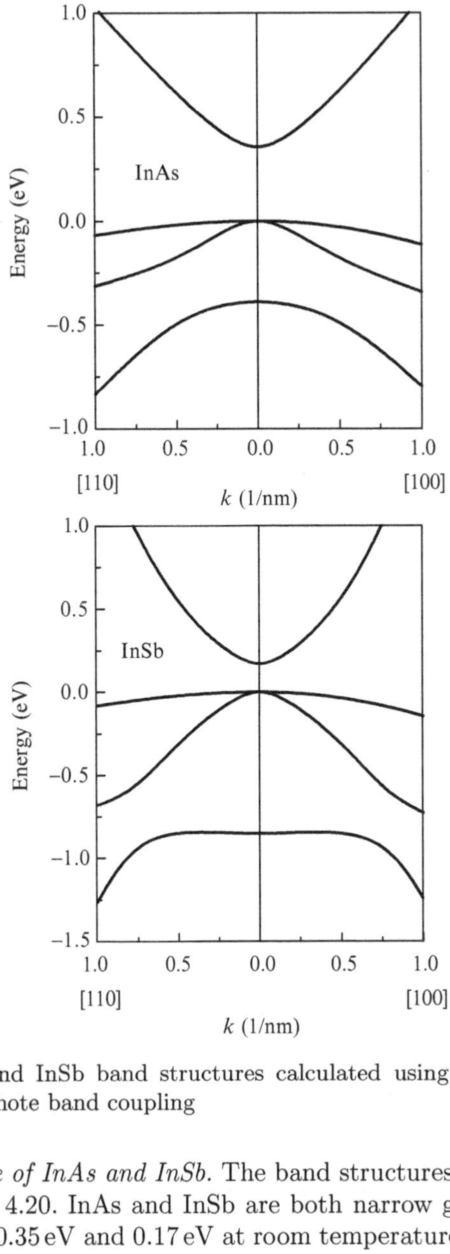

Fig. 4.20. InAs and InSb band structures calculated using Kane's model with consideration of remote band coupling

Band structure of InAs and InSb. The band structures for InAs and InSb are shown in Fig. 4.20. InAs and InSb are both narrow gap semiconductors with band gap of 0.35 eV and 0.17 eV at room temperature, respectively. The split-off energies for InAs and InSb are 0.38 and 0.85 eV, respectively. For InAs, the split-off energy is comparable to the band gap, while for InSb, the split-off energy is much larger than the band gap, thus in many cases, a six-band model including only the conduction and the $J = 3/2$ bands is employed for the InSb bands structure. The HH and LH band have no essential difference from those of Si, Ge, and GaAs.

The electron effective mass for InAs and InSb is extremely small. It is about $0.024m_0$ for InAs and $0.014m_0$ for InSb. However, due to the narrow gap nature, the electron effective mass value is only valid at the Γ point. Away from the Γ point, the nonparabolic character of the conduction band is very strong. In high electron energy range, the energy is nearly linear to k. The nonparabolicity can be described by one first-order nonparabolicity parameter α and a second-order parameter β, and the energy dispersion takes the form (we call this the empirical expression)

$$\frac{\hbar^2 k^2}{2m_\Gamma^*} = E(1 + \alpha E + \beta E^2), \tag{4.164}$$

where α and β are related to the band gap as

$$\alpha = \frac{1}{E_g}\left(1 - \frac{m_\Gamma^*}{m_0}\right)^2, \tag{4.165a}$$

$$\beta = -\frac{2}{E_g^2}\frac{m_\Gamma^*}{m_0}\left(1 - \frac{m_\Gamma^*}{m_0}\right). \tag{4.165b}$$

The values of α and β for GaAs, InAs, and InSb are shown in Table. 4.7. We show the plots for the conduction bands of InAs and InSb computed by the parabolic approximation using the effective mass at the Γ point, by the eight-band Kane's model with remote band coupling and the empirical expression in Fig. 4.21. At low-energy range, the three different models show no big difference. The three energy dispersion lines coincide very well around the Γ point. Far away from the Γ point, the results of the latter two models show great nonparabolicity.

Table 4.7. Nonparabolicity parameters for GaAs, InAs and InSb

Material	α (eV^{-1})	β (eV^{-2})
GaAS	0.62	−0.059
InAs	2.7	−0.34
InSb	5.6	−0.93

4.11 DENSITY OF STATES AND CONDUCTIVITY MASS

Density of States. In solids, electronic states are distinguished by two quantum numbers, the band number n and the wave vector k. The E–k diagram is a good and traditional way to show a band structure. However, its 2D nature cannot provide a detailed description of the electronic state distribution with energy. To outline the energy state situation in solids, the concept of "density

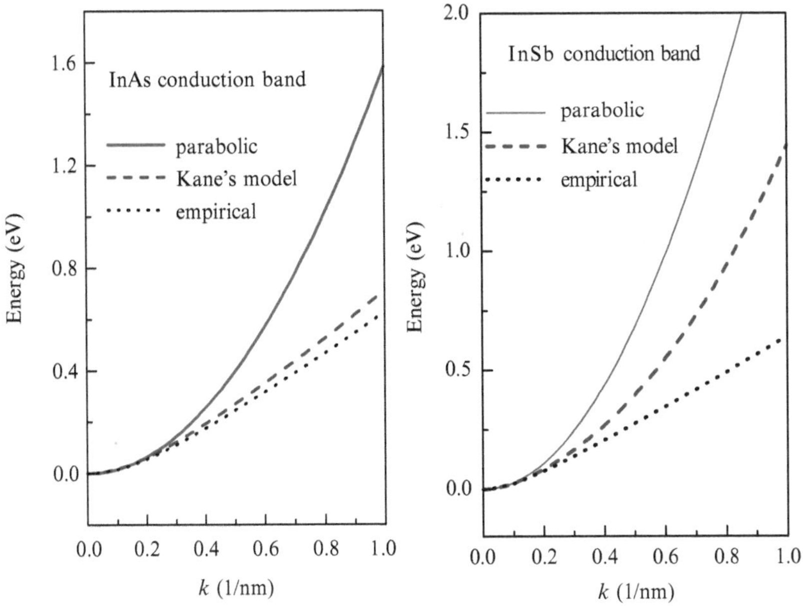

Fig. 4.21. Conduction band for InAs and InSb calculated with parabolic approximation, using Kane's model, and by empirical equation

of states" (DOS) is introduced. Considering the number of electronic states ΔZ in the energy range

$$E \to E + \Delta E, \tag{4.166}$$

DOS is defined as

$$N(E) = \lim_{\Delta E \to 0} \frac{\Delta Z}{\Delta E}. \tag{4.167}$$

ΔZ is actually the number of states between the equi-energy surfaces of energy E and $E + \Delta E$ in the three-dimensional k space. Since the distribution of k is even with a density of $\frac{V}{(2\pi)^3}$, where V is the volume of the solid, thus

$$\Delta Z = \frac{V}{(2\pi)^3} \int_S \mathrm{d}S\mathrm{d}k, \tag{4.168}$$

where $\mathrm{d}k$ represents the distance between the two energy surfaces, and $\mathrm{d}S$ is an infinitesimal surface element on the energy surface, as shown in Fig. 4.22. Since

$$\Delta E = |\nabla_k E|\mathrm{d}k, \tag{4.169}$$

thus

$$\Delta Z = \left(\frac{V}{(2\pi)^3} \int_S \frac{\mathrm{d}S}{|\nabla_k E|} \right) \Delta E. \tag{4.170}$$

Therefore, we obtain the equation for calculating the DOS

$$N(E) = \frac{V}{(2\pi)^3} \int\limits_S \frac{\mathrm{d}S}{|\nabla_k E|}. \tag{4.171}$$

If the band structure $E(\mathbf{k})$ is known, the DOS can be obtained by solving the above equation. If there is no magnetic field applied, spin states are normally not split in nonmagnetic semiconductors. Taking into account the spin degeneracy, we then have

$$N(E) = \frac{V}{4\pi^3} \int_S \frac{dS}{|\nabla_k E|}. \tag{4.172}$$

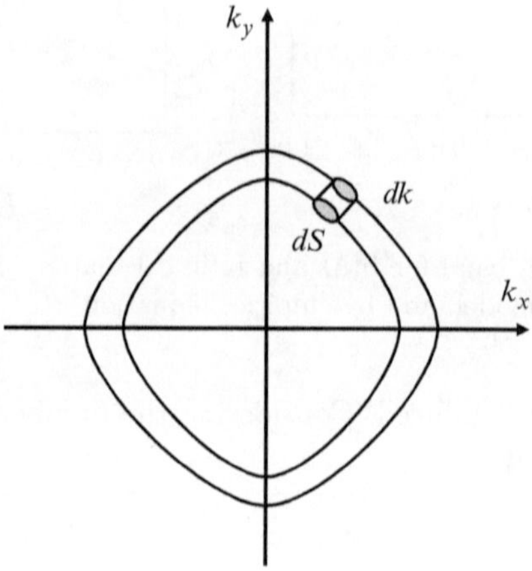

Fig. 4.22. Illustration of a unit volume enclosed by equi-energy shift in the k-space. DOS is the number of states enclosed by the equi-energy surface shift by a unit energy in the k-space

DOS Effective Mass. For free electrons,

$$E(\mathbf{k}) = \frac{\hbar^2 k^2}{2m_0} \tag{4.173}$$

is parabolic and only depends on the magnitude of $\mathbf{k}$, and thus the equi-energy surface is spherical. Obviously, the volume for a sphere with radius k in k-space is

$$Z(E) = 2\frac{V}{(2\pi)^3}\frac{4\pi}{3}k^3 = 2\frac{V}{(2\pi)^3}\frac{4\pi}{3}\left(\frac{2m_0 E}{\hbar^2}\right)^{3/2}. \tag{4.174}$$

The DOS is then given by

$$N(E) = \frac{dZ(E)}{dE} = \frac{V}{2\pi^2}\left(\frac{2m_0}{\hbar^2}\right)^{3/2} E^{1/2}. \tag{4.175}$$

Certainly, for a parabolic band with effective mass m^*, the DOS is given by the same equation with m_0 replaced by m^*. We can see that this effective mass is the inverse of the curvature of the band,

$$\frac{1}{m^*} = \frac{1}{\hbar^2} \frac{d^2E}{dk^2}. \tag{4.176}$$

Thus, this mass is also called the DOS effective mass. When the DOS effective mass is used in context, the parabolic approximation is already implicitly assumed. We can see this in a very simple instance. For parabolic energy dispersion, d^2E/dk^2 is the same at any k. However, as we discussed earlier, the conduction bands of the narrow gap semiconductors show linear relation with k at relatively large k, where $d^2E/dk^2 \sim 0$, and thus the DOS mass is infinite if we use the same equation, which is obviously not true. In such cases, only (4.171) can be used to calculate the DOS. The density of states in a unit of volume for InSb conduction band is shown in Fig. 4.23, where the DOS with the parabolic approximation is compared to that with nonparabolicity taken into account by assuming $\alpha = 5.6\, eV^{-1}$, and $\beta = 0$.

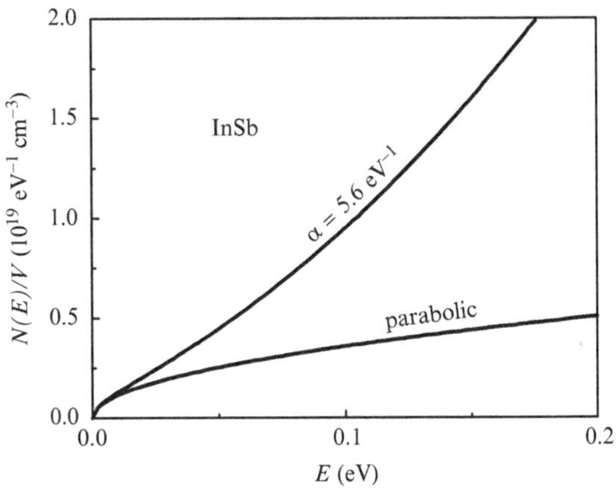

Fig. 4.23. DOS of InSb conduction band by assuming a parabolic band and by using the empirical expression which considers the nonparabolicity

The DOS of a parabolic band shows $E^{1/2}$ dependence and is proportional to $m^{3/2}$. For anisotropic bands such as the HH and LH bands, a DOS effective mass can still be assigned to each band, by assuming the DOS at energy E equal to an isotropic spherical band with effective mass m^*. To make this DOS effective mass approximation hold, the energy E needs to be close to the valence band edge so that the bands are still parabolic, though not spherical.

Conductivity mass. Do not confuse the DOS effective mass with the conductivity effective mass. The conductivity effective mass is actually the momentum effective mass. In solids, an electron is quantum mechanically a wave packet consisting of different wave vectors, following a Gaussian distribution, with center wave vector **k**. Only when the wave packet distribution range is much larger than the size of the primitive cell (thus the momentum distribution of the wave packet is much less than the size of the Brillouin zone) can the electron be considered as a quasi-classical particle. When considering the quasi-classical motion of a electron with wave vector **k** in solids, its velocity is

$$\mathbf{v_k} = \frac{1}{\hbar}\nabla_{\mathbf{k}} E. \tag{4.177}$$

In MOSFETs when the bias between the source and drain is large, the carrier velocity will reach a saturation value. This takes place because the hole carriers have large energies and are away from the band edge. With large k, the band is no longer parabolic, but almost linear to k, and thus the velocity is a constant. There is also another important factor accounting for the velocity saturation, which will be discussed in Section 6.7.5.

When put in an electric field, an electron will experience a force **F**. In a time period dt, the field exerts work on the electron with value

$$\mathbf{F}\cdot\mathbf{v_k}\mathrm{d}t. \tag{4.178}$$

With the work on the electron, its energy must change. Its energy depends on the wave vector **k**, as discussed in the band structure calculations. Suppose the wave vector changes by $d\mathbf{k}$, then we obtain

$$\mathrm{d}\mathbf{k}\cdot\nabla_{\mathbf{k}} E = \mathbf{F}\cdot\mathbf{v_k}\mathrm{d}t. \tag{4.179}$$

With the electron velocity given by 4.177, we obtain

$$\left(\hbar\frac{d\mathbf{k}}{dt} - \mathbf{F}\right)\cdot\mathbf{v_k} = 0. \tag{4.180}$$

Thus, it is proven that along the $\mathbf{v_k}$ direction, the change of $\hbar\frac{\mathrm{d}\mathbf{k}}{\mathrm{d}t}$ is equal to **F**. When the electric field is perpendicular to $\mathbf{v_k}$, it can also be shown that in the direction perpendicular to the velocity, the change of $\hbar\frac{\mathrm{d}\mathbf{k}}{\mathrm{d}t}$ is also equal to **F**. Therefore, we obtain

$$\frac{\mathrm{d}}{\mathrm{d}t}(\hbar\mathbf{k}) = \mathbf{F}. \tag{4.181}$$

This equation dictates how the electronic states change under an external force. It exactly resembles the Newton's second law, with the momentum replaced by $\hbar\mathbf{k}$. When dealing with the transport problems in solids, $\hbar\mathbf{k}$ is considered as the momentum of the electrons, which already takes into effects of the periodic crystal potentials and describes how an electron moves under both the crystal potential and external forces.

On the other hand, the electron momentum can also be written as

$$m_c\mathbf{v_k} = m_c\frac{1}{\hbar}\nabla_{\mathbf{k}} E, \tag{4.182}$$

where m_c is the momentum effective mass. By equating $m_c \mathbf{v_k}$ to $\hbar\mathbf{k}$, we obtain the relation of the momentum effective mass with the electron energy $E(\mathbf{k})$,

$$\frac{1}{m_{c,i}} = \frac{1}{\hbar^2 k_i} \frac{\mathrm{d}E}{\mathrm{d}k_i}, \tag{4.183}$$

where $i = x, y, z$. This momentum effective mass is the conductivity mass in carrier transport in solids. We can see that for a parabolic band, the conductivity mass is equal to the DOS mass, while for a linear band, the conductivity mass is proportional to k.

For narrow gap semiconductor conduction bands, since the nonparabolicity effect is strong, the conductivity mass increases with energy quickly. Using the empirical expression (4.164) for the conduction band and only taking into account the first-order nonparabolicity correction (i.e., assume $\beta = 0$), combining with (4.183), we obtain a linear dependence of the conductivity mass on electron energy

$$m_c = m_\Gamma^*(1 + 2\alpha E). \tag{4.184}$$

The electron conductivity mass as a function of the electron energy of InAs and InSb is shown in Fig. 4.24.

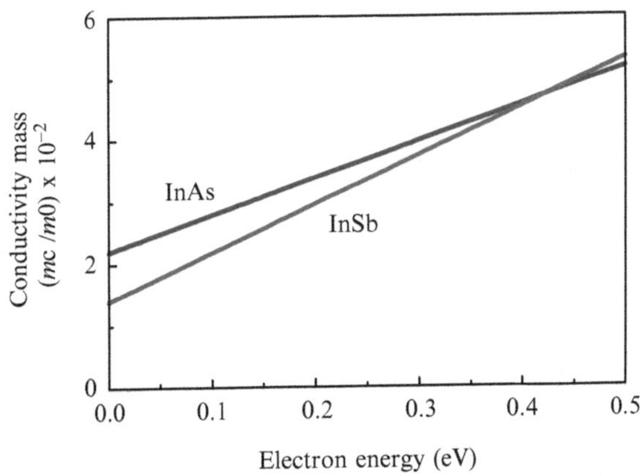

Fig. 4.24. Conductivity effective mass vs. electron energy for InAs and InSb by taking into account the nonparabolicity

Because the conductivity mass depends on electron energy, it also depends on the electron concentration. The electron conductivity mass of InSb and InAs vs. electron concentration is shown in Fig. 4.25.

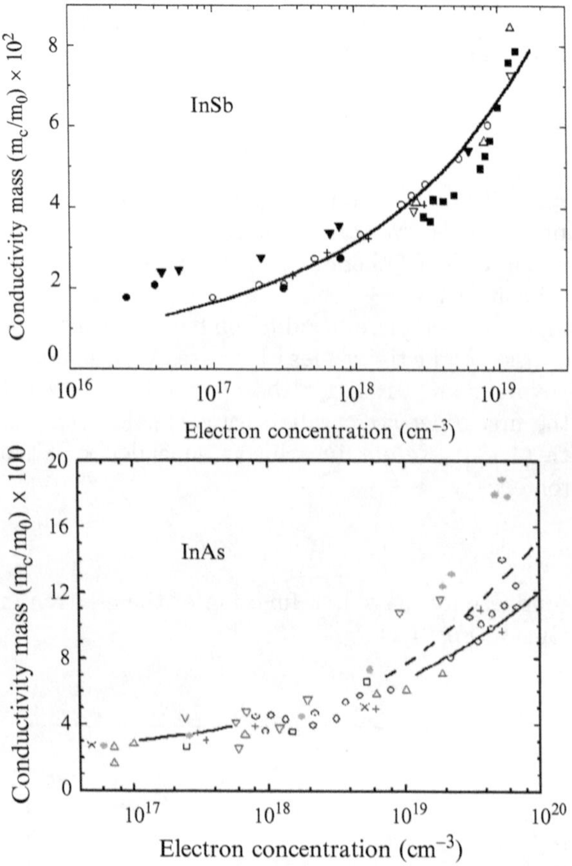

Fig. 4.25. Experimental results of electron effective mass vs. electron concentration for InSb and InAs. See Ref (Zawadski, 1974) for InSb and see (Semikolenova et al, 1978) for InAs

4.12 PIKUS–BIR STRAIN HAMILTONIAN

In this section we introduce the method to treat the effect of a homogeneous strain, i.e., the Pikus–Bir Hamiltonian, under the $\mathbf{k}\cdot\mathbf{p}$ framework, and explain how it is constructed.

One may think to treat the strain-induced effect as a perturbation, thus the strain-induced band structure change can be obtained through perturbation theory as we have discussed earlier. However, there are two difficulties for directly adopting the perturbation theory:

1. The strain-induced potential difference is not really small.
2. Strain changes the periodicity of the crystal.

In the uniformly deformed crystal, the potential is still periodic except that the function $V(\mathbf{r}')$ is a different potential from the undeformed potential $V_0(\mathbf{r})$. Even for an infinitesimal strain, the difference $V_0(\mathbf{r}) - V(\mathbf{r})$ at one same position can be of the order of $V_0(\mathbf{r})$, since at sufficient distance, the relative displacement of the two crystal systems will be of the order of the lattice constant. The other point is that in perturbation theory, the wave function of the perturbed system is always expressed as a superposition of the wave functions of the unperturbed system, thus these two sets of wave functions satisfy the same boundary conditions, which are set by the lattice periodicity. However, now strain changes the lattice periodicity, and consequently the periods of the lattice periodic functions $u_k(\mathbf{r})$ in the Bloch waves. To avoid these difficulties, Pikus and Bir used a coordinate transformation to make the deformed and undeformed crystals have the same boundary conditions, and then brought the strain effects into the perturbation formalism.

This transformation involves writing the coordinates of the deformed crystal in terms of those of the undeformed crystal. Under strain, which is represented by the strain tensor $\bar{\bar{\varepsilon}}$, the coordinates of the deformed and undeformed crystal are linked by the transformation

$$r_i' = r_i + \sum_j \varepsilon_{ij} r_j, \tag{4.185}$$

where $r_i, r_j = x, y, z$. Written in the vector form, it is

$$\mathbf{r}' = (1 + \bar{\bar{\varepsilon}}) \cdot \mathbf{r}. \tag{4.186}$$

For small strain, the inverse transformation is given by

$$\mathbf{r} = (1 - \bar{\bar{\varepsilon}}) \cdot \mathbf{r}'. \tag{4.187}$$

Correspondingly, the transformation between the reciprocal vectors is given by

$$\mathbf{k} = (1 + \bar{\bar{\varepsilon}}) \cdot \mathbf{k}'. \tag{4.188}$$

$V(\mathbf{r}')$ as a function of $\mathbf{r}'$ does not have the same periodicity of the original lattice. However, $V(\mathbf{r}')$ can be expressed as the function of $\mathbf{r}$ through the above transformation and thus converted to a function with the same boundary conditions with $V_0(\mathbf{r})$. Under the above transformation, we have

$$V(\mathbf{r}') = V[(1 + \bar{\bar{\varepsilon}}) \cdot \mathbf{r}] = V_0(\mathbf{r}) + \sum_{i,j} V_{ij} \varepsilon_{ij}, \tag{4.189}$$

where

$$V_{ij} = \lim_{\varepsilon \to 0} \frac{V[(1 + \bar{\bar{\varepsilon}}) \cdot \mathbf{r}] - V_0(\mathbf{r})}{\varepsilon_{ij}} = \frac{\partial V}{\partial \varepsilon_{ij}}. \tag{4.190}$$

Clearly, V_{ij} is related to deformation potentials. As discussed in the tight-binding model, the crystal potential $V(\mathbf{r})$ is assumed as the sum of the potentials of the individual ions, and deformation of the lattice causes only a displacement of lattice site $\mathbf{R_n}$, without distorting their potentials. Thus,

$$
\begin{aligned}
V[(1+\bar{\bar{\varepsilon}})\cdot\mathbf{r}] - V_0(\mathbf{r}) &= \sum_n V_a[(1+\bar{\bar{\varepsilon}})\cdot(\mathbf{r}-\mathbf{R_n})] - V_a(\mathbf{r}-\mathbf{R_n}) \\
&= \sum_n \sum_{i,j} \frac{\partial V_a(\mathbf{r}-\mathbf{R_n})}{\partial r_i}\varepsilon_{ij}(\mathbf{r}-\mathbf{R_n})_j,
\end{aligned}
\tag{4.191}
$$

and consequently,

$$
V_{ij} = \frac{1}{2}\sum_n \left[\frac{\partial V_a(\mathbf{r}-\mathbf{R_n})}{\partial r_i}(\mathbf{r}-\mathbf{R_n})_j + \frac{\partial V_a(\mathbf{r}-\mathbf{R_n})}{\partial r_j}(\mathbf{r}-\mathbf{R_n})_i\right], \tag{4.192}
$$

where $V_a(\mathbf{r}-\mathbf{R_n})$ is the atomic potential for an ion located at $\mathbf{R_n}$. It is appropriate to assume a spherical symmetry for $V_a(\mathbf{r})$, and thus the derivative of it with respect to r_i is an odd function of r_i.

Since

$$
\frac{\partial}{\partial r'_i} = \sum_j \frac{\partial r_j}{\partial r'_i}\frac{\partial}{\partial r_j} = \frac{\partial}{\partial r_i} - \sum_j \varepsilon_{ij}\frac{\partial}{\partial r_j}, \tag{4.193}
$$

we have

$$
\mathbf{p}' = \mathbf{p}\cdot(1-\bar{\bar{\varepsilon}}) \tag{4.194}
$$

and

$$
p'^2 = p^2 - 2\sum_{i,j} p_i\varepsilon_{ij}p_j. \tag{4.195}
$$

The Hamiltonian of the deformed system becomes

$$
H' = \frac{p'^2}{2m_0} + V(\mathbf{r}') = \frac{p^2}{2m_0} + V_0(\mathbf{r}) + H_\varepsilon, \tag{4.196}
$$

where

$$
H_\varepsilon = \sum_{i,j}\left(-\frac{1}{m_0}p_ip_j + V_{ij}\right)\varepsilon_{ij}. \tag{4.197}
$$

Next, we inspect the Bloch function in the deformed system,

$$
\begin{aligned}
&\psi_{n\mathbf{k}'}(\mathbf{r}') = \psi_{n\mathbf{k}'}[(1+\bar{\bar{\varepsilon}})\cdot\mathbf{r}] \\
&= e^{i\mathbf{k}'\cdot\mathbf{r}'}u_{n\mathbf{k}'}(\mathbf{r}') = e^{i\mathbf{k}'\cdot(1+\bar{\bar{\varepsilon}})\cdot\mathbf{r}}u_{n\mathbf{k}'}[(1+\bar{\bar{\varepsilon}})\cdot\mathbf{r}] = e^{i\mathbf{k}\cdot\mathbf{r}}u'_{n\mathbf{k}}(\mathbf{r}),
\end{aligned}
\tag{4.198}
$$

where $u'_{n\mathbf{k}}(\mathbf{r})$ is used to represent the periodic modulation part of the Bloch function for the strained system. If we substitute the above expression into the Schödinger equation in the deformed system

$$
H'\psi_{n\mathbf{k}'}(\mathbf{r}') = E_n(\mathbf{k}')\psi_{n\mathbf{k}'}(\mathbf{r}'), \tag{4.199}
$$

we obtain

$$\left[H_0 + \frac{\hbar}{m_0}\mathbf{k}\cdot\mathbf{p} + \frac{\hbar^2 k^2}{2m_0} + H_\varepsilon + H_{\varepsilon k}\right] u'_{n\mathbf{k}}(\mathbf{r}) = E_n(\mathbf{k}) u'_{n\mathbf{k}} \tag{4.200}$$

with

$$H_0 = \frac{\hbar^2 k^2}{2m_0} + V(\mathbf{r}), \tag{4.201a}$$

$$H_{\varepsilon k} = -\frac{2\hbar}{m_0}\sum_{i,j} k_i \varepsilon_{ij} p_j. \tag{4.201b}$$

The Hamiltonian in (4.200) is just the normal $\mathbf{k}\cdot\mathbf{p}$ Hamiltonian with the strain contribution. We can use the same perturbation procedure to treat the $\mathbf{k}\cdot\mathbf{p}$ term together with the strain term as the perturbation and expand the lattice periodic function $u'_{n\mathbf{k}}$ using the eigenfunctions of H_0. Basically, there is no essential difference from the procedure for treating the normal $\mathbf{k}\cdot\mathbf{p}$ Hamiltonian. However, as we mentioned earlier, we need to treat the $\mathbf{k}\cdot\mathbf{p}$ term up to the second order in k, while we only need to seek the first-order energy correction due to strain.

Next, we inspect the strain effects on a single band and one set of degenerate bands. In either case, the $\mathbf{k}\cdot\mathbf{p}$ term has the same Hamiltonian elements as in discussions in Sects. 4.7.3 and 4.7.4. Therefore, we only need to study the Hamiltonian elements of the strain terms.

To the first order of strain, the matrix element of $H_{\varepsilon k}$ is zero for a single s band such as the GaAs conduction band $|n\rangle$ due to its even parity, only leaving the nonvanishing elements of H_ε, which can be parameterized by

$$\begin{aligned}\langle n|H_\varepsilon|n\rangle &= \langle n|\sum_{i,j}\left(-\frac{1}{m_0}p_i p_j + V_{ij}\right)\varepsilon_{ij}|n\rangle \\ &= \langle n|\sum_{i=j}\left(-\frac{1}{m_0}p_i p_j + V_{ij}\right)\varepsilon_{ij}|n\rangle \\ &= a_c(\varepsilon_{xx} + \varepsilon_{yy} + \varepsilon_{zz}),\end{aligned} \tag{4.202}$$

where a_c is the conduction band deformation potential. Because of the isotropic nature of the s state and the symmetry property of V_{ij} by inspecting (4.192), $\langle n|V_{ij}|n\rangle = 0$ for $i \neq j$. Thus, for a singly degenerate band with edge located at the Γ point, the energy dispersion with strain is given by

$$E_n = E_{n0} + \frac{\hbar^2 k^2}{2m^*} + a_c(\varepsilon_{xx} + \varepsilon_{yy} + \varepsilon_{zz}). \tag{4.203}$$

For degenerate bands, the first-order perturbation theory is to expand the perturbation only in the space spanned by the basis states, which we choose as the six valence band states as defined in (4.142) as an example. Now H_0 also contains the spin–orbit coupling term, which we assume to be independent on strain. Similar to the single band case, the matrix elements of $H_{\varepsilon k}$ vanish, due to the parity consideration. Next, we inspect the matrix elements of H_ε. We have

$$\begin{aligned}\langle X|H_\varepsilon|X\rangle &= \sum_{ij}\langle X|(-\frac{1}{m_0}p_ip_j + V_{ij})\varepsilon_{ij}|X\rangle \\ &= \langle X|(-\frac{1}{m_0}p_xp_x + V_{xx})|X\rangle\varepsilon_{xx} \\ &+ \langle X|(-\frac{1}{m_0}p_yp_y + V_{yy})|X\rangle\varepsilon_{yy} \\ &+ \langle X|(-\frac{1}{m_0}p_zp_z + V_{zz})|X\rangle\varepsilon_{zz} \\ &= l\varepsilon_{xx} + m(\varepsilon_{yy} + \varepsilon_{zz}),\end{aligned} \tag{4.204}$$

and

$$\begin{aligned}\langle X|H_\varepsilon|Y\rangle &= \sum_{ij}\langle X|(-\frac{1}{m_0}p_ip_j + V_{ij})\varepsilon_{ij}|Y\rangle \\ &= \langle X|(-\frac{1}{m_0}p_xp_y + V_{xy})|Y\rangle\varepsilon_{xy} \\ &= n\varepsilon_{xy},\end{aligned} \tag{4.205}$$

where l, m, and n are three different deformation potentials.

The matrix elements of H_ε can be expressed by identifying a straightforward relation

$$k_{ij} \leftrightarrow \varepsilon_{ij} \tag{4.206}$$

and defining corresponding deformation potentials

$$\frac{\hbar^2\gamma_1}{2m_0} \leftrightarrow a_v, \tag{4.207a}$$

$$\frac{\hbar^2\gamma_2}{2m_0} \leftrightarrow -\frac{b}{2}, \tag{4.207b}$$

$$\frac{\hbar^2\gamma_3}{2m_0} \leftrightarrow -\frac{d}{2\sqrt{3}}, \tag{4.207c}$$

where a_v is the Pikus–Bir hydrostatic deformation potential for the valence band edge, and b and d are two Pikus–Bir shear deformation potentials. We need to note that the convention of the sign of a_v in some different literatures is different. One easy way to determine whether the positive or negative value is to be used is that the hydrostatic tensile strain tends to shift the valence band edge up. These Pikus–Bir deformation potentials relate to l, m, and n by

$$a_v = -\frac{2}{3}(l + 2m), \tag{4.208a}$$

$$b = \frac{2}{3}(l - m), \tag{4.208b}$$

$$d = \frac{n}{\sqrt{3}}. \tag{4.208c}$$

Thus, the strain Hamiltonian has an identical form as the $\mathbf{k} \cdot \mathbf{p}$ Hamiltonian (4.146), with the strain counterparts defined by

$$P_\varepsilon = a_v(\varepsilon_{xx} + \varepsilon_{yy} + \varepsilon_{zz}), \tag{4.209a}$$

$$Q_\varepsilon = -\frac{b}{2}(\varepsilon_{xx} + \varepsilon_{yy} - 2\varepsilon_{zz}), \tag{4.209b}$$

$$S_\varepsilon = -d(\varepsilon_{xz} - i\varepsilon_{yz}), \tag{4.209c}$$

$$R_\varepsilon = \frac{\sqrt{3}}{2} b(\varepsilon_{xx} - \varepsilon_{yy}) - id\varepsilon_{xy}. \tag{4.209d}$$

Table 4.8. Deformation potentials for selected semiconductors (eV)

	Ξ_d	Ξ_u	a_c	a_v	b	d
Si	−6.0	7.8		−2.46	−2.1	−4.8
Ge	−9.1	15.9		−1.24	−2.9	−5.3
GaAS			−7.17	−1.16	−2.0	−4.8
InAs			−5.08	−1.00	−1.8	−3.6
InSb			−6.94	−0.36	−2.0	−4.7

All values are in unit of eV. Ξ_d and Ξ_u are for the lowest indirect conduction bands of Si and Ge, and a_c is for the conduction bands of direct gap semiconductors.

Pikus and Bir also studied the strain effects using the theory of invariants based on symmetry. In summary, the strain Hamiltonian is also determined by symmetry the same as the $\mathbf{k} \cdot \mathbf{p}$ Hamiltonian. The number of deformation potentials needed in the strain Hamiltonian is the same as the number of independent effective mass parameters in the $\mathbf{k}{\cdot}\mathbf{p}$ Hamiltonian. At an arbitrary point in the Brillouin where there is no other symmetry operation except for the identity, the strain energy is

$$E_\varepsilon = D_{xx}\varepsilon_{xx} + D_{yy}\varepsilon_{yy} + D_{xx}\varepsilon_{zz}. \tag{4.210}$$

For Si and Ge conduction band edges, the total Hamiltonian with strain is written as

$$H = \frac{\hbar^2(k_l - k_0)^2}{2m_l^*} + \frac{\hbar^2 k_t^2}{2m_t^*} + D_{\shortparallel}\varepsilon_{\shortparallel} + D_\perp\varepsilon_\perp \tag{4.211}$$

where the subscripts l and $\shortparallel$ indicate values along the longitudinal direction and t and $\perp$ for the transverse directions. Because of the symmetry, both Si and Ge conduction bands need two effective masses and two deformation potentials to describe the band under strain. Following Herring and Vogt's notation, the energy shift of valley i for a homogeneous deformation described by a strain tensor $\bar{\bar{\varepsilon}}$ can be expressed as (de Walle, 1989)

$$\Delta E_c^i = (\Xi_d \bar{\bar{1}} + \Xi_u \{\hat{\mathbf{a}}_i \hat{\mathbf{a}}_i\}) : \bar{\bar{\varepsilon}}, \tag{4.212}$$

where $\bar{\bar{1}}$ is the unit tensor, $\hat{\mathbf{a}}_i$ is a unit vector parallel to the $\mathbf{k}$ vector of valley i, and $\{\}$ denotes a dyadic product. Ξ_d and Ξ_u are the dilation and uniaxial deformation potentials at the conduction band edges, related to $D_{\parallel}$ and $D_{\perp}$ in (4.211). The shift of the mean energy of the conduction band edges is

$$\Delta E_{c,av} = (\Xi_d + \frac{1}{3}\Xi_u)\bar{\bar{1}} : \bar{\bar{\varepsilon}} = (\Xi_d + \frac{1}{3}\Xi_u)Tr(\bar{\bar{\varepsilon}}), \tag{4.213}$$

where $Tr(\bar{\bar{\varepsilon}})$ stands for the trace of the strain tensor, which represents the crystal volume change. Thus, the quantity $(\Xi_d + \frac{1}{3}\Xi_u)$ corresponds to the hydrostatic deformation potential for the conduction band. Split of the conduction band edges is caused by the shear strain.

The conduction and valence band structures under strain for various semiconductors are discussed in the next section.

4.13 STRAINED BAND STRUCTURES

4.13.1 Conduction Band

Strain affects the conduction bands of Si, Ge, and direct gap III–V semiconductors differently due to the positions of their band edges in the Brillouin zone. For the direct gap semiconductors, only hydrostatic strain shifts the band edge according to (4.203). Shear strain terms do not have any effects in shifting or warping the conduction band located at the Γ point. However, this is not really true, since in (4.203), we neglected the coupling from the other bands. If strain shifts the other bands, the energy distance between them and the conduction band will change, thus from (4.116), we can see that the coupling strength changes. Also, if strain changes the symmetry of the band structure, and consequently the electron wave function along different directions, the momentum matrix element along different directions will also be altered. This may cause anisotropy in the strained conduction band, which is originally isotropic. We can show this using Kane's model for InAs conduction band as in Fig. 4.26, where we put the two conduction band edges together to show the difference. The strain is induced by 1 GPa compression along the [110] direction, which also results in a ~29meV band edge shift upward. The effective mass along [110] changes by about 10%. Aspnes and Cardona (Aspnes and Cardona, 1978) investigated the strain dependence of the electron effective masses for GaAs, and the results are reproduced in Fig. 4.27. Generally, the effect is very small. The largest change shown is less than 5% under 10 Kbar (1 GPa) tetragonal stress.

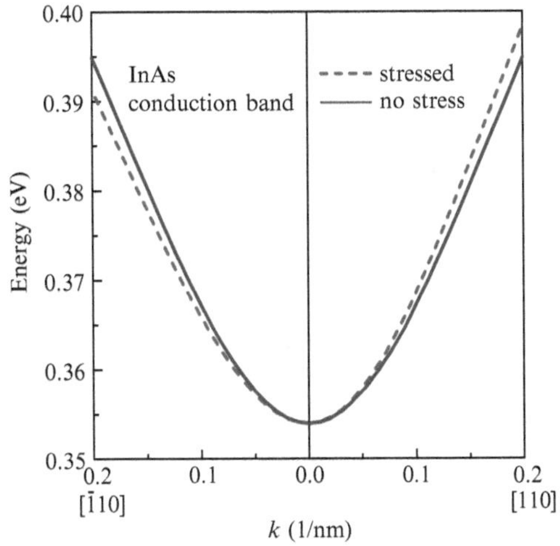

Fig. 4.26. InAs conduction band with and without 1 GPa [110] uniaxial compressive stress. Calculated using Kane's model

Stress Type		**Mass change**
Hydrostatic (330 MPa)		2.87%
Tetragonal (1 GPa)	longitudinal	–0.758%
	transverse	4.69%
Trigonal (1 GPa)	longitudinal	3.81%
	transverse	2.41%

Fig. 4.27. Electron effective mass changes under various stress conditions. After Aspnes and Cardona (Aspnes and Cardona, 1978)

For conduction bands with edges located not at the Γ point as is the case for Si and Ge, except for the net shift of the weight of the conduction band caused by the hydrostatic strain, which is given by

$$\Delta E_{c,av} = (\Xi_d + \frac{1}{3}\Xi_u)Tr(\bar{\bar{\varepsilon}}), \tag{4.214}$$

where Ξ_d is the hydrostatic deformation potential and Ξ_u is the shear deformation potential, shear strain may lift the star degeneracy. The conduction band minima along Δ, which is the case for Si, are split by uniaxial stress along $\langle 001\rangle$ or $\langle 110\rangle$. The splitting of the bands with respect to the mean energy is given by

$$\Delta E_c^{001} = \frac{2}{3}\Xi_u^{\Delta}(\varepsilon_{zz} - \varepsilon_{xx}), \tag{4.215a}$$

$$\Delta E_c^{100,010} = -\frac{1}{3}\Xi_u^{\Delta}(\varepsilon_{zz} - \varepsilon_{xx}). \tag{4.215b}$$

However, uniaxial stress along ⟨111⟩ does not split the Δ valleys. For conduction band minima at the L point, which is the case for Ge, ⟨001⟩ stress does not have effect. Uniaxial stress along ⟨110⟩ or ⟨111⟩ split the L bands with the band edge energy given by

uniaxial stress along ⟨110⟩:

$$\Delta E_c^{111,11\bar{1}} = \frac{2}{3}\Xi_u^L \varepsilon_{xy}, \tag{4.216a}$$

$$\Delta E_c^{\bar{1}11,1\bar{1}1} = -\frac{2}{3}\Xi_u^L \varepsilon_{xy}. \tag{4.216b}$$

uniaxial stress along ⟨111⟩ (in this case, $\varepsilon_{xy} = \varepsilon_{yz} = \varepsilon_{zx}$):

$$\Delta E_c^{111} = \frac{2}{3}\Xi_u^L \varepsilon_{xy}, \tag{4.217a}$$

$$\Delta E_c^{\bar{1}11,1\bar{1}1,11\bar{1}} = -\frac{1}{3}\Xi_u^L \varepsilon_{xy}. \tag{4.217b}$$

The Si and Ge conduction valley shifts and splitting are shown schematically in Figs. 4.28 and 4.29. From these results, we can see that the conduction band shift and splitting are precisely consistent with the symmetry analysis and tight-binding discussions.

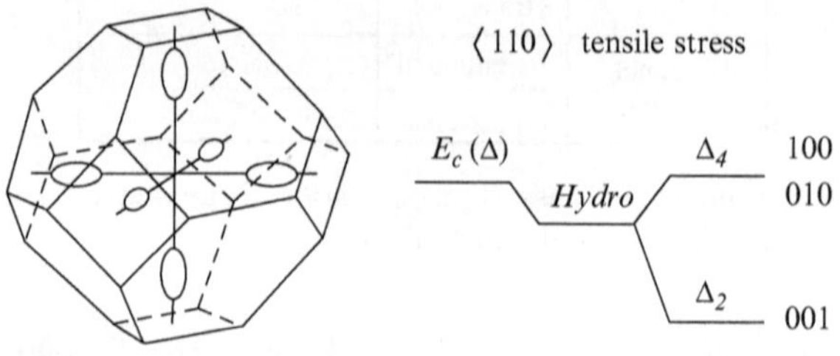

Fig. 4.28. Si conduction band under ⟨110⟩ uniaxial tensile strain

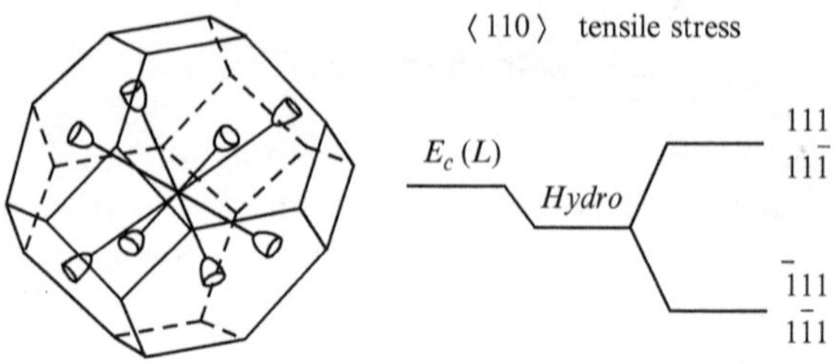

Fig. 4.29. Ge conduction band under ⟨110⟩ uniaxial tensile strain

4.13.2 Analytical Results for Valence Bands with 4 × 4 Hamiltonian

First we study the valence bands with strain using the 4×4 model without split-off band coupling, since it can provide analytical results. The six-band model will be discussed later because of its complexity.

The 4×4 Hamiltonian is given by

$$H_{4\times4}(k=0) = \begin{bmatrix} -P-Q & S & -R & 0 \\ S & -P+Q & 0 & -R \\ -R & 0 & -P+Q & -S \\ 0 & -R & -S & -P-Q \end{bmatrix}. \tag{4.218}$$

With strain, P, Q, R, and S are the sum of both the k terms and the strain terms,

$$P = P_k + P_\varepsilon, \tag{4.219a}$$

$$Q = Q_k + Q_\varepsilon, \tag{4.219b}$$

$$S = S_k + S_\varepsilon, \tag{4.219c}$$

$$R = R_k + R_\varepsilon, \tag{4.219d}$$

where P_k, Q_k, S_k, and R_k are defined in Eq. (4.147) and P_ε, Q_ε, S_ε, and R_ε are defined in Eq. (4.209). Similar to 4.149, diagonalizing Hamiltonian (4.218) gives the HH and LH energy expression with strain

$$E(k) = -P \pm \sqrt{|Q|^2 + |S|^2 + |R|^2}, \tag{4.220}$$

but here this equation has both contributions from second-order k terms and strain. Strain has no effects on spin, thus these two energy bands are each doubly degenerate.

The first thing to notice is that P_ε appears at every diagonal term, even in the 6×6 Hamiltonian. Obviously it corresponds to the hydrostatic strain effect and shifts the entire valence bands uniformly. Thus, it has no effects on band splitting and warping. We will come back later to P_ε when we discuss strain-altered bandgap.

The splitting at the zone center is given by

$$\Delta E(k=0) = 2\sqrt{|Q_\varepsilon|^2 + |S_\varepsilon|^2 + |R_\varepsilon|^2}. \tag{4.221}$$

Since the strain terms are just constants under a particular strain situation, the band curvatures are still determined by the k terms. Under strain, we still call the bands with the diagonal terms containing $-P_k - Q_k$ the HH bands and bands with diagonal terms containing $-P_k + Q_k$ the LH bands. Q_ε can be positive (for tensile strain) or negative (for compressive strain). Because Q_k is always negative along the [001] axis, which is the quantization axis, then if

$Q_\varepsilon < 0$, in (4.220), the "+" sign corresponds to the HH bands, and the "−" sign corresponds to the LH bands. However when $Q_\varepsilon > 0$, one single sign may correspond to two different band characters, which change at the k point with $Q_k = |Q_\varepsilon|$. Suppose $S = R = 0$, we investigate the "+" sign in (4.220). When $|Q_k| < Q_\varepsilon$, $Q > 0$, then for the "+" sign, $E_k = -P + Q$, which corresponds to LH band, and when $|Q_k| > |Q_\varepsilon|$, $Q < 0$, $E_k = -P - Q$, which corresponds to HH band. This also takes place for the "−" sign. Thus, when $Q_\varepsilon > 0$, we can expect that the HH and LH bands will exchange band character at a certain k point. This takes place when two bands cross, as shown in Fig. 4.30. If we

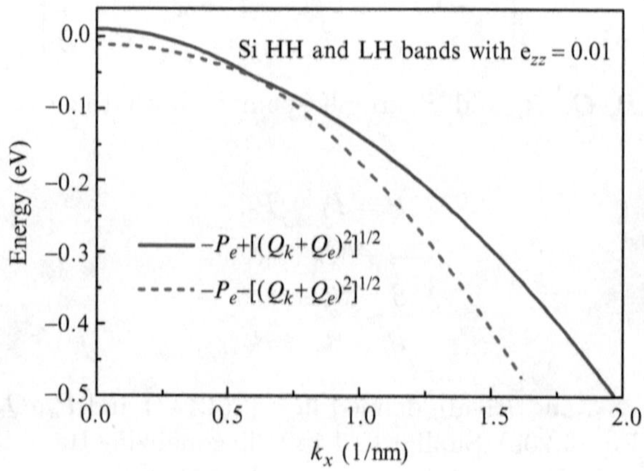

Fig. 4.30. Band curvature characteristic of the upper band under certain types of strain. At certain point, the band curvature characteristic of the upper and lower bands swaps

only concentrate on the zone center properties, then for $Q_\varepsilon > 0$ case, the "+" sign corresponds to the LH bands, and the "−" sign corresponds to the HH bands. Along the other directions, because the HH and LH states vary, the situation shall be reanalyzed. This seems complicated; however, when coming to a particular strain case, it is not as difficult as it seems like. As a matter of fact, because Q_k and Q_ε are always together in the total Hamiltonian, then from the sign of Q_ε, we can see how HH and LH bands shift at the zone center.

Next, we study the splitting and band dispersion of two strain cases based on Hamiltonian (4.218).

Biaxial Stress. For biaxial stress, $\varepsilon_{ij} = 0$ for $i \neq j$, and $\varepsilon_{xx} = \varepsilon_{yy} \neq \varepsilon_{zz}$, and thus $R_\varepsilon = S_\varepsilon = 0$. Then at the zone center, the band splitting is $2Q_\varepsilon$. Since the off-diagonal strain terms vanish, then at the Γ point, the HH and LH effective masses are not affected along the [001] direction. The shear strain shifts the HH and LH bands oppositely in energy. We need to note that Q_ε along [100] or [010] is different from that along [001]. Along [001],

$Q_\varepsilon = -b(\varepsilon_{xx} - \varepsilon_{zz})$, and along [100] or [010], $Q_\varepsilon = b/2(\varepsilon_{xx} - \varepsilon_{zz})$. Therefore, the trend of relative band shift is opposite between [001] and [100] or [010], but is the same for [100] and [010]. For tensile strain ($\varepsilon_{xx} - \varepsilon_{zz} > 0$), HH bands shift down, and LH bands shift up along [001], while HH bands shift up, and LH bands shift down along [100] or [010]. Therefore, we may see the band character mixing in one single band under strain.

From the 4×4 Hamiltonian while neglecting the coupling from the split-off band, the energy dispersion in the small k limit ($\hbar^2k^2/2m_0 \ll |Q_\varepsilon|$) for the biaxial stress case is

$$E(k) = -P_e \mp Q_e - \frac{\hbar^2}{2m_0}[(\gamma_1 \pm \gamma_2)k_t^2 + (\gamma_1 \mp 2\gamma_2)k_z^2], \tag{4.222}$$

where k_t is the in-plane k vector. Then, for biaxial stress, the top and lower bands are both ellipsoids with the energy contours in the x-y plane being circles. The top band under biaxial tensile strain corresponds to the lower sign, and the in-plane bands are HH-like with effective mass given by

$$\frac{m_t^{\text{hh}}}{m_0} = \frac{1}{\gamma_1 - \gamma_2}, \tag{4.223}$$

and the out-of-plane band is LH-like with effective mass given by

$$\frac{m_t^{\text{lh}}}{m_0} = \frac{1}{\gamma_1 + 2\gamma_2}. \tag{4.224}$$

The second band corresponds to the upper sign, and the in-plane bands are LH-like with effective mass given by

$$\frac{m_t^{\text{lh}}}{m_0} = \frac{1}{\gamma_1 + \gamma_2}, \tag{4.225}$$

and the out-of-plane band is HH-like with effective mass given by

$$\frac{m_t^{\text{hh}}}{m_0} = \frac{1}{\gamma_1 - 2\gamma_2}. \tag{4.226}$$

Uniaxial Stress along $\langle 110\rangle$. For [110] uniaxial stress, $\varepsilon_{xx} = \varepsilon_{yy} \neq \varepsilon_{zz}$, and $\varepsilon_{xy} \neq 0$. In this case, $S_\varepsilon = 0$ and $R_\varepsilon \neq 0$. The two shear strain terms are the diagonal term ($\varepsilon_{xx} + \varepsilon_{yy} - 2\varepsilon_{zz}$) and the off-diagonal term ε_{xy}. The band splitting at the zone center is $2\sqrt{Q_\varepsilon^2 + R_\varepsilon^2}$. At the Γ point, the HH bands and LH bands are coupled by R_ε. Under [110] uniaxial compression ($\varepsilon_{xx} < 0$ and $\varepsilon_{zz} > 0$), the HH-like band rises and the LH-like band lowers in energy along [001], which is mainly determined by the diagonal shear strain term Q_ε. Because of the coupling between the HH and LH bands at the Γ point, the LH effective mass becomes larger and the HH effective mass becomes smaller along [001]. The band structure in the x-y plane is a little complicated. We can expand the energy at small k limit, and define $\alpha = \varepsilon_{xx} - \varepsilon_{zz}$, and $\beta = \varepsilon_{xy}$.

Then the dispersion expression for the top two bands under [110] uniaxial stress is

$$\begin{aligned} E(k) = & - P_e \pm \sqrt{b^2\alpha^2 + d^2\beta^2} - \frac{\hbar^2 k^2}{2m_0}\gamma_1 \mp \frac{\hbar^2 k_1^2}{2m_0} \frac{(\sqrt{3}\gamma_3 d\beta + \gamma_2 b\alpha)}{\sqrt{b^2\alpha^2 + d^2\beta^2}} \\ & \pm \frac{\hbar^2 k_2^2}{2m_0} \frac{(\sqrt{3}\gamma_3 d\beta - \gamma_2 b\alpha)}{\sqrt{b^2\alpha^2 + d^2\beta^2}} \pm \frac{2\gamma_2 b\alpha k_z^2}{\sqrt{b^2\alpha^2 + d^2\beta^2}}, \end{aligned} \tag{4.227}$$

where k_1 is the k vector along [110] and k_2 is the k vector along $[\bar{1}10]$. The energy dispersion with [110] uniaxial stress also depends on the compliance parameters of the particular semiconductor besides the band parameters through the relation between α and β. At the small k limit, both the top and lower bands are ellipsoids, with the three axes along $[\bar{1}10]$, [110], and z. Under uniaxial compression, the energy contour for the top band in the x-y plane is an ellipse with the major axis along $[\bar{1}10]$ and the minor axis along [110].

4.13.3 Valence Bands of Strained Semiconductors with Split-Off Band Coupling

In the 4×4 model, the coupling from the split-off bands is totally neglected. If there is no strain, then at the zone center, because all the k-related terms are very small, the HH and LH band dispersion is negligibly affected. However, when there is finite strain, even at the zone center, there are nonvanishing coupling of the HH and LH bands (the $|J = 3/2\rangle$ bands at the valence band edge) with the split-off bands (the $|J = 1/2\rangle$ bands). Both the band splitting between the HH and LH bands and the band curvature of them will be altered. Because the strain terms are generally not small, the error of neglecting the coupling from the split-off band can be quite large even for relatively large split-off energy. For this reason, the six-band model must be used for accurate description. The total Hamiltonian with strain is in the same form of Hamiltonian (4.146), but with P, Q, R, and S replaced by those defined in (4.219). Qualitatively, there is no essential difference between the results of the four-band and six-band models. The difference is only about the numeric. We can have some comparisons in the following.

At the zone center, the valence band splitting is given by the eigenenergies of the following Hamiltonian,

$$H(k{=}0) = \begin{bmatrix} -P_\varepsilon - Q_\varepsilon & S_\varepsilon & -R_\varepsilon & 0 & S_\varepsilon/\sqrt{2} & -\sqrt{2}R_\varepsilon \\ S_\varepsilon^+ & -P_\varepsilon + Q_\varepsilon & 0 & -R_\varepsilon & \sqrt{2}Q_\varepsilon & -\sqrt{3/2}S_\varepsilon \\ -R_\varepsilon^+ & 0 & -P_\varepsilon + Q_\varepsilon & -S_\varepsilon & -\sqrt{3/2}S_\varepsilon^+ & -\sqrt{2}Q_\varepsilon \\ 0 & -R_\varepsilon^+ & -S_\varepsilon^+ & -P_\varepsilon - Q_\varepsilon & \sqrt{2}R_\varepsilon^+ & S_\varepsilon^+/\sqrt{2} \\ S_\varepsilon^+/\sqrt{2} & \sqrt{2}Q_\varepsilon & -\sqrt{3/2}S_\varepsilon & \sqrt{2}R_\varepsilon & -P_\varepsilon - \Delta & 0 \\ -\sqrt{2}R_\varepsilon^+ & -\sqrt{3/2}S_\varepsilon^+ & -\sqrt{2}Q_\varepsilon & S_\varepsilon/\sqrt{2} & 0 & -P_\varepsilon - \Delta \end{bmatrix}. \tag{4.228}$$

Particular strain situation will reduce the complexity of the above matrix. For biaxial strain, $R_\varepsilon = S_\varepsilon = 0$, the situation is almost the same to the four band model, with the only difference that the LH band now coupled with the split-off band by Q_ε. The zone center energies then are given by:

$$E_1(k=0) = -P_\varepsilon - Q_\varepsilon, \tag{4.229a}$$

$$E_2(k=0) = -P_\varepsilon + \frac{1}{2}\left(Q_\varepsilon - \Delta + \sqrt{9Q_\varepsilon^2 + 2\Delta Q_\varepsilon + \Delta^2}\right), \tag{4.229b}$$

$$E_{\mathrm{SO}}(k=0) = -P_\varepsilon + \frac{1}{2}\left(Q_\varepsilon - \Delta - \sqrt{9Q_\varepsilon^2 + 2\Delta Q_\varepsilon + \Delta^2}\right). \tag{4.229c}$$

We label the energies using the subscripts "1" and "2" instead of "HH" and "LH" here, because the former two bands are actually mixed in different directions, as we discussed in the four-band model. Compared to the zone center splitting in the four-band model, where $\Delta E(k=0) = 2Q_\varepsilon$, the splitting in the six-band model apparently depends on the value Q_ε/Δ. For Si which has a small split-off energy, the zone center splitting between the four-band model and six-band model could be appreciable. For example, if $Q_\varepsilon = \Delta$, the splitting difference is $(\sqrt{3}-1)Q_\varepsilon/2Q_\varepsilon = 37\%$. The band shift and warping follows qualitatively the same way as in the four-band model. First look of the Hamiltonian shows that there is no strain coupling between the HH and split-off bands, thus the HH effective mass shall not be affected. Because of coupling between the LH and split-off bands, their effective masses shall vary oppositely. The band dispersion is difficult to obtain, since even we can decompose the 6×6 Hamiltonian into two identical block matrices using basis transformation according to the cubic symmetry, the solution to an equation with E^3 is still tedious. Hasegawa (Hasegawa, 1963) found that under biaxial tensile strain, the LH mass becomes larger and the inverse of it varies linearly with stress due to the mixing with the split-off band, and the HH mass does not change to the first order with strain. The Si HH and LH dispersion with 1 Gpa biaxial tensile stress is shown in Fig. 4.31 to illustrate the band splitting

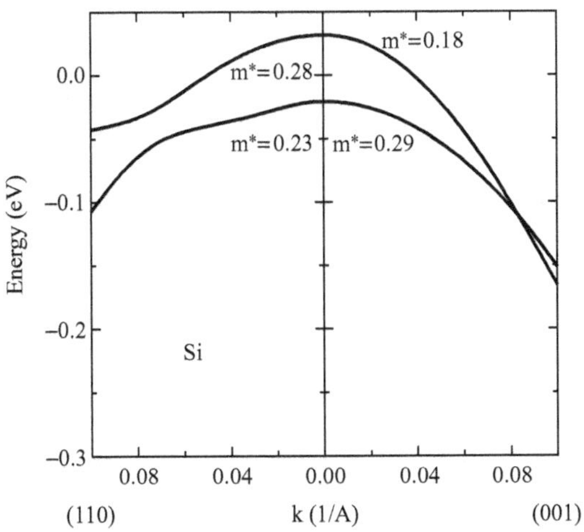

Fig. 4.31. Si valence bands under biaxial tensile stress

and warping. Under biaxial tensile stress, the LH shifts up and HH shifts down along the [001] direction, and the HH shifts up and LH shifts down along the [110] direction.

The case for a uniaxial stress along [110] is more complicated. Because of nonvanishing R_ε, the HH, LH, and split-off bands are all coupled together. The band shifts and warping also follow closely to the discussions in the four-band model. We shall not go into the detailed analytical expressions for the energy dispersions here, since even the analytical expressions for the zone center eigenenergies are very complex. The split top and second valence bands are both ellipsoids at small energy range, with major and minor axes along the diagonal direction in the x-y plane. The numerical E–k relations for 1 Gpa uniaxial stressed Si HH and LH bands are shown in Fig. 4.32. Under uniaxial

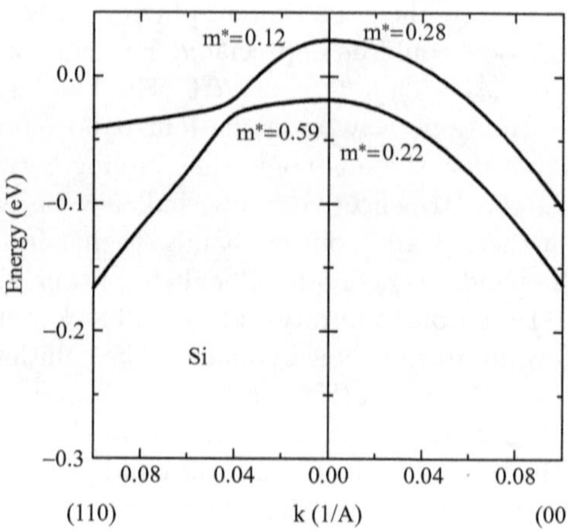

Fig. 4.32. Si valence bands under uniaxial compressive stress

compressive stress, the HH shifts up and LH shifts down along the [001] direction, and the HH shifts down and LH shifts up along the [110] direction. These results support the earlier tight-binding discussions.

Strain affects tetrahedral semiconductor valence bands similarly. For a comprehensive understanding, we show the 3D equi-energy surfaces for the top valence band for Si, Ge, and GaAs with no stress, 1 Gpa biaxial stress and 1 Gpa uniaxial stress in Fig. 4.33. One conspicuous difference between the biaxial stressed and uniaxial stressed energy surface is the symmetry in the x-y plane. At the energy under study, the 2D energy contour for biaxial stress is circle-like, whereas the 2D energy contour for uniaxial stress is ellipse-like. In the latter case, the effective mass along [110] is much smaller than that along [$\bar{1}$10]. For [110] channel p-MOSFETs, this will result in a small conductivity mass while still keeping a relatively large DOS.

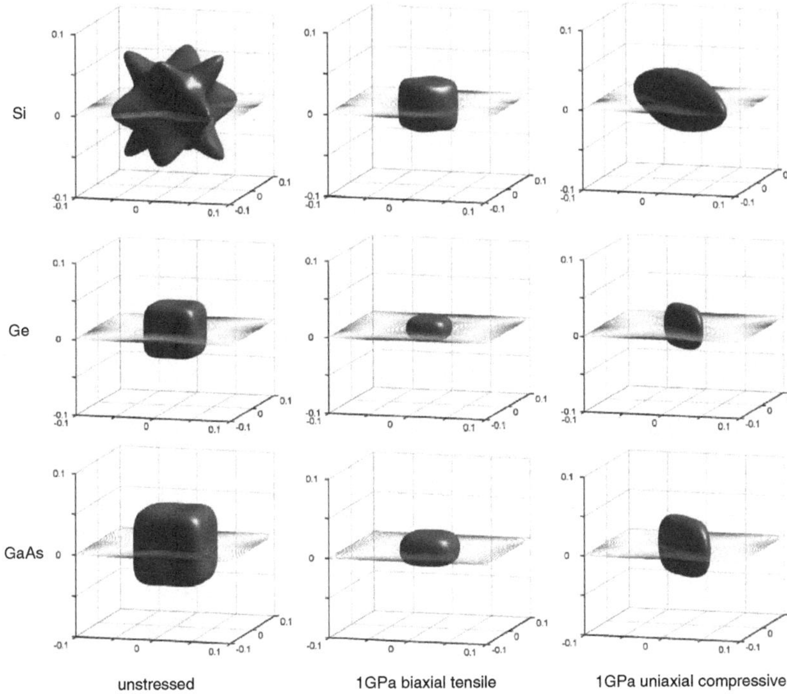

Fig. 4.33. The 3D energy surfaces of the top valence band for Si, Ge, and GaAs under no stress, 1 GPa biaxial tensile stress, and 1 GPa uniaxial compressive stress

4.13.4 Band Gap Shift with Strain

Band gap plays an important role in optical absorption, electron–hole pair recombination, and off-state leakage in semiconductor electronic devices, and so on. Band gap is altered by the strain as a result of strain-shifted conduction and valence band edges. If strain is introduced by applied external stress, the band gap can even be continuously tuned.

Band gap is not only altered by hydrostatic strain, which shifts the weight of the bands, but by the shear strain as well, which splits the valence bands and conduction bands not located at the Γ point. We may choose the biaxial strain as an example to illustrate how it affects the band gap, because the shift due to biaxial strain can be analytically obtained. However, it is not difficult to obtain the bang gap shift due to the other types of strain numerically.

For valence bands, $-P_\varepsilon$ denotes the shift of the valence band weight under a tensile hydrostatic strain. Here, we use the band characters along the [001] direction to label the bands, i.e., in (4.229), the band with zone center energy $-P_\varepsilon - Q_\varepsilon$ is the HH bands, and the bands with zone center energy $-P_\varepsilon + \frac{1}{2}\left(Q_\varepsilon - \Delta + \sqrt{9Q_\varepsilon^2 + 2\Delta Q_\varepsilon + \Delta^2}\right)$ is the LH bands. For biaxial tensile strain, $Q_\varepsilon > 0$, and the valence band edge is the LH bands. For compressive stress, $Q_\varepsilon < 0$, and the band edge is the HH bands. Thus, for tensile or compressive stress, the band edge energy is given by different energy expressions, so generally at the zero strain point, the first derivative of band gap with strain is not continuous.

The band gap is the energy distance between the top valence band and the lowest conduction band. For Si, the Δ_2 valleys shift with strain faster than the Δ_4 valleys, and both Δ_2 and Δ_4 valley energies depend linearly on strain. For GaAs, there is only one conduction valley, the Γ valley, and its shift is only caused by the hydrostatic strain. For the valence bands, the HH band depends linearly on strain, but the LH band shows a nonlinear relation with compressive strain, especially for the LH of Si. This nonlinearity is caused by the split-off band coupling. For GaAs with a relatively large split-off energy, the nonlinearity is not remarkable at small strain range. The conduction and valence band shifts of Ge with biaxial strain shall be similar to GaAs, since first the split-off energies of them are comparable, and second, under biaxial stress, the conduction valleys of Ge shift just due to hydrostatic strain. Shifts and splitting for uniaxial stress can also be obtained numerically through $\mathbf{k}\cdot\mathbf{p}$ method.

Calculated conduction band and valence band edge shifts and splitting for Si and GaAs are shown in Fig. 4.34 as functions of biaxial strain ($\varepsilon_{xx} = \varepsilon_{yy}$, ε_{zz} can be obtained through $\varepsilon_{zz} = (2\varepsilon_{xx}S_{12}/(S_{11} + S_{12}))$). We plot the band edge energies of both the conduction and valence bands for Si, Ge and GaAs

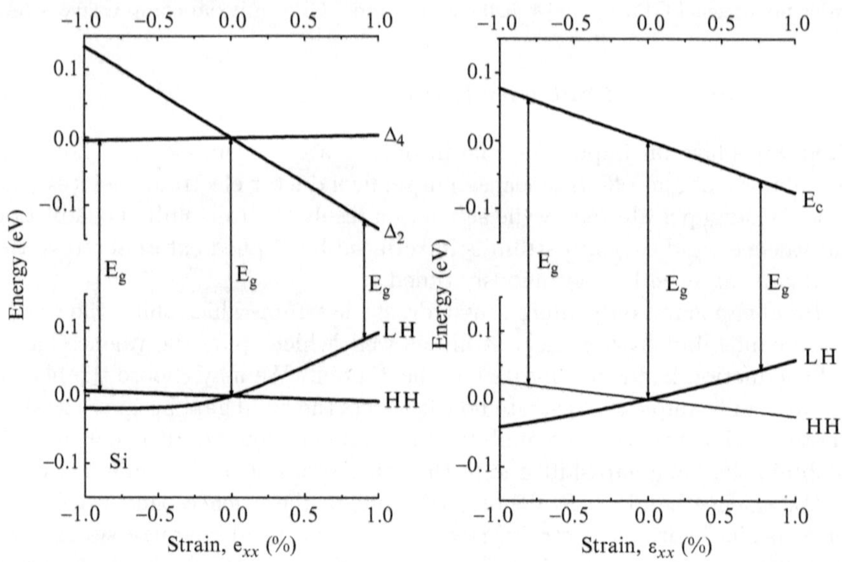

Fig. 4.34. The conduction and valence band shift for Si and GaAs with biaxial strain. E_g is shown as the energy gap between the lowest conduction band the highest valence band

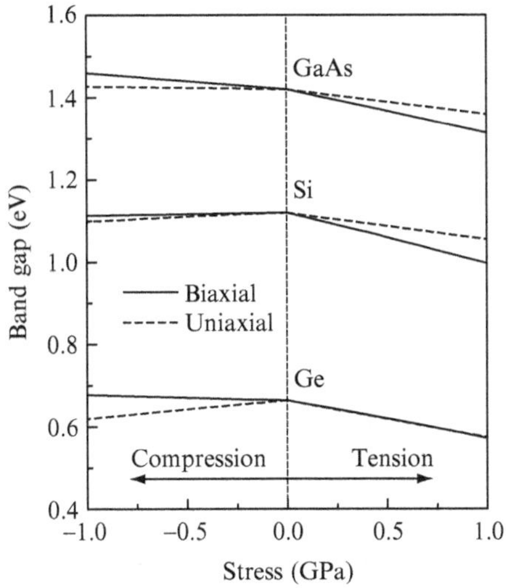

Fig. 4.35. Band gap as a function of stress for GaAs, Si, and Ge

in Fig. 4.34 as functions of biaxial strain ($\varepsilon_{xx} = \varepsilon_{yy}$, ε_{zz} can be obtained through $\varepsilon_{zz} = (2\varepsilon_{xx}S_{12}/(S_{11} + S_{12})))$.

The band gap shifts of Si, Ge and GaAs as a function of stress for both biaxial and uniaxial stress are shown in Fig. 4.35. It can be seen that the band gap change under compression is smaller than under tension for both types of stress. For Si and Ge, which have degenerate conduction valleys, uniaxial stress reduced the band gap, for either compression or tension.

5

Low-Dimensional Semiconductor Structures

5.1 INTRODUCTION

In the last chapter, we have introduced the theories for calculating the band structures of bulk semiconductors and discussed the corresponding strain effects. However, carriers in most modern electron devices such as FETs are electrically or spatially confined. Confinement interrupts the periodic potential where an electron in a bulk semiconductor has. Quantum mechanically, electric confinement and spatial confinement have no essential difference. They alter the potential term $V(\mathbf{r})$ in the single electron Schröinger equation (3.1) such that in the confinement direction(s), the potential loses the periodic characteristic the bulk crystals possess. This has significant effects on carrier transport properties, since electron systems are distinguished by this very potential term $V(\mathbf{r})$, as discussed in Chap. 3. Typical confined electron systems include quantum wells, quantum wires, and quantum dots, which are 2D, 1D, and 0D structures, respectively. A MOSFET is a symbolic 2D electron system, which invention is probably the most important event in the history of modern semiconductor industry, and has been the driving force for the technology development for over half a century. The operation of a MOSFET is based on the control of the electronic behavior through tunable external electric field to create a 2D electron gas (2DEG) layer close to the semiconductor/oxide interface. Structural 2D systems such as quantum wells extend the means of 2DEG generation and present extensive research interest and electrical and optical applications as well, especially when the crystal growth techniques such as molecular beam epitaxy are very mature nowadays. Quantum wells can be created by growing a layer of semiconductor sandwiched by two semiconductor layers whose band gaps are larger than the former, or formed in a heterojunction such as GaAs/$Ga_{1-x}Al_xAs$. The former sandwich structures are extensively found in quantum well lasers. The GaAs/$Ga_{1-x}Al_xAs$ heterojunction is the kernel part for a high electron mobility field effect transistor (HEMFET) since the similar lattice constant and coefficient of expansion of $Ga_{1-x}Al_xAs$ permits the growth of high mobility

Y. Sun et al., *Strain Effect in Semiconductors: Theory and Device Applications*, DOI 10.1007/978-1-4419-0552-9_5,

thin $Ga_{1-x}Al_xAs$ film on a GaAs substrate. With aggressive scaling of modern planar MOSFET devices, short channel effects become increasingly grave. Other device architectures such as fully depleted silicon-on-insulator (SOI), double-gate MOSFET, and FinFET devices are currently under intense investigation. Conceptual devices using carbon nanotubes and Si-nanowires also present promising application prospects. A Si-nanowire with transverse cross section length scale in nanometer scale is typically 1D structure. The motion of electrons is restricted along the nanowire and quantized in the transverse directions. In these nontraditional semiconductor structures, both electric and spatial confinement are present and often interact with each other.

The application of these low-dimensional structures makes use of their unique properties determined by their unique electronic structures. Under confinement, one energy band in the corresponding 3D structure splits into a series of subbands in the confining direction. The subband structure has vast difference from the bulk 3D band structures. Strain in these low-dimensional structures has significantly different effects on the band structures and consequently on transport properties, compared to bulk materials. In this chapter we will explore the electronic properties of low-dimensional systems and discuss their strain effects. At first, we will give a quick overview of the electronic properties of low-dimensional semiconductor structures and briefly introduce their strain effects. Then we will adopt the triangular potential well approximation, which is simple and can give analytical results and has been vastly employed to a variety of 2D systems, to provide a simple quantitative picture for 2D and 1D band structures, particularly for MOSFET devices. A more complicated and believed-to-be more systematic and precise treatment for low-dimensional electronic systems is introduced afterward. In Sect. 5.4, we introduce some mathematical techniques, including the self-consistent procedure and finite difference method to quantitatively study the subband structures. After this, results for Si MOSFETs are presented in Sect. 5.5 as an example. The subband splitting, carrier distribution under inversion, and comparison with some III–V channels materials are discussed. The effects of strain on the low-dimensional subband structures are discussed lastly based on theoretical computations. In this chapter, we are not trying to give a comprehensive review for all low-dimensional structures, but rather to focus on some physics properties and to provide a formalism for theoretical study. The detailed discussion of strain effects on some particular device structures such as MOSFETs, strain sensors, and quantum well lasers will be investigated in the other separate chapters.

5.2 OVERVIEW: LOW-DIMENSIONAL SEMICONDUCTOR STRUCTURES

Before we address their electronic properties of the low-dimensional semiconductor structures, first we introduce some typical low-dimensional semiconductor structures. According to the different formation of the 2D semiconductor systems, we can ascribe them into three types: MOS structures, heterojunctions, and square quantum wells. In the following we introduce how these structures are formed and how the carriers are confined in these structures. Following the 2D structures, we introduce the Si-nanowire as an example of the 1D structures.

5.2.1 MOS Structure and MOSFET Channel

A typical Si MOS inversion device is shown in Fig. 5.1. The MOS device is fabricated on a p-type or n-type silicon substrate, which is called the body and labeled B in the figure. On top of the semiconductor body, an insulating silicon dioxide layer is thermally grown forming the gate oxide, followed by a metal layer or gate electrode (labeled G in the figure); this metal-oxide-semiconductor structure is used to apply an electric field through the oxide to the silicon. According to the conduction carrier type, the 2D carrier system in a MOS structure can be 2D electron gas or hole gas. For the MOS device shown in the figure, the body or substrate is p-type and the source (S) and drain (D) regions are n-type. Studies on the 2D MOS system have been concentrated on the Si MOS inversion electron gas, since the Si technology is mature, and the density of Si–SiO_2 interface state can be very low (e.g., $10^8 - 10^9/\text{cm}^2$ eV). Furthermore, the physical properties of SiO_2 are ideal for insulation. Its breakdown electric field can reach 10^6 V/cm, and the band gap is > 9 eV so that the potential barrier between Si and SiO_2 is very high (3.1 eV), and thus the carriers are difficult to penetrate into the SiO_2 insulating layer. These beneficial conditions make it easy to control the 2D charge density, which is either in an accumulation layer or an inversion layer, according to the transverse electric field applied through the gate. As in the figure, in the inversion mode, the applied voltage shall be high enough so that the conduction band edge is below the Fermi level, thus the carrier of the opposite type to that of the substrate can accumulate just beneath the gate oxide and form an inversion layer. In Fig. 5.1, the inversion charge type is electron. The electrons are strongly confined to a Si–SiO_2 surface layer by the potential induced by band bending and form a so-called conducting channel between the source and drain. The electrons can only move in the plane parallel to the surface. The motion in the transverse direction is quantized. Similarly, modification of the doping and bias can induce p-type channel, where holes are confined to the surface layer. Reversing the gate bias can cause the same type of carriers as in the substrate to accumulate in the Si–SiO_2 surface to form an accumulation layer. However, the accumulation layer is not well defined and carriers are extended in the substrate, and thus the quantization for an accumulation layer is not strong. In this book, we only concentrate on the inversion layer of the MOS devices, which is typically the operational mode of most logical devices.

The SiO_2 layer grown on Ge is not stable, and the density of the interface states in the Ge–SiO_2 surface is quite high. These have limited the study and application of the Ge MOS devices.

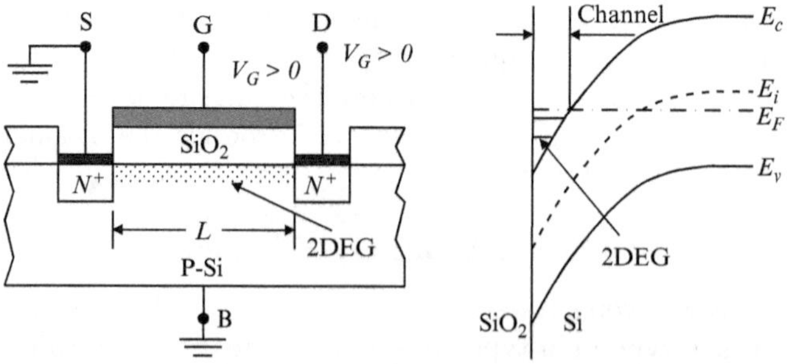

Fig. 5.1. A typical structure of a Si MOSFET, and the band bending profile of the Si substrate at the Si/SiO_2 interface. Channel is formed by bending the band to create an inversion layer

III–V semiconductors are very attractive since they normally have high electron mobilities. However, the semiconductor–insulator interface has been a historical issue. The III–V semiconductors do not have a good native oxide, and for a variety of lab-grown dielectrics, the interface contains a high density of interfacial states, e.g., $10^{12} - 10^{13}/\text{cm}^2\cdot$ eV (Schulz and Klansmann, 1979; Shiue and Sah, 1979; Suzuki et al, 1978; Chang and Coleman, 1978; Mimura et al, 1979; Nishizawa and Shiota, 1980). When the density of the interface states is as high as $10^{13}/\text{cm}^2\cdot\text{eV}$, the surface Fermi level will be pinned relative to the band edge, and the applied gate voltage can no longer control the surface band bending, and thus inversion cannot be induced. Recently, Bell laboratories (Passlack et al, 1995, 1997; Hong et al, 1997; Wang et al, 1999) and Freescale (Rajagopalan et al, 2006, 2007) reported a high-quality Ga_2O_3 film on GaAs with unpinned Fermi level in the surface channel and successful enhancement and depletion mode operation of GaAs MOSFET.

5.2.2 Heterojunction

When two semiconductors with different band gaps and Fermi energies are put together and made to contact, after reaching the equilibrium state, they will form a heterojunction. The electron affinity and work function of each constituent materials determine the band bending and alignment. The electron affinity and work function are the energy needed to excite an electron from the conduction band edge and the Fermi level to the vacuum level, respectively. The electron affinity is an intrinsic property of a semiconductor. The work function depends on the the doping type and concentration. Heterojunctions can be formed using two intrinsic materials or by doping one or both materials either n- or p-type. Therefore, a large variety of junctions can be formed (i.e., p–n, n–n, n–i, p–p, etc.). For purposes of illustration, let us consider the formation of an i-n GaAs/AlGaAs heterojunction using Fig. 5.2. The band gap of $Al_xGa_{1-x}As$ is greater than GaAs and depends on Al component x. When $x < 0.41$, $Al_xGa_{1-x}As$ is a direct gap semiconductor; when $x > 0.41$, it is an indirect gap semiconductor with the conduction valley located at the X point. When $x < 0.41$, its band gap is given by $E_g = 1.424 + 1.247x\,\text{eV}$, and the conduction band offset to GaAs is given by $\Delta E_c = 0.79x\,\text{eV}$. Let

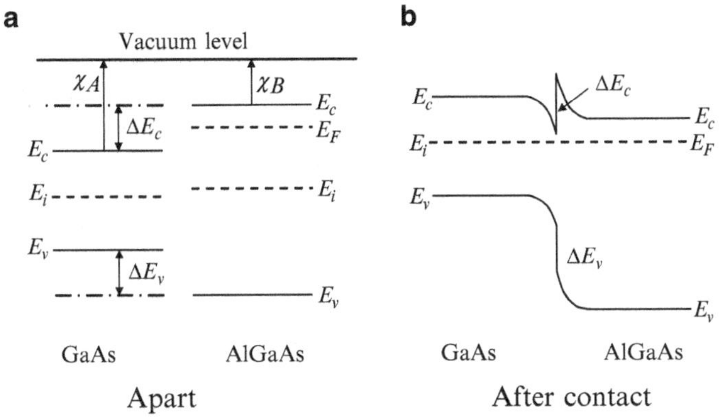

Fig. 5.2. Schematic to show how a GaAs/AlGaAs heterojunction is formed. (**a**) The band alignment when the two materials are apart. (**b**) The band alignment when GaAs and AlGaAs are brought together to have contact and have reached the equilibrium state

us write the electron affinities for GaAs and AlGaAs as χ_A and χ_B, respectively. The position of the band edges of one semiconductor can be referred by electron affinities as shown in the figure. It is always regarded as truth that the vacuum level is the same for all materials when they are apart. Using the electron affinities and band gaps, the conduction and valence band offsets ΔE_c and ΔE_v are found. The fermi energy of each material is determined by doping. The equilibrium band diagram of the junction follows from the application of the two rules that a) the Fermi level is flat everywhere and b) far from the junction the bulk-like properties of the materials are recovered. The built-in potential between the well and barrier is easily found to be the difference between the work functions of the constituent materials. As we can see, due to the band bending at the interface, a thin confined layer (sometimes is also called a triangular potential well) is formed in the GaAs side. Typical thickness of this confined layer is around 10 nm. Because of the band bending, a high concentration of electrons can be created in the confined layer, and on the AlGaAs side at the junction the electrons are depleted.

For GaAs/AlGaAs heterostructures, modulation doping is often adopted, i.e., the barrier layer AlGaAs is doped while the well layer GaAs is intentionally left undoped. This technique was first demonstrated by Dingle in 1978 (Dingle et al, 1978). Sometimes an undoped thin AlGaAs spacer is grown in between to further isolate the intrinsic GaAs and doped AlGaAs layer. Modern devices require high carrier concentration to enhance conductivity, which requires high density of dopant impurities, which degrades the carrier mobility significantly due to impurity scattering. However, in a band alignment as in Fig. 5.2, the electrons transfer from the AlGaAs layer into GaAs layer, thus the electrons and impurities are spatially separated. This modulation doping offers two great advantages over the traditional doping techniques: a) The carrier concentration in the GaAs layer can be greatly increased without introducing dopants in GaAs layer itself; and b) Because of the high purity, the mobility of the carriers in the GaAs layer is greatly enhanced. The first high electron mobility transistor (HEMT), or heterostructure FET (HFET), was created using GaAs/AlGaAs structure in 1980 (Mimura et al, 1980).

The large mobility in GaAs/AlGaAs HEMT is possible due to also another important factor: the excellent heterointerface quality. Both GaAs and AlGaAs are zinc-blende crystals, and the lattice constants of AlGaAs and GaAs are very well matched. Ultralow density of interface states diminishes the interface scattering. $Ga_{0.53}In_{0.47}As/InP$ is the other lattice-matched heterojunction system.

Because of the transfer of electrons from AlGaAs barrier to GaAs well layer, the band in the GaAs side at the interface bends downward. At equilibrium state the Fermi energy is constant across both sides. Electronic states in the GaAs side close to the interface are quantized due to the electric confinement as a result of the band bending as illustrated in Fig. 5.3.

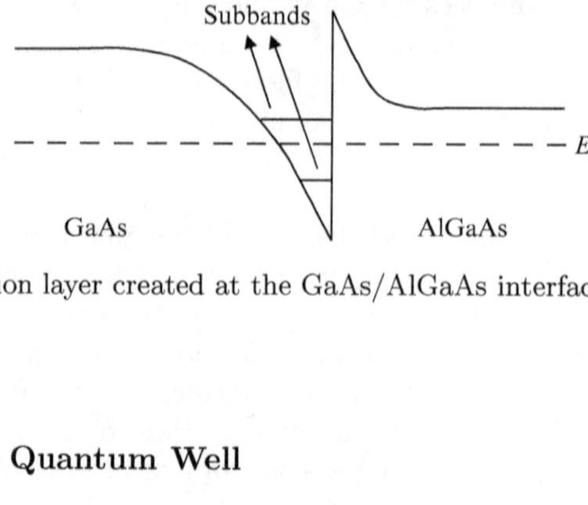

Fig. 5.3. Inversion layer created at the GaAs/AlGaAs interface by band bending

5.2.3 Square Quantum Well

Not any two semiconductors put together will form a quantum well by band bending. Both the conduction and valence bands may have a continuous transition at the interface as the valence band of the i-n GaAs/AlGaAs heterostructure. However, we can form a square quantum well using a sandwiched structure, if the constituent semiconductors have different band gaps. According to the band alignment determined by the intrinsic semiconductors with no free charges that can move, the sandwiched quantum wells are classified into two types: the type-I and type-II quantum wells, as shown in Fig. 5.4. Writing the well material in the front and the barrier material in the back like

A/B, the type-I quantum well includes GaAs/$Ga_{1-x}Al_xAs$, $Ga_{1-x}In_xAs$/InP, GaSb/AlSb, etc., where the B is the barrier for both electrons and holes. The typical type-II quantum well includes InAs/GaSb and Ge/Si, where one material acts as a well for electrons but a barrier for holes. Among these quantum

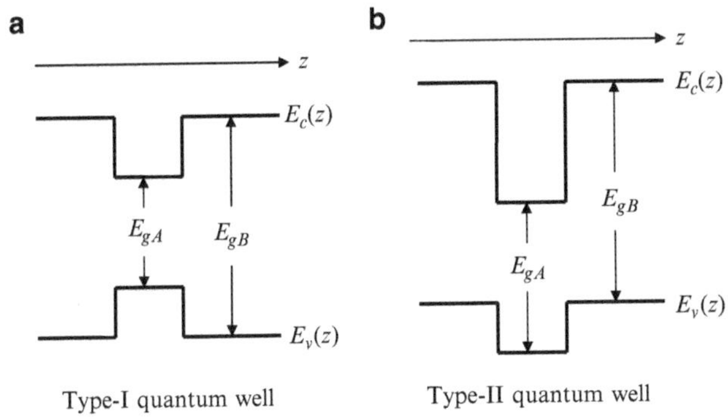

Fig. 5.4. Type-I and Type-II square quantum wells

well structures, some of them are lattice-matched while some of them are not. For thin films grown on substrates, people usually take a pseudomorphic approximation that the lattice mismatch is accommodated by strain in the thin film in order to achieve a common in-plane lattice constant.

Charge transfer takes place when the Fermi energies are not the same for well and barrier in doped square quantum wells. The potential profile is also determined by other factors besides the built-in potential such as the doping type and concentration, as we discussed for the heterostructure case. If the well layer is thick, then the quantum well is nothing but two separate heterojunctions. However, when the well layer is thin and the electronic wave function of one of the two heterojunctions can penetrate to the other, the two heterojunctions then form a single quantum well. Therefore, the thickness of the well layer is typically in nanometer scale.

The type I quantum wells play an important role in semiconductor lasers. In an n-i-p AlGaAs/GaAs/AlGaAs quantum well shown in Fig. 5.5, the electrons and holes can be injected from the electrodes connected with n- and p-type AlGaAs barrier layer, and then drop and are confined in the GaAs well and recombine there to emit photons. Light tends to propagate in regions with a higher refractive index. This can be viewed as total internal reflection of light within the medium with a higher refractive index. A small band gap is usually associated with a higher refractive index, and the refraction index of GaAs is indeed larger than AlGaAs. Therefore, the light emitted in the GaAs layer is also confined in the GaAs layer. The electron–hole recombination and light transmission occur in the same region in this case, resulting in a very efficient stimulated photon emission.

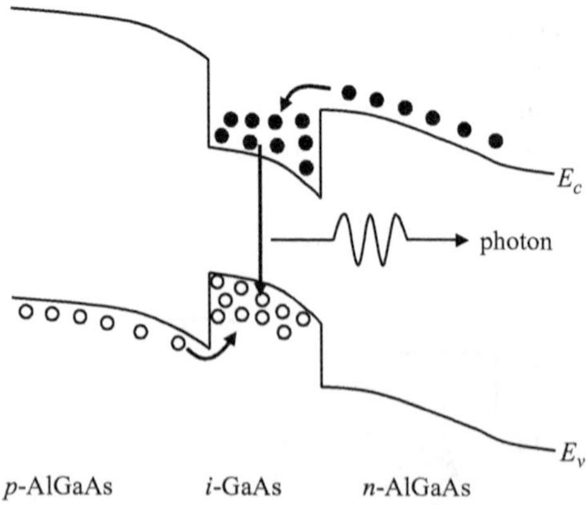

Fig. 5.5. Diagram of a quantum well structure used for quantum well laser application

5.2.4 Nanowire

In aggressively scaled MOSFETs, the off-state leakage current is the primary factor that degrades the device performance. In order to reduce the substrate leakage, the Si-on-insulator (SOI) architecture is proposed. In an SOI structure, the substrate is "floating," and thus there is no substrate leakage. If we consider this architecture to restrict the thickness dimension of the bulk substrate, one more aggressive idea might be to also restrict the width dimension. For a Si channel with both thickness and width greatly reduced, say, to nanometer scale, it becomes a nanowire. If we say that in a traditional MOSFET, the carriers are only confined in the normal-to-the-channel direction, but are free to move in the lateral dimension, in a nanowire FET, they can only move in one direction, the longitudinal direction.

A typical Ge nanowire FET is shown in Fig. 5.6. The ends of the nanowire are connected to metal electrodes that act as source and drain. The gate is the Si layer that is called a back gate in this situation. Fin-type or coaxial gate is also used in some cases. Though not mature, nanowire FET already demonstrated substantial advantages over conventional planar FETs. The hole mobility of a Ge/Si nanowire FET can reach ten times higher than that of a silicon planar pMOSFET (Cui et al, 2003; Xiang et al, 2006).

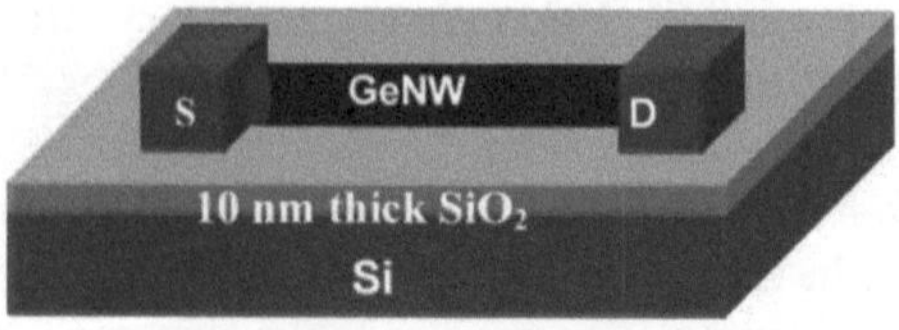

Fig. 5.6. A back-gate nanowire transition using Ge nanowire as channel

Except for the applications in transistors, p-type Si nanowire is found to have much larger piezoresistance than bulk Si (~40 times maximum), and it increases with decreasing nanowire transverse area (He and Yang, 2006). This effect can be employed to replace bulk Si to make strain sensors with high sensitivity.

5.3 ELECTRONIC PROPERTIES OF LOW-DIMENSIONAL STRUCTURES

For a qualitative description of the electronic structures of the MOS devices, we can employ the triangular potential well approximation. For square quantum wells and quantum wires, their electronic structures may be understood from an infinite quantum well model. In bulk semiconductors, the wave vector k is considered a variable rather than an operator, but due to spatial confinement, the crystal potential periodicity vanishes in some directions, and thus the corresponding wave vector in those directions become operators. The effective mass theory evolves into the envelope function theory in low-dimensional semiconductor systems, and it is the theoretical basis for the calculations of the subband structures. In the following section, we first introduce the envelope function theory and then apply it to some quantum structures.

5.3.1 Envelope Function Theory

In a uniform space, motions of one electron have the plane wave characteristic, with well-defined wave vector k_x, k_y, and k_z. In a quantum well, if we define the confining direction as the z direction, the motion of carriers is quantized in the z direction, represented by a series of discrete energy levels. In describing the states of the carriers along z, k_z no longer has significant meaning. But wave vectors k_x and k_y are still good quantum numbers. Inside the x-y plane, their motion is represented by plane waves. Similarly, if the electron is confined in two directions, the electronic motion is quantized along the two directions. One simplest analogy for the energy levels along the confining directions for the quantum well and nanowire could be the 1D and 2D infinite square quantum wells.

The realistic low-dimensional systems are not simple infinite quantum wells, since the electrons in these structures also experience potentials from the localized ions, the other electrons, etc., in the semiconductors. However, even though these low-dimensional systems are quite complicated at the first glance, the energy structure and computation formalism are quite similar to the simplest quantum wells, based on the introduction of the envelope function theory. The envelope function formalism is basically very similar to the effective mass theory we introduced for the bulk semiconductors. The difference is that in a 2D system, the wave vector along the confining direction is replaced by an operator. In bulk materials, the probability of finding a carrier is equal in space. It is not the case for confined systems. The envelope function is used to describe the variation of the electron distribution probability with space. There are also the other formalisms to solve the quantum well problems, such as the tight-binding method and pseudopotential method. In the tight-binding method, a heterostructure is approximated as a bulk material with very large unit cell, and the wave function is built atom by atom in the super cell. In the pseudopotential method, the heterostructure is considered as a perturbation to the bulk material, and the band structure is found similarly as for a bulk semiconductor. Both these formalisms give an entire zone band structure. However, compared with the envelope function formalism, both suffer from complexity, and sometimes it is hard to trace the physics of the numerical results. The simplicity and satisfactory accuracy make the envelope function formalism extensively used.

When applying the envelope function formalism to low-dimensional semiconductor systems, the effective mass theorem is central. As an approximation, we can still expand the wave function using the bulk eigenstates at the band edges in low-dimensional systems, and the wave function is a multiplication of a function of position z, i.e., the envelope function, and the periodic band edge functions. The point of the effective mass theorem is to replace the wave vector, say, k_z, in the effective mass Hamiltonian for a perfect bulk semiconductor using the differential operator, $-\mathrm{i}\frac{\partial}{\partial z}$, in the perturbed Hamiltonian, to seek a solution for envelope functions. The perturbation could be a quantum well, barrier, impurity, or superlattice. Writing the perturbation as $\delta V(\mathbf{r})$, and the envelope function as $F_i(\mathbf{r})$ for the ith band, we need to solve a second-order differential equation to obtain the energy levels,

$$\left[-\frac{\hbar^2\nabla^2}{2m^*}+\delta V(\mathbf{r})\right]F_i(\mathbf{r})=EF_i(\mathbf{r}), \tag{5.1}$$

and find the wave function in real space.

A a simple proof of the effective mass theorem can be found in Appendix. The unperturbed bulk semiconductor can serve as a limiting case of the effective mass theorem. For example, for a single parabolic band with energy dispersion in the form of $E(\mathbf{k})=\hbar^2k^2/2m^*$, the Schrödinger equation for the envelope function becomes

$$-\frac{\hbar^2\nabla^2}{2m^*}F(\mathbf{r})=EF(\mathbf{r}). \tag{5.2}$$

Under a periodic boundary condition, this equation directly gives the solution of $F(\mathbf{r})=\frac{1}{\sqrt{\Omega}}\mathrm{e}^{\mathrm{i}\mathbf{k}\cdot\mathbf{r}}$ and energy of $E(\mathbf{k})=\hbar^2k^2/2m^*$, and the coefficient $\frac{1}{\sqrt{\Omega}}$ is introduced to normalize the wave function to a δ-function. The total wave function is

$$\psi(\mathbf{r})=F(\mathbf{r})u_{n\mathbf{k}_0}(\mathbf{r}) \tag{5.3}$$

with $u_{n\mathbf{k}_0}(\mathbf{r})$ the band edge basis function. It can be seen that in the effective mass theorem, the effective mass is explicitly used in the Schrödinger equation. It already takes into account the effect of the periodic potential of the bulk semiconductors, and thus it greatly reduces the complexity of the procedures to solve the Schrödinger equation described in the last chapter. If the confinement is in one direction, say, the z direction, in a quantum well, then the envelope function may be written as a product of an in-plane and an out-of-plane function,

$$F(\mathbf{r}) = \mathrm{e}^{(ik_x x + ik_y y)} \times f(z). \tag{5.4}$$

Substituting this equation into (5.1), we obtain an equation for $f(z)$,

$$\left(-\frac{\hbar^2}{2} \frac{\partial}{\partial z} \frac{1}{m_z^*} \frac{\partial}{\partial z} + \delta V(z) \right) f(z) = \left[E - \frac{\hbar^2 (k_x^2 + k_y^2)}{2m^*} \right] f(z), \tag{5.5}$$

where we write the effective m_z^* between two differentials to ensure that the whole operator is Hermitian, since the effective mass in the z direction in a quantum well structure also depends on space, which we need to take into account in the equation.

In degenerate case such as for the valence bands, replacing k_z with the differential operator $-\mathrm{i}\frac{\partial}{\partial z}$ in the Luttinger Hamiltonian results in an equation set with six second-order differential equations. Solving these coupled equations gives a series of subbands along z, which also have intricate in-plane structure.

Since the low-dimensional systems are composed of different materials, the boundary condition at the interface is an important issue to determine the solutions of the envelope equation (5.5). One common boundary condition adopted is to ensure the continuity of the probability flux at the interface. This boundary condition at the interface is then

$$f(0_\mathrm{A}) = f(0_\mathrm{B}); \ \frac{1}{m_\mathrm{A}} \frac{\mathrm{d}f(z)}{\mathrm{d}z}\bigg|_{z=0_\mathrm{A}} = \frac{1}{m_\mathrm{B}} \frac{\mathrm{d}f(z)}{\mathrm{d}z}\bigg|_{z=0_\mathrm{B}}, \tag{5.6}$$

where the subscripts A and B represent the different sides of the interface. Since the derivative of position is essentially the momentum, then this boundary condition requires the velocity to be the same at both sides to conserve the current.

For a system confined in two directions (quantum wire) or even three directions (quantum dot), the formulation of the problem follows exactly the same way as discussed earlier for one-direction confinement under the envelope function theory. The wave vectors in the confining directions are replaced by the differential operator $-\mathrm{i}\frac{\partial}{\partial r}$, where r is the coordinate in the confining direction, then energies and envelope functions are solved. For example, in a quantum wire, the equation for the envelope function is

$$\left[-\frac{\hbar^2}{2m^*} \left(\frac{\partial^2}{\partial x^2} + \frac{\partial^2}{\partial y^2} \right) + V(x,y) \right] f_{m,n}(x,y) = E_{m,n} f_{m,n}(x,y), \tag{5.7}$$

where m, n label the subbands due to the confining from the x and y directions. If the envelope function $f_{m,n}(x,y)$ is solved, the total wave function is given by

$$\psi(\mathbf{r}) = f_{m,n}(x,y)\exp(\mathrm{i}k_z z), \tag{5.8}$$

and the energy is given by

$$E = E_{m,n} + \frac{\hbar^2 k_z^2}{2m^*}. \tag{5.9}$$

5.3.2 Triangular Potential Well Approximation

In the case of electric confinement, the band bending profile determines the $V(\mathbf{r})$, which is central for an electron system. The band bending profile of the MOS structure is a well-studied subject. It is related to the charge density distributions through the Poisson's equation,

$$\frac{\mathrm{d}^2 V(z)}{\mathrm{d}z^2} = \frac{e}{\epsilon_\mathrm{s}}[p(z) - n(z) + N_\mathrm{d}^+(z) - N_\mathrm{a}^-(z)], \tag{5.10}$$

where $V(z)$ is the z-dependent confining electric potential, $p(x)$ and $n(x)$ are hole and electron concentrations, respectively, and $N_\mathrm{d}^+(z)$ and $N_\mathrm{a}^-(z)$ are ionized donor and acceptor impurity concentrations, respectively. For an n-channel MOS device on a p-doped substrate, a) assuming Boltzmann distribution and b) using the effective DOS for the conduction and valence bands, solving the Poisson's equation in a classic limit gives (Taur and Ning, 1998)

$$\frac{\mathrm{d}V(z)}{\mathrm{d}z} = \sqrt{\frac{2k_BTN_\mathrm{a}}{\epsilon_\mathrm{s}}}\left[\left(e^{-V(z)/k_BT} + \frac{V(z)}{k_BT} - 1\right) + \frac{n_i^2}{N_\mathrm{a}^2}\left(e^{V(z)/k_BT} - \frac{V(z)}{k_BT} - 1\right)\right]. \tag{5.11}$$

Together with the boundary condition of the band bending at the surface that is determined by the gate bias, and the band bending vanishing in the deep substrate, this equation can be solved numerically. However, though this classic result depicts well of the potential due to charge depletion, it works poorly for the inversion by predicting most charge density at the interface. In the inversion layer, normally the inversion electrons are degenerate, and thus the Fermi-Dirac distribution function shall be used instead of the Boltzmann distribution. Furthermore, the electronic states are quantized, and the DOS is energy-dependent, distinguishing significantly from the effective DOS. Therefore, the classic model is not adequate for charges in the inversion layers. This can be illustrated by a qualitative quantum mechanical solution of the electronic structure for the MOS device channels based on a triangular potential well approximation, which assumes that the electric field in the channel is constant in the transverse direction, and thus the semiconductor has a linear band bending profile. Under the triangular potential well approximation, the insulator barrier and the linear confining potential form a triangular potential well, which is described by

$$V(z) = \begin{cases} eFz & z > 0 \\ \infty & z \leq 0, \end{cases} \tag{5.12}$$

where F is the effective field along the z-direction. The Schrödinger equation (5.5) then becomes

$$\begin{cases} \dfrac{\mathrm{d}^2 f(z)}{\mathrm{d}z^2} + \dfrac{2m_z^*}{\hbar^2}(E - eFz)f(z) = 0 & z > 0 \\ f(z) = 0 & z \leq 0, \end{cases} \tag{5.13}$$

where we neglected the in-plane dispersion at first. Defining

$$r = \left(\frac{2m_z^* eF}{\hbar^2}\right)^{1/3}\left(z - \frac{E}{eF}\right), \tag{5.14}$$

for $z > 0$, (5.13) becomes

$$\frac{\mathrm{d}^2 f(r)}{\mathrm{d}z^2} - rf(r) = 0. \tag{5.15}$$

This is the Airy equation. Its finite solution is called the Airy function, i.e., $f(r) = A(r)$. $A(r)$ has an analytical expression as

$$A(r) = \frac{1}{\pi}\int_0^\infty \cos\left(\frac{1}{2}x + rx\right)\mathrm{d}x. \tag{5.16}$$

The Airy function is plotted in Fig. 5.7. The roots for $A(-r) = 0$ are discrete. The lowest ones are such as $r_0 = 2.338$, $r_1 = 4.087$, $r_2 = 5.520$,

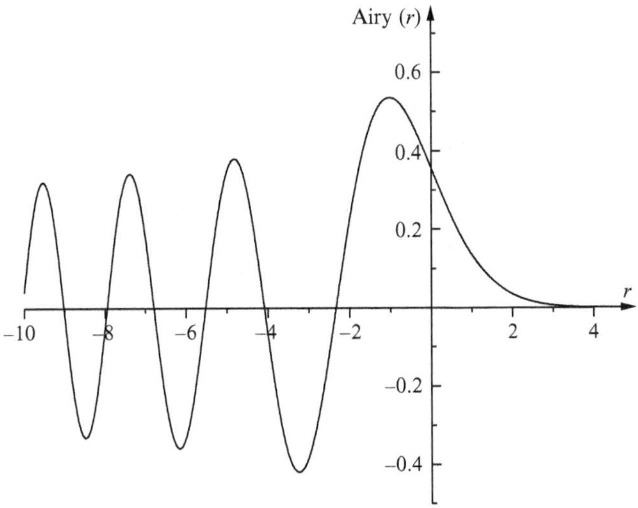

Fig. 5.7. Airy function

$r_3 = 6.787, \ldots$. The energies are obtained by matching the boundary conditions for the $f(r)$ at $z = 0$,

$$f(z)|_{z=0} = A(r)|_{z=0} = A\left[-\left(\frac{2m_z^* eF}{\hbar^2}\right)^{1/3} \frac{E}{eF}\right] = 0. \tag{5.17}$$

Thus, through solving the equation

$$\left(\frac{2m_z^* eF}{\hbar^2}\right)^{1/3} \frac{E}{eF} = r_n, \qquad n = 0, 1, 2, 3, \ldots, \tag{5.18}$$

the energies are given by

$$E_n = \left(\frac{\hbar^2}{2m_z^*}\right)^{1/3} (eF)^{2/3} r_n. \tag{5.19}$$

When $n > 4$, r_n can be approximated pretty well by

$$r_n = \left[\frac{3}{2}\pi\left(n + \frac{3}{4}\right)\right]^{2/3}, \tag{5.20}$$

hence for $n > 4$, the energy level can be approximated by

$$E_n = \left(\frac{\hbar^2}{2m_z^*}\right)^{1/3} \left[\frac{3}{2}\pi eF\left(n + \frac{3}{4}\right)\right]^{2/3}. \tag{5.21}$$

After we obtain the energies, the envelope functions are given by

$$f_n(z) = C \times A\left[\left(\frac{2m_z^* eF}{\hbar^2}\right)^{1/3} \left(z - \frac{E_n}{eF}\right)\right], \tag{5.22}$$

where C is the normalization constant. The envelope functions must be normalized in the confining direction. In the triangular potential well approximation, the integration of $f_n(z)$ from $z = 0$ to $z = \infty$ shall be 1. The lowest three energy levels and the corresponding wave functions for Si electrons with effective mass $m_z = 0.19m_0$, which is the effective mass along the minor axis of the Si conduction band ellipsoid, are shown in Fig. 5.8. For any subband, the probability of finding charges at the interface is zero, in contrast with the classic result. Larger effective mass results in lower eigenenergies, and consequently the electrons are concentrated closer to the surface.

After finding the energy levels in the z-direction, the subband energy can be written as

$$E_n(k) = E_n + \frac{\hbar^2 k_x^2}{2m_x} + \frac{\hbar^2 k_y^2}{2m_y}. \tag{5.23}$$

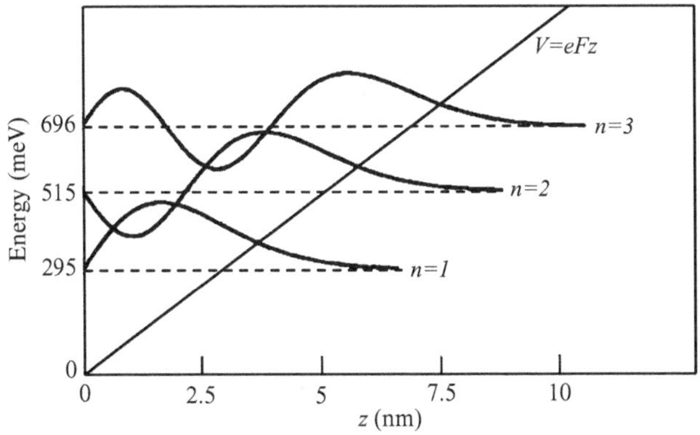

Fig. 5.8. The triangular potential well and the wave function for electrons with $m_z = 0.19m_0$, in an electric field of $F = 1\,MV/cm$

For the Δ_2 valley, the energy can be written as

$$E_n(k) = \left(\frac{\hbar^2}{2m_z}\right)^{1/3}\left[\frac{3}{2}\pi eF\left(n+\frac{3}{4}\right)\right]^{2/3} + \frac{\hbar^2 k_t^2}{2m_t}, \tag{5.24}$$

where $k_t^2 = k_x^2 + k_y^2$. In the quantum limit (electrons only occupy the ground state), the average width of the electronic distribution in the z-direction is given by

$$\bar{z} = \frac{\int_0^\infty z|f_0(z)|^2 \mathrm{d}z}{\int_0^\infty |f_0(z)|^2 \mathrm{d}z} = \frac{2E_0}{3eF}. \tag{5.25}$$

We need to point out that the electric field F used in the equation is not the field at the Si/SiO_2 interface. It is in fact the effective field and related to the depletion and inversion charge density by an empirical equation

$$F_{\text{eff}} = \frac{1}{\epsilon_s}\left(|Q_d| + \eta|Q_i|\right), \tag{5.26}$$

where ϵ_s is the semiconductor permittivity. The coefficient η is normally chosen as $\frac{1}{2}$ for Si n-channel MOS devices and $\frac{1}{3}$ for Si p-channel MOS devices. The value of η might be different for different materials.

Similarly, for a qualitative description, the subband structure in the GaAs/AlGaAs heterojunction can also be calculated using the triangular potential well approximation, just as in the Si/SiO_2 system, to assume that the potential barrier between GaAs and AlGaAs is infinite.

5.3.3 Quantum Well and Quantum Wire Band Structures

Next, we only use quantum wire as an example to illustrate how to calculate their subband structures. Procedures for quantum wells are similar.

Since in a nanowire the confinement is 2D, the electronic subband is discrete in the two confining directions, say x and y, and quasi-continuous in the channel direction, z. This is in contrast with the 2D charge system, where the electronic states are quasi-continuous in-plane. The qualitative electronic subband structure can be solved by approximating the nanowire with an infinite square quantum well, and the subband structure is approximately given by:

$$E = \frac{\pi^2\hbar^2}{2m^*}\left(\frac{n_1^2}{a^2} + \frac{n_2^2}{b^2}\right), \quad n_1, n_2 = 1, 2, 3, \cdots \tag{5.27}$$

where a and b are the widths of the cross-section of a square nanowire. For a spherical nanowire with diameter D, the subband energy is

$$E = \frac{2\pi^2\hbar^2(n+1)^2}{m^* D^2}, \quad n = 0, 1, 2, \cdots \tag{5.28}$$

In a Si n-type nanowire grown on the (100) surface, we have two effective masses in each valley along x and y. Then in each valley, the subband is given by

$$E = \pi^2\hbar^2\left(\frac{n_1^2}{2m_t^* a^2} + \frac{n_2^2}{2m_l^* b^2}\right), \quad n_1, n_2 = 1, 2, 3, \cdots \tag{5.29}$$

where we assume that the transverse direction of the equi-energy ellipsoid is along the nanowire transection side with width a and the longitudinal direction is along the nanowire transection side with width b. For quantum wire transverse dimension is significantly small, e.g., $< 3\,\text{nm}$; it is believed that the effective mass theory fails to capture the unique electronic properties under this condition. Si nanowire has a significant change from indirect bandgap to direct bandgap, shown by tight-binding (Luisier et al, 2006; Shiri et al, 2008) and ab initio calculations (Gnani et al, 2007; Leu et al, 2008). This reflects the limitation of the effective mass theory, which focuses on a small range of the Brillouin and fails to capture the whole band picture.

5.3.4 P-Type Structures

Qualitatively, the subband structure of p-type low-dimensional structures can be analogized to an n-type Si structure, which has two series of subbands originated from the two different effective masses. In the valence bands, the HH and LH bands will both split into a series of subbands under confinement. However, one important difference is that the valence bands are generally anisotropic and highly nonparabolic. It means that, in calculating the splitting, the effective masses that shall be employed in (5.5) are not well defined. The strong coupling between the HH and LH bands also affects the splitting between subbands. To obtain a quantitative band structure, we need to solve coupled equations using the Luttinger Hamiltonian, as described in Sect. 5.3.1. To illustrate this, we inspect the band splitting between the ground and the second hole states for a confinement along [001] with inversion hole density of $10^{13}/\mathrm{cm}^2$ for a Si p-channel MOS device. If we use the bulk HH and LH mass values along the [001] direction in the triangular potential well approximation, i.e., $0.29m_0$ and $0.20m_0$, they give a result of $E_g = 256\,\mathrm{meV}$ and $E_1 = 290\,\mathrm{meV}$, and the splitting is 34 meV. Using the Luttinger Hamiltonian and numerically solving the coupled equation under the triangular well potential approximation gives $E_g = 209\,\mathrm{meV}$ and $E_1 = 229\,\mathrm{meV}$, and the splitting is only 20 meV.

5.3.5 2D and 1D Density of States

In a 2D system, the motion of the electrons in the x-y directions is similar to that of the corresponding bulk solid, which can be treated by the conventional single electron approximation, and its energy dispersion can be expressed by the respective effective masses, which approximately takes into account the effect of the periodic potential in the x-y plane. The total electron energy is the superimposition of the quasi-plane wave motion in the x-y plane and the motion under the confinement in the z-direction and can be expressed as,

$$E_i(k_x, k_y) = E_{i,z} + E_\perp(k_x, k_y) = E_{i,z} + \frac{k_x^2 + k_y^2}{2m^*}, \tag{5.30}$$

which is also the result of the effective mass theorem (5.5). $E_{i,z}$ is the energy level obtained through the procedures we discussed earlier using the triangular potential well approximation or self-consistent procedures. The energy levels in the z-direction in a Si nMOS and the energy dispersions in the x-y plane are illustrated in Fig. 5.9. For each subband, at the point $k_\perp = 0$, the energy is exactly the energy $E_{i,z}$, thus each subband is a 2D parabolic band with energy beginning from $E_{i,z}$.

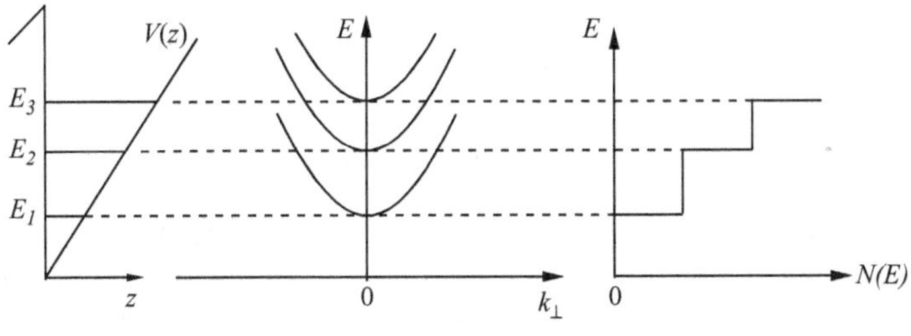

Fig. 5.9. Subband, in-plane dispersion and DOS of 2DEG

The state density in the 2D k-space is $1/(2\pi)^2$. For each subband, we have

$$D(E)\mathrm{d}E = \frac{1}{(2\pi)^2} 2\pi k_\perp \mathrm{d}k_\perp. \tag{5.31}$$

Since $\mathrm{d}E/\mathrm{d}k_\perp = \hbar^2/m^*$, then

$$D(E) = \frac{m^*}{2\pi\hbar^2} \tag{5.32}$$

for each spin state. Including the spin degeneracy, and considering that the energy states begin from $E_{i,z}$ for each subband, then the DOS for subband E_i can be written as

$$\begin{aligned} D_i(E) &= \tfrac{m_D}{\pi\hbar^2}, \quad && E > E_i \\ &= 0, && E < E_i, \end{aligned} \tag{5.33}$$

where m_D is the 2D DOS mass. In a isotropic parabolic band, the effective mass equals the DOS mass. In a anisotropic band such as in Si conduction valleys, m_x is not necessarily equal to m_y, and $m_D = \sqrt{m_x m_y}$. For a nonparabolicity band with nonparabolicity parameter α, the DOS is also dependent on $E_\perp$ through the relation

$$D_i(E) = \begin{cases} \frac{m_{D0}}{\pi\hbar^2}(1 + 2\alpha E), & E > E_i \\ 0, & E < E_i, \end{cases}$$

where m_{D0} is the DOS mass at the point $k_\perp = 0$.

Under parabolic approximation, the contribution to the 2D charge density from the subband E_i is given by

$$N_i = \int_{E_i}^{\infty} D_i(E) f_D(E) \mathrm{d}E = \frac{m_D k_B T}{\pi\hbar^2} \ln\left[1 + \exp\left(\frac{E_F - E_n}{k_B T}\right)\right], \tag{5.34}$$

where $f_D(E)$ is the Fermi-Dirac distribution function. Then the 2D charge density (e.g., the "surface" charge density in MOS structure) is given by

$$N = \frac{m_D k_B T}{\pi\hbar^2} \sum_{i=1}^{\infty} \ln\left[1 + \exp\left(\frac{E_F - E_n}{k_B T}\right)\right]. \tag{5.35}$$

For nonparabolic bands, the 2D charge density can only be solved numerically.

In a 1D system such as in a Si nanowire, as its energy is written in (5.9), each subband is a 1D E-k line in the k_z direction with the bottom energy given by $E_{m,n}$. The DOS of a 1D system can be similarly obtained as in the 2D system, and it is found to be proportional to $E^{-1/2}$. The DOS for one subband is

$$D_i(E) = \begin{cases} \frac{1}{\pi\hbar}\sqrt{\frac{2m^*}{E - E_{m,n}}}, & E > E_{m,n} \\ 0, & E < E_{m,n}. \end{cases}$$

In the 1D case, the unit for the DOS is the reverse of length and energy. At each subband bottom, the DOS is divergent. The diagram of the 1D DOS for a GaAs quantum wire is shown in Fig. 5.10

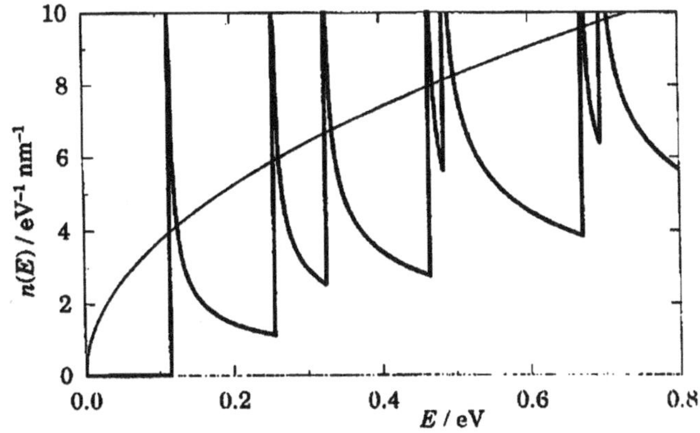

Fig. 5.10. 1D DOS for a GaAs quantum wire. From Davies (Davies, 1998)

5.4 SELF-CONSISTENT CALCULATION

The validity of the triangular well approximation is an important issue. We need to consider several factors before we can safely apply it. For a MOS device, first, we shall understand that the triangular well approximation is not appropriate for the strong inversion situation. Under strong inversion, the inversion charge significantly alters the electric field induced by depletion, which can be assumed constant near surface and is the basis for the triangular approximation. According to the electron distribution probability in Fig. 5.8, when the inversion charge density is high, it apparently induces strongly space-dependent field near the surface and thus changes the linear potential profile. Second, the normal electric field at the surface in the triangular well approximation is from an empirical expression

$$F_{\text{eff}} = \frac{1}{\epsilon_s}\left(|Q_d| + \eta|Q_i|\right), \tag{5.36}$$

where ϵ_s is the semiconductor permittivity. The coefficient η is normally chosen as $\frac{1}{2}$ for Si electron MOS devices and $\frac{1}{3}$ for Si hole MOS devices. These values are from the empirical fit and obviously smooth out some fine features of the electric field, and there is no reason that they shall be kept constant for all channel charge densities. For materials other than Si, the value for η shall obviously be different. Also, the infinity barrier assumption is quite good for Si-based MOS devices, but may not be proper for some MIS structures and some heterostructures such as the GaAs/AlGaAs system, which has relatively lower barrier height. In a GaAs/$Al_{0.3}Ga_{0.7}As$ heterojunction with a surface charge density of $5 \times 10^{11}/\text{cm}^2$ and depletion charge density of $5 \times 10^{10}/\text{cm}^2$, the average distance of the free electron from the interface is 2 nm more away using the triangular potential well approximation than the self-consistently obtained result. This is a significant discrepancy. Accurate determination of confining potential profile and carrier distribution in strong inversion (or accumulation) layer in MOS devices or in most MIS structures and heterostructures can be obtained by solving coupled Schrödinger and Poisson equations.

5.4.1 Self-Consistent Procedure

The coupled Schrödinger–Poisson equations can generally be solved only through self-consistent procedures. Stern (Stern, 1972) took one year to obtain the self-consistent subband structures in Si inversion layers in 1973 by using computers at that time. Now a person computer can solve the same problem in minutes.

To introduce the self-consistent procedure, first we do not include the in-plane band structures into consideration, but just consider the energy subbands in the confining direction. This is a one-dimensional (1D) problem. Let us consider an n-channel Si MOSFET first. The Poisson equation is given by:

$$\frac{\mathrm{d}^2V(z)}{\mathrm{d}z^2} = \frac{e}{\epsilon_\mathrm{s}}[p(z) - n(z) + N_d^+(z) - N_a^-(z)], \tag{5.37}$$

where $V(z)$ is the z-dependent confining electric potential, $p(x)$ and $n(x)$ are hole and electron concentrations, respectively, and $N_\mathrm{d}^+(z)$ and $N_\mathrm{a}^-(z)$ are ionized donor and acceptor impurity concentrations, respectively. In n-doped (p-doped) semiconductors, the acceptor (donor) doping density is zero. Under the deep depletion or accumulation condition, the minor carrier concentration can also be neglected. For instance, in an n-type inversion MOSFET device, the right-hand side of (5.37) is simplified to $\frac{e}{\epsilon_\mathrm{s}}[n(z) - N_\mathrm{a}^-(z)]$, i.e., the variation of the electric field is induced by both the depletion of the holes and accumulation of the electrons in the channel. The 1D Schrödinger equation in the effective mass approximation is

$$-\frac{\hbar^2}{2}\frac{\partial}{\partial z}\frac{1}{m_z^*}\frac{\partial}{\partial z}\psi(z) + [E - eV(z)]\,\psi(z) = 0, \tag{5.38}$$

here $V(z)$ is the confining potential in (5.37). Solving (5.38) gives the electronic wave function in the z direction, and the carrier density $n(z)$ (for nMOS) or $p(z)$ (for pMOS) is proportional to $|\psi(z)|^2$. Then substitute the carrier density into (5.37) to obtain a new $V(z)$. Iterate this process until satisfactory results are obtained. This self-consistent procedure is shown in Fig. 5.11, where $0 < \alpha < 1$ is a number that corrects the original potential profile after each process cycle.

In a 2D electron system where only the electron occupation in the ground state (or also the first excited states) is important, the ground state wave function can be obtained conveniently by a variational method. But when the band structure of the system considered is complicated and degenerate, or we have to consider multiple subbands, the variational method becomes complicated. In such cases, we can employ the finite difference method, where the wave functions and potential profile $V(z)$ are evaluated on a grid where these discrete values are used to approximate the continuous functions. In the following, we briefly introduce the variational and finite difference methods.

5.4.2 Variational Method

In many cases, if the analytical expression for the wave functions is available, the determination of the properties of a quantum system becomes very convenient. The variational method is an approach that may be used to estimate the ground state energy and also gives an analytical wave function. We already used the variational method in Chap. 4 to obtain the energy band structure under a tight-binding framework. Now we briefly introduce the principle of the variational method again. The wave function of the ground state, ψ_1, must satisfy the Schrödinger equation,

$$H\psi_1 = E_1\psi_1. \tag{5.39}$$

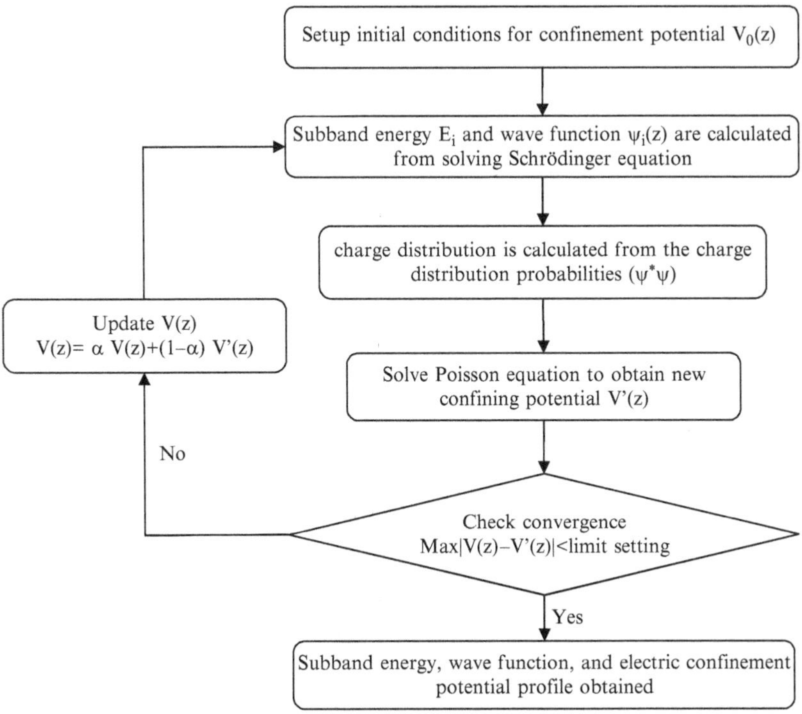

Fig. 5.11. Procedures for Schrödinger–Poisson self-consistent calculations

Then from this equation the energy E_1 can be written as

$$E_1 = \frac{\int \psi_1^* H \psi_1}{\int \psi_1^* \psi_1}. \tag{5.40}$$

Any other wave function ψ instead of ψ_1 in the above equation will result in an energy $E \geq E_1$,

$$E_1 \leq E = \frac{\int \psi^* H \psi}{\int \psi^* \psi}. \tag{5.41}$$

For the ground state of a 2D system, we can first guess the form of the wave function with a small number of adjustable parameters, then adjust these parameters to minimize the ground state energy, and through this procedure, the values of the parameters are determined. The form of the guessed wave function is very important in the implantation of the variational method. For Si and GaAs/AlGaAs 2D electron systems, Fang–Howard function (Fang and Howard, 1966) is extensively adopted as the guess function for the ground state, where

$$\psi(z) = Az\exp(-\frac{b}{2}z), \tag{5.42}$$

where A is the normalization coefficient and b is the adjustable parameter. If we assume that for a 2D electron system $\psi(z) = 0$ at $z = 0$, then normalization of $\psi(z)$

$$\int_0^\infty \psi(z)^2 = 1 \tag{5.43}$$

gives $A = (b^3/2)^{1/2}$, and thus

$$\psi(z) = \left(\frac{b^3}{2}\right)^{\frac{1}{2}} z\exp(-\frac{b}{2}z). \tag{5.44}$$

Next, we apply the variational method to a triangular well potential to check the consistency between the variational method and the exact solution. We substitute $\psi(z)$ into (5.41) to evaluate the numerator,

$$\begin{aligned}&\int_0^\infty \psi(z)^* H\psi(z)\mathrm{d}z \\ &= \int_0^\infty \frac{b^3}{2} z\exp(-\frac{b}{2}z)\left[-\frac{\hbar^2}{2m}\frac{\mathrm{d}^2}{\mathrm{d}z^2} + eFz\right] z\exp(-\frac{b}{2}z)\mathrm{d}z \\ &= \frac{\hbar^2 b^2}{8m} + \frac{3eF}{b}.\end{aligned} \tag{5.45}$$

This indicates that the ground state energy

$$E_1 \leq E = \frac{\hbar^2 b^2}{8m} + \frac{3eF}{b}. \tag{5.46}$$

The minimization of E with respect to b then gives

$$b = 6\frac{2meF}{\hbar^2}. \tag{5.47}$$

Finally, the ground state energy set by this variational function is given by

$$E = 2.476\left(\frac{\hbar^2}{2m}\right)^{1/3}(eF)^{2/3}, \tag{5.48}$$

compared with the exact energy found previously for the triangular well potential

$$E_1 = 2.338 \left(\frac{\hbar^2}{2m} \right)^{1/3} (eF)^{2/3}. \tag{5.49}$$

This result is amazingly good for using only one adjustable parameter in the variational function.

For GaAs/AlGaAs heterojunction system, the wave function has a non-vanishing amplitude in the AlGaAs layer due to low barrier height, then the variational function shall have a little different form to take into account this effect. Ando (Ando, 1982b,a) proposed a piecewise function according to different regions:

$$\psi(z) = \begin{cases} Bb^{1/2}(bz+\beta)\exp(-bz/2), & z > 0 \\ B^{'}b^{'1/2}\exp(b^{'}z/2), & z < 0. \end{cases} \tag{5.50}$$

where b, $b^{'}$, β, B, and $B^{'}$ are variational parameters. Among these parameters, B, $B^{'}$, and β can be expressed by b and $b^{'}$ through boundary conditions stated in (5.6) and normalization. Figure 5.12 shows the comparison of the ground state wave functions obtained by Fang-Howard function, Ando's trial function, and wave function calculated numerically for a GaAs/$Al_0.3Ga_{0.7}As$ heterostructure with a 5-nm-thick undoped AlGaAs spacer. The wave function

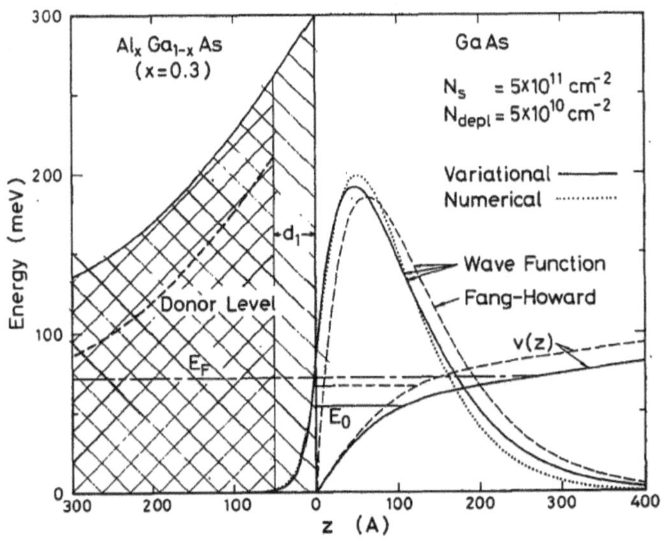

Fig. 5.12. Band profile and electron wave functions calculated by variational and numerical methods. After Ando (Ando, 1982a)

given by the variational method coincides with the numerical one well for small z, while it tends to give larger amplitude for large z. However, overall they are pretty consistent. Thus, the variational method is expected to be sufficient for calculation of the subband energy and transport properties, but only if the ground subband occupation is dominant.

5.4.3 Finite Difference Method

If the band structure is complicated, which cannot be approximated parabolic, or there are many subbands involved, the variational method is not as efficient. Rather, we seek numerical solutions for subband energies and wave functions. Numerically, the wave function can only be evaluated on a series of discrete points, i.e., a grid. The differentiations to the wave function then can be approximated by finite difference method. In many cases, the finite difference method can implicitly take into account the boundary conditions in a 2D system Hamiltonian, without producing spurious solutions. This method has great advantages in a self-consistent procedure to find the confining potential profile along the confining direction, especially if there is external electric field as in the MOS structures. The finite difference method is easily implemented, but sometimes the self-consistent procedure can be quite time-consuming.

The finite difference method is a numerical method to solve the differential equations. It turns the differential into finite differences and evaluates them at a series of grid points. To illustrate this method, we can use a 1D infinite potential well problem for an example.

Problem: Find the energy levels and wave functions for a particle with a mass m in an infinite 1D box with width a shown in Fig. 5.13(a).

Normal procedure: This is a well-known quantum mechanical problem. Solving the Schrödinger equation

$$\frac{\mathrm{d}^2}{\mathrm{d}x^2}\psi(z) + \frac{2m}{\hbar^2}E\psi(z) = 0 \tag{5.51}$$

with boundary condition

$$V(z) = \begin{cases} 0 & 0 < z < a \\ \infty & z < 0 \quad z > a \end{cases} \tag{5.52}$$

gives the energy level

$$E = E_n = \frac{\hbar^2\pi^2 n^2}{2ma^2}, \quad n = 1, 2, 3, \cdots \tag{5.53}$$

and the wave functions

$$\psi_n(z) = \sqrt{\frac{2}{a}}\sin\left(\frac{n\pi}{a}z\right). \tag{5.54}$$

The exact energy levels and the wave functions are shown in Fig. 5.13b.

Finite difference method: To numerically solve the Schrödinger equation, first we break the continuous z-axis into grid points as in Fig. 5.14 with

$d = a/(N+1)$. We evaluate the wave function $\psi(z)$ at these grid points, and thus the continuous wave function $\psi(z)$ is now converted to a vector $\{\psi_i\}$ with N components, where N is the number of the grid points between $z = 0$ and $z = a$.

The first-order derivative of ψ to z can be approximately expressed as

$$\left.\frac{\mathrm{d}\psi}{\mathrm{d}z}\right|_{z_{i-\frac{1}{2}}} \simeq \frac{\psi^i - \psi^{i-1}}{d}, \quad \left.\frac{\mathrm{d}\psi}{\mathrm{d}z}\right|_{z_{i+\frac{1}{2}}} \simeq \frac{\psi^{i+1} - \psi^i}{d}. \tag{5.55}$$

Then the second-order derivative of ψ at z_i is approximately

$$\left.\frac{d^2\psi}{dz^2}\right|_{z_i} \simeq \frac{\psi^{i+1} - 2\psi^i + \psi^{i-1}}{d^2}. \tag{5.56}$$

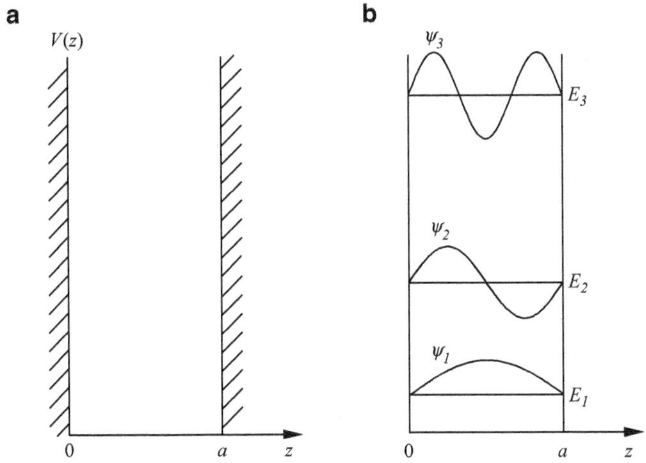

Fig. 5.13. (**a**) An infinite potential well, and (**b**) energies and wave functions for a particle in an infinite potential well

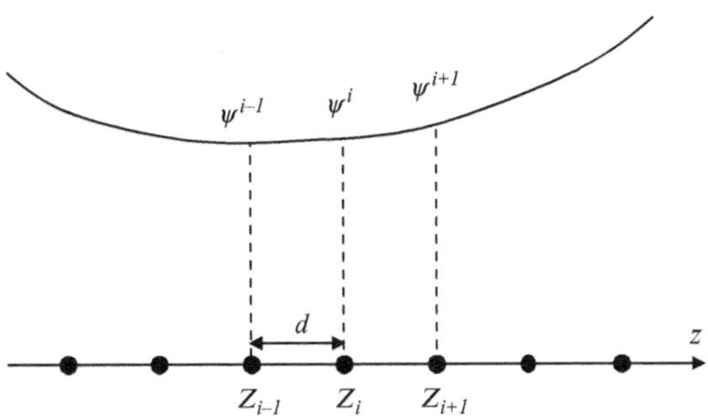

Fig. 5.14. 1D grid on the z axis. An arbitrary function is evaluated on these grid points in the finite difference method

The second-order derivative of ψ at one point also requires its values at the adjacent points. Equation (5.56) can be written in a matrix form on the grid

$$\frac{d^2\psi}{dz^2} \rightarrow H_D\{\psi_i\}, \tag{5.57}$$

where

$$H_D = \begin{array}{c|cccccccccc} & 1 & 2 & \cdots & i-1 & i & i+1 & \cdots & N-1 & N \\ \hline 1 & 2h_0 & -h_0 & \cdots & 0 & 0 & 0 & \cdots & 0 & 0 \\ 2 & -h_0 & 2h_0 & \cdots & 0 & 0 & 0 & \cdots & 0 & 0 \\ \cdots & \cdots & \cdots & \cdots & \cdots & \cdots & \cdots & \cdots & \cdots & \\ i-1 & 0 & 0 & \cdots & 2h_0 & -h_0 & 0 & \cdots & 0 & 0 \\ i & 0 & 0 & \cdots & -h_0 & 2h_0 & -h_0 & \cdots & 0 & 0 \\ i+1 & 0 & 0 & \cdots & 0 & -h_0 & 2h_0 & \cdots & 0 & 0 \\ \cdots & \cdots & \cdots & \cdots & \cdots & \cdots & \cdots & \cdots & \cdots & \\ N-1 & 0 & 0 & \cdots & 0 & 0 & 0 & \cdots & 2h_0 & -h_0 \\ N & 0 & 0 & \cdots & 0 & 0 & 0 & \cdots & -h_0 & 2h_0 \end{array}, \tag{5.58}$$

with $h_0 = -1/d^2$. Since the second-order derivative at grid point z_i depends on the function values at z_{i-1}, z_i, and z_{i+1}, so the matrix is in a block diagonal form with $H_{i,i-1} = H_{i,i+1} = -H_{i,i}/2$ and other elements equal to zero.

A special care should be taken with respect to the boundary conditions. In finite difference method, the boundary conditions are automatically taken care of by setting specific values for a few matrix elements at the boundary points. There are typically two types of boundary conditions:

1. *Periodic boundary condition.* In this case, the grid point 1 is also the point $N+1$. This boundary condition is typically taken for bulk semiconductors and superlattices. In this case, the last element in the first row and the first element in the last row are also $-h_0$.
2. *Infinite well boundary condition.* This is what we applied in matrix (5.58), where the wave function vanishes at both ends and thus at the boundary the matrix elements are all zero in matrix 5.58.

In matrix (5.58), we set grid points 1 and N next to the ends, and the well width a is divided into $N+1$ equal intervals with length $d = \frac{a}{N+1}$. Using the expression (5.58), the Schrödinger equation (5.51) becomes

$$H_S\{\psi_i\} = E\{\psi_i\} \tag{5.59}$$

where H_S has the same form as H_D, with h_0 now changing into $\frac{\hbar^2}{2md^2}$. The eigenvalue problem of (5.59) can be solved easily using a computer.

The number of eigenvalues obtained equals the dimension of the matrix, i.e., the number of the grid points in the range we study. The number of grid points also affects the accuracy of the obtained energy levels through the finite difference method. Obviously, increasing the number of the grid points can enhance the accuracy of the numerical results since the derivatives can be better simulated. However, increasing the number of the grid points also greatly increases the computation labor. The trade-off between the accuracy and computation labor depends on the property of the problem under investigation. In Fig. 5.15 the eigenenergies for an electron in the 1D infinite well with width $a = 10\,nm$ obtained by exact analytical solution (5.53) and by finite difference method with $N = 20$, $N = 50$, and $N = 100$ are compared. At higher energy, the numerical results show big deviation from the exact

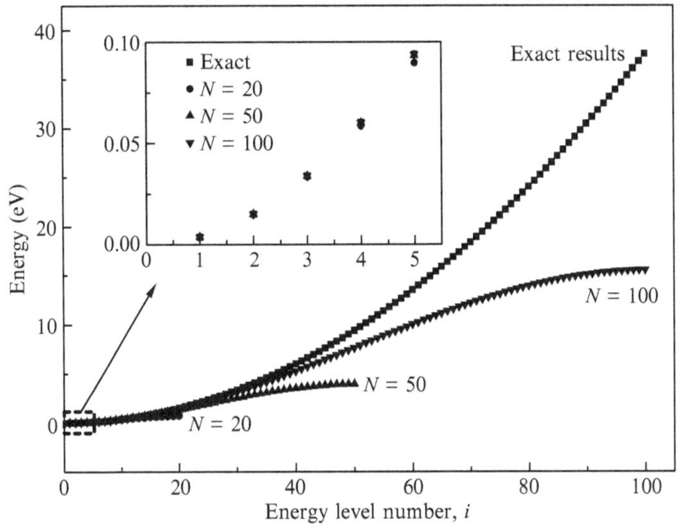

Fig. 5.15. Comparison between the eigenenergies obtained by finite difference method and the exact values

solution. This is because that the wave function oscillates rapidly for states with elevated quantum number, and thus its derivative cannot be accurately simulated using a finite grid. However, for the lowest several energy levels, the energies are very close to the exact value. Three different grid point numbers give about the same ground state energy. For the fourth ($i = 5$) excited state, the energy obtained by setting $N = 20$ is only 4.6% different from the exact value, and the difference is diminishingly 0.2% for $N = 100$. In most cases, only the lowest several energy levels are of interest, where the finite difference method is most accurate.

In a multiband case such as in Hamiltonian (4.146) or (4.159), every Hamiltonian element spans on the grid to become an $N \times N$ block. If the element does not contain k_z, then the $N \times N$ block is just a constant matrix. If the element contains k_z or k_z^2, then the $N \times N$ block follows the same form as in (5.58). Thus the 6×6 Luttinger Hamiltonian on a 1D grid with N points becomes a $6N \times 6N$ matrix. Diagonalizing this Hamiltonian gives $6N$

eigenenergies and corresponding wave functions, each of which is composed of six components (the Luttinger basis), which are spanned on the N grid points.

One important case is that the barrier height is not so large that the wave function penetrates into the barrier for a finite depth. In such a case, the application of the finite difference method shall also take into account this wave penetration to set up the infinite wall at a far enough distance from the barrier/well interface. Together with this setup is the mass (or mass parameter) difference between the barrier and well layer. This mass difference must be considered following the rule in (5.5). The mass difference greatly affects the wave penetration probabilities. In the application of finite difference method in Luttinger Hamiltonian, the masses are represented by the Luttinger parameters.

5.5 SUBBAND STRUCTURES OF 2D ELECTRON/HOLE GAS

5.5.1 Self-Consistent Confining Potential

Employing the self-consistent procedure introduced in Fig. 5.11, and the finite difference method introduced earlier for the second-order differential equation, the coupled Schrödinger and Poisson equations are solved for an Si n-channel MOS device. From the surface to 50 nm deep into the substrate, 1,000 grid points have been evenly chosen. For an electron density of $1 \times 10^{13}/\text{cm}^2$ in the inversion layer, the confining potential $V(z)$ is shown in Fig. 5.16 together

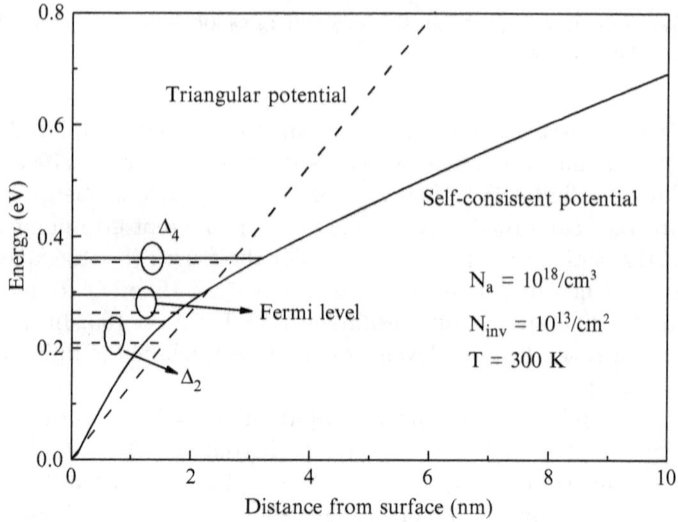

Fig. 5.16. Band bending profile and the ground subband energies for the Δ_2 and Δ_4 valleys obtained by triangular potential well approximation and self-consistent procedures

with the results from the triangular potential well approximation for a comparison. In this figure, the potential at the interface is selected as zero. Above the Fermi energy, the self-consistent confining potential quickly turns flat. In a MOS device, although the gate voltage is an important device parameter, it is however an extrinsic parameter, depending on various other parameters such as the gate oxide thickness, substrate bias, etc. The early MOSFET gate voltage was as large as 10 V, and the gate voltage for the current state-of-the-art ultrasmall size MOSFET devices is at about 1 V. Even though the gate voltage changes about ten times, the charge density in the channel always keeps almost constant at around $1.5--2 \times 10^{13}/\text{cm}^2$. This is mainly determined by the breakdown electric field of the gate oxide. The channel charge density is an intrinsic device parameter that determines the operation of the MOS devices. Thus, in the self-consistent procedure, the channel charge density rather than the gate voltage shall be the input and fixed device parameter.

In Fig. 5.17, the channel electron distribution profile from the surface to substrate is shown for triangular potential well approximation and self-consistent procedure. It clearly shows that under triangular potential well approximation the electrons are more closely concentrated near the surface, which then will result in larger gate capacitance and stronger surface scattering than from the fully numerical method, and consequently affect the results of electron transport calculations in MOS devices.

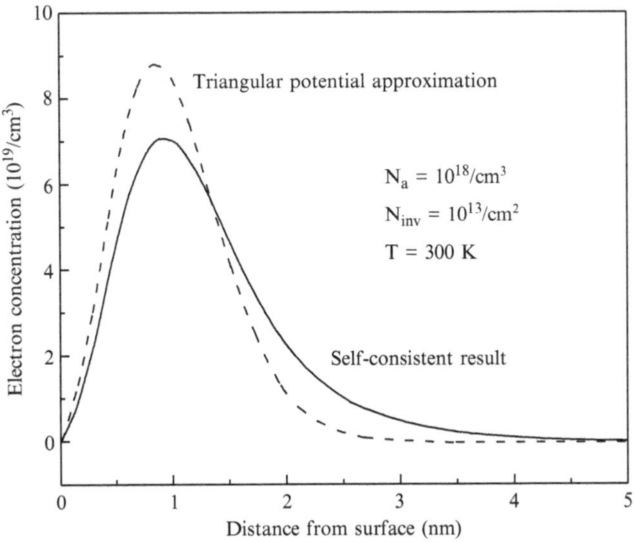

Fig. 5.17. Electron concentration distribution for a Si MOSFET obtained by triangular potential well approximation and self-consistent procedures

5.5.2 Charge Distribution vs. Material

The inversion charge distribution computed quantum mechanically is significantly different from that calculated classically where the charge concentrates at the surface. Quantum mechanically, the charge concentration peaks some distance away from the surface, and at the surface, the concentration is actually zero if we assume that the barrier height is infinite. From either the triangular well potential model or the infinite box model, we can see that for the same quantum number labeling the energy level, smaller effective mass results in higher eigenenergy. This indicates one important quantum mechanical conclusion in normal device structures that smaller effective mass results in weaker confinement. For Si whose electron effective mass is $0.19m_0$ for in-plane valleys and $0.92m_0$ for the out-of-plane valleys, the confining field approaches zero at about 25 nm. For semiconductors with smaller electron effective masses, the charge distribution can peak much deeper than in Si. In Fig. 5.18, the electron concentration as a function of the distance from the surface is shown for Si, GaAs, InAs, and InSb. The surface barrier is assumed infinite for all materials. This repulsion of electrons from the barrier results in an increase of the distance between the opposite charge layers of the gate, which is classically the gate oxide thickness. Take a Si n-type MOS capacitor as an example. If the average distance of the electrons from the surface (the inversion layer depth) is t_{inv}, then the effective oxide thickness can be written as

$$t_{\mathrm{ox,eff}} = t_{\mathrm{ox,phy}} + \frac{\epsilon_{\mathrm{ox}}}{\epsilon_{\mathrm{Si}}} t_{\mathrm{inv}}, \tag{5.60}$$

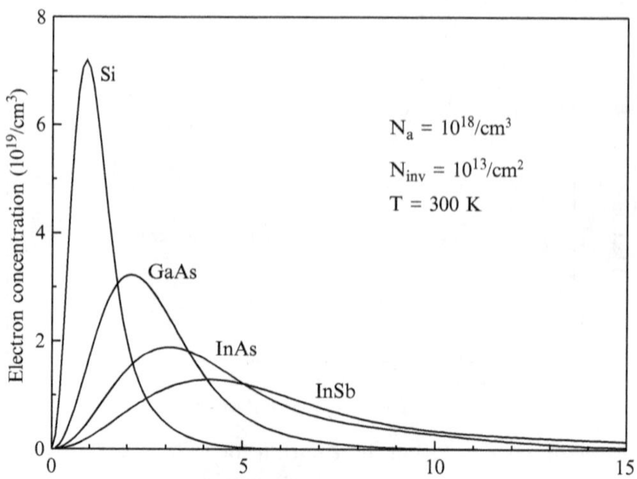

Fig. 5.18. Electron concentration distribution vs. distance from the surface for Si, GaAs, InAs, and InSb under strong inversion condition

where $t_{\mathrm{ox,phy}}$ is the physical oxide thickness. The gate capacitance

$$C_{\mathrm{ox}} = \frac{\epsilon_{\mathrm{ox}}}{t_{\mathrm{ox,eff}}} \tag{5.61}$$

is reduced by this quantum effect. The inversion layer depths of Si, GaAs, InAs, and InSb for two typical inversion electron densities, $1 \times 10^{12}/\mathrm{cm}^2$ and $1 \times 10^{13}/\mathrm{cm}^2$, are listed in Table 5.1. For Si MOS devices, this quantum effect

Table 5.1. Inversion layer depths in units of nm for Si, GaAs, InAs, and InSb channels for inversion electron densities of $1 \times 10^{12}/\mathrm{cm}^2$ and $1 \times 10^{13}/\mathrm{cm}^2$ at T = 300 K with $N_\mathrm{a} = 10^{18}/\mathrm{cm}^3$

	Si	GaAS	InAs	InSb
$1 \times 10^{12}/\mathrm{cm}^2$	1.6	3.2	5.1	~ 12
$1 \times 10^{13}/\mathrm{cm}^2$	1.3	2.8	3.9	~ 6.3

adds about 3-4 Å to the gate oxide thickness. If the gate oxide thickness is large enough, as was in the micron-scale devices, this may be neglected. But for aggressively scaled devices in the deep submicron region, the gate oxide thickness is in the nanometer scale, and this quantum effect-induced extra effective thickness can greatly alter the device behavior such as the drive current and transconductance of the devices.

The deep distribution of inversion electrons induces another problem in the nanoscale device applications of semiconductors such as GaAs, InAs, and InSb, which have small electron effective masses. With the centroid of the inversion electrons deep in the substrate, the gate capacitance is greatly reduced by addition of the effective oxide thickness. For modern state-of-the-art short channel devices, the channel length is typically around tens of nanometers. If InAs and InSb are used as channel materials, the average distance of inversion electrons is comparable to the channel length. Under this situation, control of channel electron through gate bias is difficult. The narrow band gap of InAs and InSb will also cause a severe leakage issue. The drain-induced barrier lowering effect is especially significant.

5.5.3 Subband Structure in GaAs/AlGaAs Heterostructures

The subband structure of the GaAs/AlGaAs can be calculated either using triangular well potential approximation or through Schrödinger–Poisson self-consistent procedure. Normally the AlGaAs layer is fully depleted, and thus the free electron charge in the GaAs well is known. If we choose the electron density in the GaAs well as a known parameter, then employing the triangular potential well approximation requires the relation between the effective electric field and the electron density. Assuming that the electron density per unit area at the interface is N_s, according to Gauss's law, the electric field at the interface is

$$F_{\mathrm{int}} = \frac{N_s}{\epsilon_{\mathrm{GaAs}}}. \tag{5.62}$$

The effective field F_{eff} can be approximated by F_{int} multiplied by a constant that needs to be determined by experiments. However, we shall understand that even if the AlGaAs layer is thick enough and it is not fully depleted, the doping in AlGaAs can solely determine the charge in the GaAs well (certainly, it also depends on some other external factors such as temperature). Thus, doping and thickness of the AlGaAs layer are important parameters to control the transport of the GaAs/AlGaAs heterostructure.

For GaAs/AlGaAs heterostructure, the triangular well potential approximation works much poorer than the Si/SiO_2 system, because the barrier height for GaAs/AlGaAs heterostructure is typically below 0.5 eV, compared with the conduction band discontinuity of 3.1 eV between Si and SiO_2. Thus the accurate band profile of GaAs/AlGaAs heterostructure requires self-consistent calculation. At low temperature and with relatively low doping, only the electron occupation in the ground state is predominant. The variational method works well in such a case. For relatively high temperature and high doping, several subbands can be occupied. If we are interested in only the bound states whose wave functions vanish at the edge of the AlGaAs layer (thin dielectric) and the bulk GaAs, we can assume an infinite barrier at the AlGaAs layer edge and deep in the GaAs substrate, including the barrier and well into one single semiconductor system as illustrated in Fig. 5.19. Along the growth direction (confining direction), the 1D Schrödinger equation applying the effective mass approximation is

$$-\frac{\hbar^2}{2}\frac{\partial}{\partial z}\left[\frac{1}{m_z}\frac{\partial}{\partial z}\right]\psi(z)+V(z)\psi(z)=E\psi(z), \tag{5.63}$$

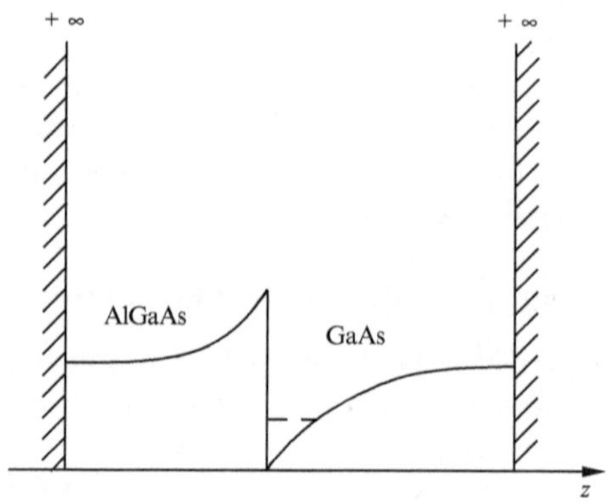

Fig. 5.19. Illustration of approximated potential profile for GaAs/AlGaAs heterojunction in finite difference method

where $1/m_z$ must be written in between the two differential operators since m_z is different in the barrier and well layers. The potential $V(z)$, is given by

$$V(z)=e\phi(z)+V_h(z), \tag{5.64}$$

where $\phi(z)$ is the electrostatic potential caused by electron transfer from the barrier to well (electrostatic potential caused by external voltage included, too, if it is applied), and $V_h(z)$ is a step function describing the potential barrier caused by the conduction band discontinuity at the interface. The electrostatic potential $\phi(z)$ is in turn determined by the electron distribution and solved by the Poisson's equation,

$$\frac{\mathrm{d}}{\mathrm{d}z}\epsilon(z)\frac{\mathrm{d}\phi(z)}{\mathrm{d}z} = e\left[n(z) + N_\mathrm{a}(z) - N_\mathrm{d}(z)\right], \tag{5.65}$$

where $\epsilon(z)$ is the z-dependent permittivity, N_a and N_d are the doping densities of the acceptors and donors, respectively, and

$$n(z) = \sum_i N_i|\psi_i(z)|^2, \tag{5.66}$$

where the sum is over interested subbands, and

$$N_i = \int_{E_i}^{\infty} \mathrm{d}E \frac{m_{i,D}}{\pi\hbar^2}\frac{1}{1+e^{\frac{E-E_F}{k_BT}}} = \frac{m_{i,D}k_BT}{\pi\hbar^2}\ln\left[1+\exp\left(\frac{E_F-E_i}{k_BT}\right)\right] \tag{5.67}$$

is the electron density occupying the energy subband E_i, where $m_{i,D}$ is the 2D density-of-states mass for the subband E_i.

Figure 5.20 (Inoue et al, 1985) shows the conduction band potential profile of a AlGaAs/GaAs/AlGaAs double-heterojunction structure under an externally applied voltage perpendicular to the layer surfaces. This is a basic structure used for selectively doped double-heterojunction FET.

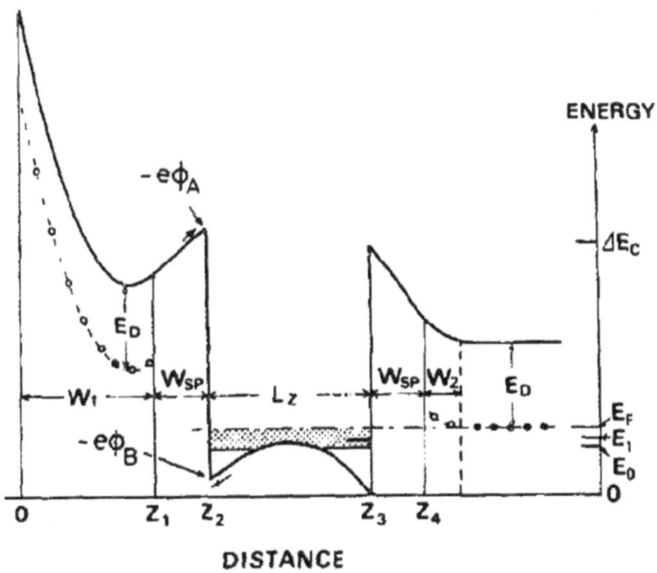

Fig. 5.20. Band alignment and band bending profile for AlGaAs/GaAs/AlGaAs double-heterojunction structure under applied external electric field. From (Inoue et al, 1985)

5.5.4 Subband Structure in Square Quantum Wells

The subband structure of GaAs/AlGaAs square quantum wells can be obtained using the same method applied for GaAs/AlGaAs heterostructures. A sqaure quantum well can also be considered as a limit case of a superlattice, where the barrier width is large enough to suppress the correlation of the wave functions in single quantum wells that together comprise a superlattice.

Here, we consider an n-type modulation-doped GaAs/$Al_{0.3}Ga_{0.7}As$ quantum well structure used by Chuang (Chuang, 1995) as shown in Fig. 5.21. The total width of the quantum well $L_w + L_b = 20\,\text{nm}$, with the well width of 10 nm. The barrier ends are doped with $N_D = 4 \times 10^{18}/\text{cm}^3$ with doping width of 1 nm at each end. The conduction band discontinuity ΔE_c is 251 meV. The self-consistent conduction band profile and the lowest two subbands are shown in Fig. 5.22a. The corresponding wave functions for these two levels are shown in Fig. 5.22b. It can be seen that the potential curves up at the middle of the well, since electron occupation probability (of the ground state) is highest at the middle region of the well, and it creates electric field pointing toward the middle region.

A Si symmetrical double-gate (SDG) MOSFET structure also forms a sqaure quantum well. The subband diagram of a SDG FET is shown in Fig. 5.23, in comparison with the single-gate (SG) MOSFET band profile. The splitting between the lowest two subbands for SDG MOSFETs is very

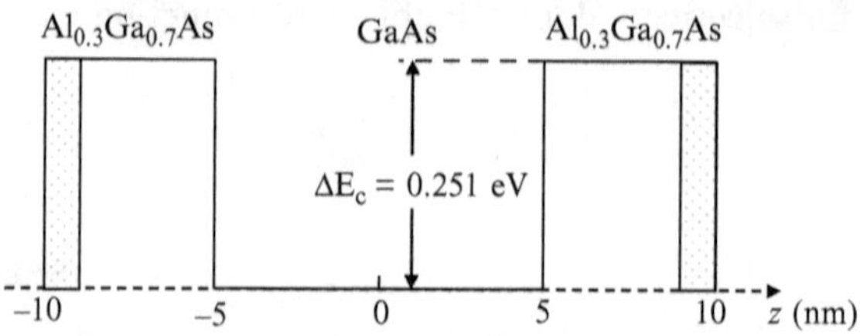

Fig. 5.21. Built-in band edge profile for a modulation doped AlGaAs/GaAs/-AlGaAs double heterojunction structure

small when the Si thickness is over 5 nm (5 meV when $t_{\text{Si}} = 5\,\text{nm}$, 3 meV when $t_{\text{Si}} = 15\,\text{nm}$). If the Si thickness is below 5 nm, the strong interaction of the two surface channels causes the subband splitting increasing drastically (i.e., 18 meV for $t_{Si} = 3\,\text{nm}$). The closely spaced energy levels $E_{1,1}$ and $E_{1,2}$ will eventually merge each other (become degenerate) when the Si thickness is large enough so that the two surface channels do not have quantum interference. Thus, these two lowest levels in the SDG structure correspond to energy level E_1 in the SG structure. The critical spacing of the two gates in SDG structure for them to strongly interfere with each other is obviously related to the charge distribution in a SG structure. When the two surface channels are brought together with Si thickness less than the critical thickness, one

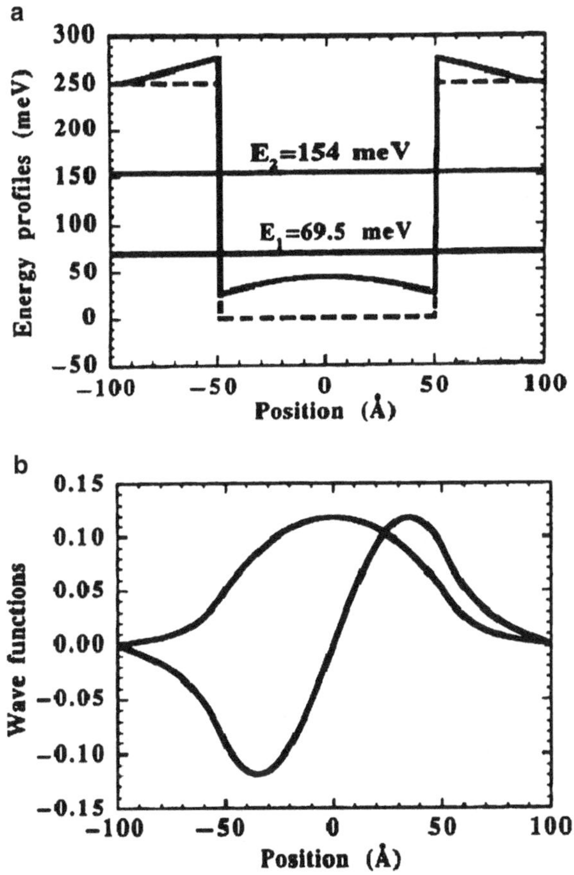

Fig. 5.22. (**a**) Built-in (*dashed*) and self-consistent band bending profile (*solid*) for the double-junction structure shown in Fig. 5.21. E_1 and E_2 are the two lowest energy levels obtained self-consistently. (**b**) The wave functions of the two corresponding energy states in (**a**). From Chuang (Chuang, 1995)

single level in the SG structure splits into two energy levels. This is exactly resembling the bonding and antibonding states formed by atomic levels when two atoms are brought together to form a molecule. The critical thickness depends on the doping density (depletion width), charge density (which in turn is controlled by the effective electric field, or gate voltage), etc. When the splitting between the lowest two levels is small, the charge tends to occupy the regions close to the gate; when the splitting becomes larger, they tend to accumulate at the center. The distribution of the charge in the SDG structure also depends on the charge density or gate voltage. Lower density of charges tend to occupy the center area between the double gates since the confinement to the surface is not strong compared with high density of charges.

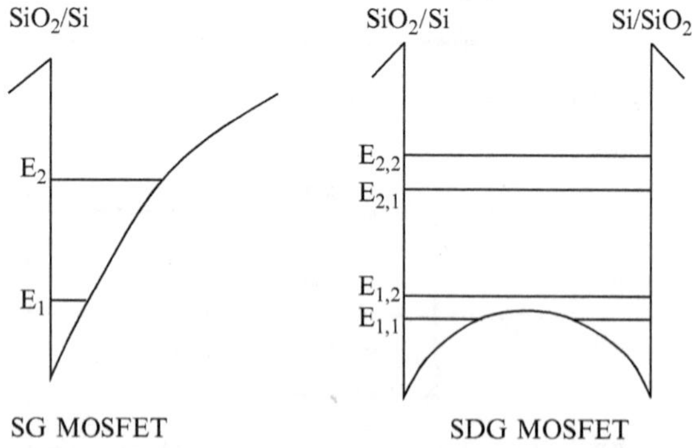

Fig. 5.23. Band diagrams for a single-gate Si MOSFET (*left*) and a double-gate Si MOSFET

In p-channel devices or n-channel devices with multiple conduction valleys, we shall notice the electronic characters of the lowest two subbands. In the SG structure, the lowest two subbands result from two types of valleys in nMOS or from HH and LH bands, respectively, in pMOS. However, since the lowest two levels in the SDG structure are actually split from one band, they have the same character.

5.5.5 Subband Energy vs. Well Width

The quantum well width measures the strength of the confinement to charge carriers. It is apparent for square quantum wells, but not obvious for a triangular well, or a well created by a self-consistent potential. In the latter cases, we can define an effective well width through the wave function overlap factor $W_{\mu\nu}$, by

$$\frac{1}{W_{\mu\nu}} = 2\pi \int_{-\infty}^{\infty} |I_{\mu\nu}(q_z)|^2 \mathrm{d}q_z, \tag{5.68}$$

where

$$I_{\mu\nu}(q_z) = \int F_\mu(z) \exp(\mathrm{i}q_z z) F_\nu(z) \mathrm{d}z, \tag{5.69}$$

where $F_\mu(z)$ is the envelope function of the wave fucntion in the μth subband. Following Price (Price, 1981), $W_{\mu\nu}$ can be expressed as

$$\frac{1}{W_{\mu\nu}} = 2\pi \int F_\mu^2(z) F_\nu^2(z) \mathrm{d}z, \tag{5.70}$$

where $W_{\mu\mu}$ represents the effective well width for the μth subband. Therefore, different subbands have different well widths.

Subband energy increases with decrease of the well width. This is straightforward from the point of Heisenberg's uncertainty relation $\Delta p \Delta z > \hbar/2$. With stronger confinement (smaller z), the momentum (and energy) becomes larger. In MOS devices, larger effective fields produce stronger confinement, and all subbands tend to have higher energy. The subband splitting generally also increases, and thus even larger effective field induces higher surface charge density, and the occupation percentage in the ground subband increases. One example for a Si p-channel MOSFET under triangular well potential approximation is shown in Fig. 5.24 (Fischetti et al, 2003).

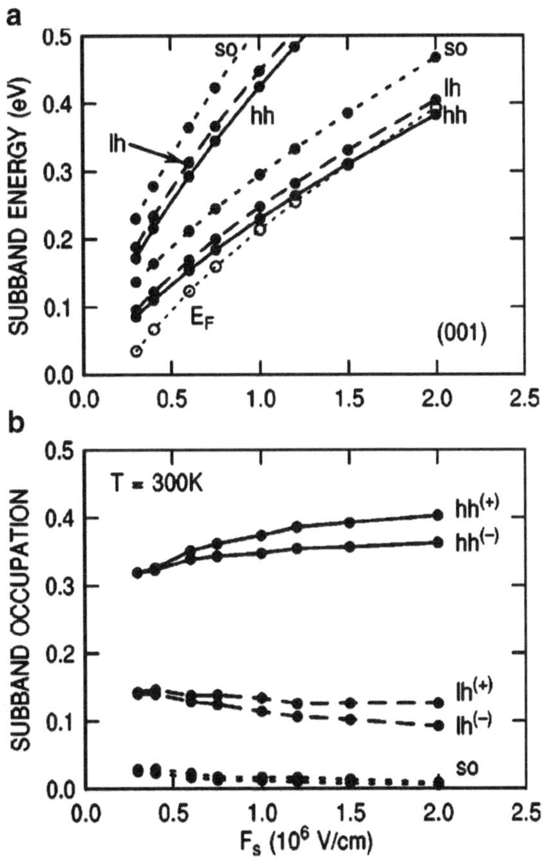

Fig. 5.24. (**a**) Lowest-lying subbands and (**b**) hole occupation at 300 K for the (001)-surface Si p-channel MOSFETs. Results obtained by triangular potential well approximation. From Fischetti (Fischetti et al, 2003)

5.5.6 In-Plane Energy Dispersion

The energy dispersion for most 2D electron systems is not difficult to obtain. For conduction bands with small nonparabolicity, such as those for Si and

GaAs, the in-plane energy band can be approximated parabolic for most cases. For strongly nonparabolic conduction bands such as that for InAs and InSb, the energy dispersion can also be written analytically using one or two nonparabolicity parameters, as in (4.164).

However, even for semiconductors with relatively large split-off energy such as GaAs, the valence band energy dispersions cannot be obtained analytically as in Chap. 4 using a 4×4 Luttinger Hamiltonian. Thus, they can be only obtained by numerical computations. The in-plane hole energy dispersion in quantum wells is strongly nonparabolic. The confinement splitting alters the coupling between the HH, LH, and split-off hole bands. For materials with small confinement splitting, the in-plane energy structure is similar to that of the corresponding bulk material. For those with large confinement splitting, the in-plane structure of the ground subband tends to become more isotropic. This can be seen from the comparison between the bulk and 2D confined in-plane energy contours for Si and GaAs, illustrated in Fig. 5.25. In

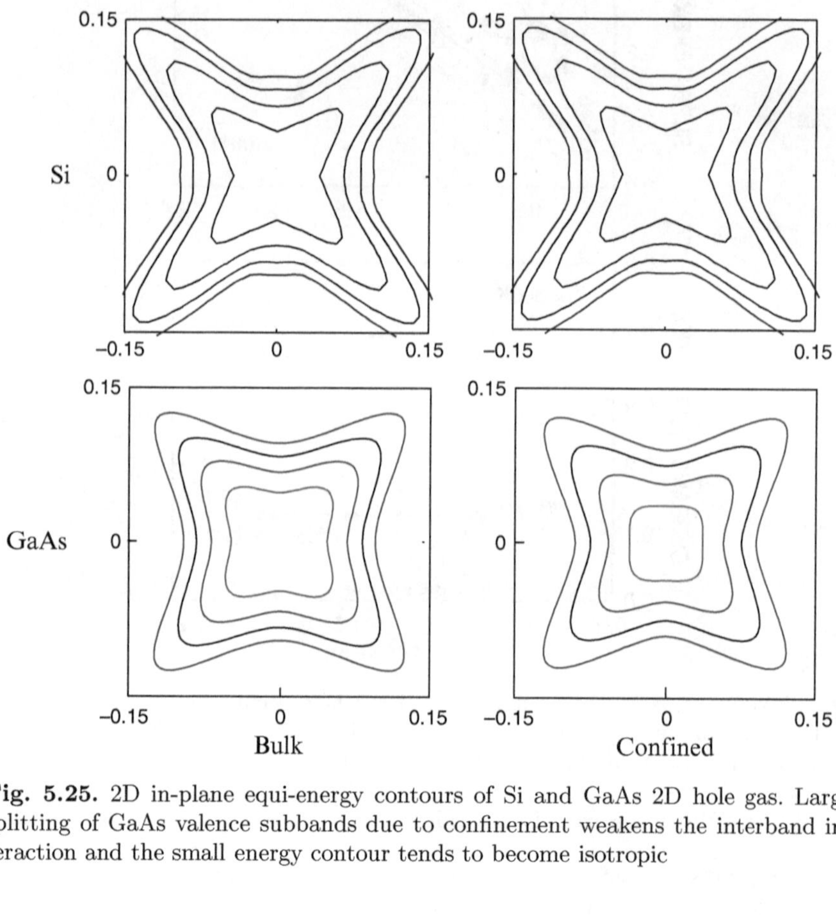

Fig. 5.25. 2D in-plane equi-energy contours of Si and GaAs 2D hole gas. Large splitting of GaAs valence subbands due to confinement weakens the interband interaction and the small energy contour tends to become isotropic

Fig. 5.26, the valence band dispersions for four GaAs/AlGaAs square wells are shown, where both the theoretical curves and experimental measurements of the energy dispersion are demonstrated (Kash et al, 1994). The complicated in-plane band structure makes the analysis of the hole transport difficult. The

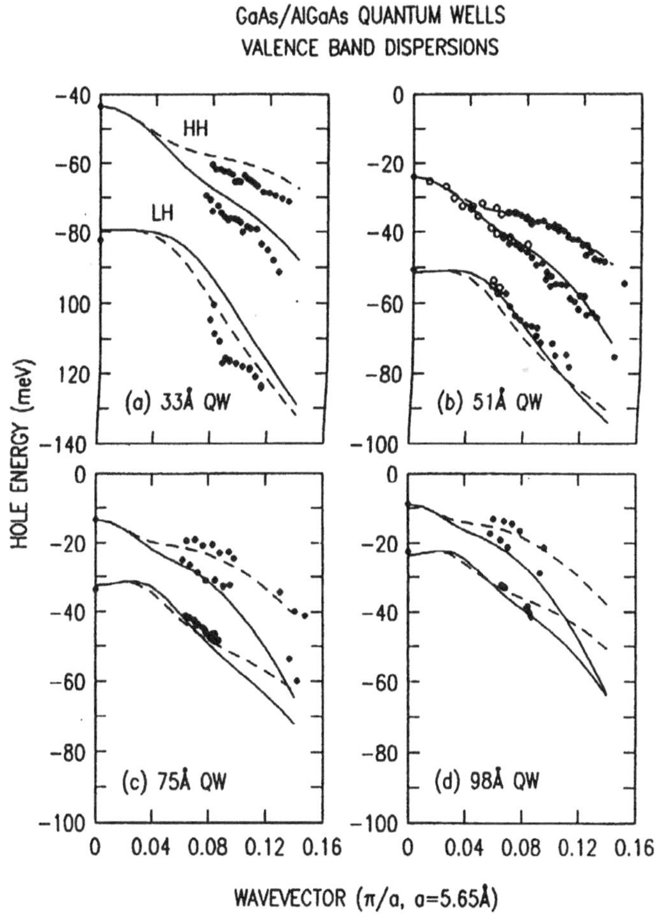

Fig. 5.26. Valence band in-plane energy dispersions for GaAs/AlGaAs quantum wells with four different well widths. Dots are experimentally obtained data points, the *solid lines* depict the dispersion along the ⟨100⟩, and *dashed lines* depict the dispersion along the ⟨110⟩ direction. From (Kash et al, 1994)

difficulties lie in: a) Effective mass values for the bulk materials cannot be used for quantum wells. Because of the band structure anisotropy, transport in one direction is different from that of the other; b) The subband DOS is not a constant, but energy-dependent. It shall be evaluated by the 2D Brillouin zone integration.

5.6 STRAIN EFFECTS ON SUBBAND STRUCTURES

Strain on low-dimensional band structures needs to be understood from its representation in bulk materials and how it transforms in low-dimensional band structure formulation. Under Pikus-Bir strain Hamiltonian framework, for a nondegenerate band (including bands with star-degeneracy), the strain energy is still a constant under confinement quantization. This means, the strain effect and confinement quantization can be separated. For degenerate bands, every matrix element in the strain Hamiltonian is diagonalized into an $N \times N$ block where N is the grid point in the finite difference formalism. The final system energy band structure is determined by both confinement quantization and strain effect. The interplay of strain and confinement quantization may be classified into three situations: the GaAs-conduction-band-like, the Si-conduction-band-like, and the valence-band-like situation.

5.6.1 GaAs Conduction Band

The conduction band of GaAs is located at the Γ point and is singly degenerate. In bulk, the strain energy is a constant, and the Hamiltonian is written as

$$E_n = E_c + \frac{\hbar^2(k_x^2 + k_y^2 + k_z^2)}{2m^*} + a_c(\varepsilon_{xx} + \varepsilon_{yy} + \varepsilon_{zz}). \tag{5.71}$$

In low-dimensional cases such as under the 1D confinement in the z-direction, k_z is replaced by the operator $-\mathrm{i}\frac{\partial}{\partial z}$. When it is converted into a matrix under a set of basis, the strain term turns into a constant matrix with the diagonal term being the strain energy. The net effect is that every subband is shifted by the strain energy. In the triangular potential well approximation, this net energy shift causes no difference, whereas since the barrier height is finite in fact, this net shift will change the potential barrier. This can be seen from (5.63), where with strain, it becomes

$$-\frac{\hbar^2}{2}\frac{\partial}{\partial z}\left[\frac{1}{m_z}\frac{\partial}{\partial z}\right]\psi(z) + (e\phi(z) + V_h(z) + H_\varepsilon(z))\psi(z) = E\psi(z). \tag{5.72}$$

When the strain energy is a constant throughout the barrier and well, it has no effect as if only the energy zero is reselected. When the strain energy is different in barrier and well, the effect is like the potential barrier at the interface is changed. Since $\phi(z)$ depends on the band potential barrier, thus strain can change the built-in potential profile.

However, we have to consider the other important strain-induced effect commonly existing in compound III–V semiconductors, especially in GaAs. That is, the strain-induced electron repopulation among different valleys such as the L and X valleys. The valley edge energies and the deformation potentials for each type of valleys are show in Table 5.2, where a_c is the hydrostatic deformation potential, equal to $(\Xi_d + \Xi_u/3)$ in L and X valleys.

Table 5.2. GaAs conduction valley energies and deformation potentials in units of eV. The Γ valley edge is selected as the energy zero

	Γ	L	X
Edge energy	0	0.29	0.45
a_c	−8.4	−2.0	1.7
Ξ_u	0	19.6	6.5

After *GaAs and related materials* by Sadao Adachi, published by World Scitific, pp. 271-288.

Because of the energetic proximity between the Γ and L valleys, and between the Γ and X valleys, the electron occupation in L and X valleys is appreciable at relatively high electron densities. This is illustrated in Fig. 5.27. The four L valleys does not split under confinement along $\langle 001 \rangle$ direction and do not spilt under biaxial stress either, but split into two doublets for uniaxial stress along $\langle 110 \rangle$ or $\langle 111 \rangle$, together with an average shift. The X valley does not split under $\langle 111 \rangle$ uniaxial stress, but splits into a singlet and a doublet under both biaxial stress and $\langle 110 \rangle$ uniaxial stress, accompanied by an average shift. For GaAs, $\Delta E_{\Gamma L} \sim 0.29\,\text{eV}$, and $\Delta E_{\Gamma X} \sim 0.45\,\text{eV}$. The stress-induced

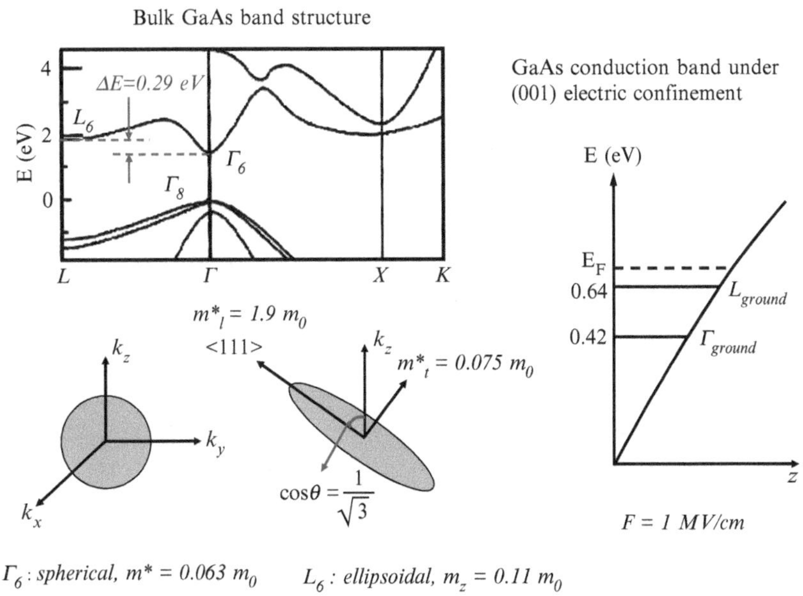

Fig. 5.27. GaAs band structure reproduced after Chelikowsky and Cohen (Chelikowsky and Cohen, 1976) and the diagrams of the conduction band Γ valley and L valley, with the diagram showing the ground states for Γ valley and L valley under an electric field of 1 MV/cm. Occupation of electrons in the L valley is considerable

splitting is sketched in Fig. 5.28, where the numbers show the deformation potentials for hydrostatic and shear strains. The stress-induced relative shifts of the Γ, L, and X valleys repopulate the electrons between valleys.

First we assume that there is negligible wave penetration into the barrier layer, which is probably the case for GaAs-channel MOSFETs, and assume a 2D electron density $10^{13}/\mathrm{cm}^2$ in the channel which has a low doping of $1 \times 10^{15}\, cm^3$, the electron population in each type of valley is shown in Fig. 5.29.

GaAs/AlGaAs is a lattice-matched system. The similar InGaAs/GaAs system is a lattice-mismatched system. Strain between InGaAs and GaAs layers

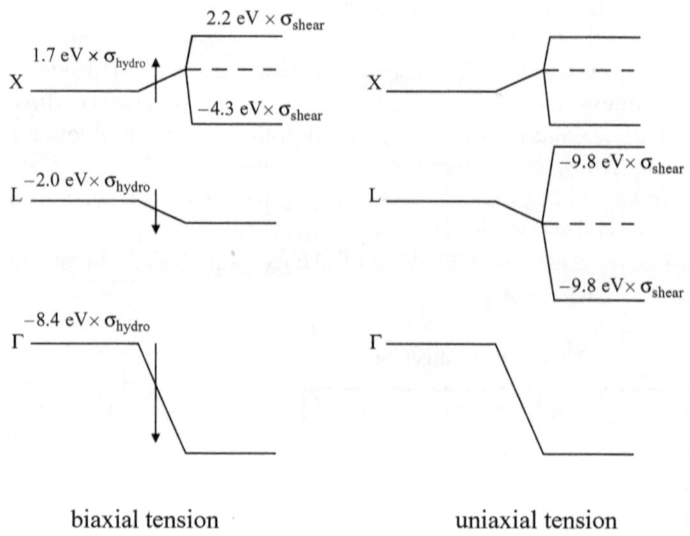

Fig. 5.28. Valley splitting for GaAs under biaxial tensile and uniaxial tensile stress. σ_{hydro} represents the hydrostatic stress, and σ_{shear} represents the shear stress. Numbers are the deformation potentials in unit of eV, and the *up and down arrows* describe the splitting and shifts of different valleyss under corresponding stress

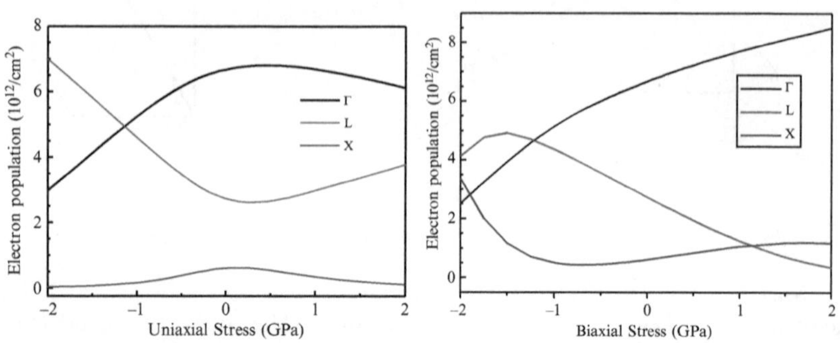

Fig. 5.29. Electron occupation variance among the Γ, L, and X valleys in GaAs with stress for biaxial (*left*) and uniaxial (*right*) tensile stress, respectively

can be very large. Very large piezoelectric effect can take place for growth direction other than ⟨100⟩. The strain-induced large electric field in the InGaAs quantum well alters the potential profile significantly. Readers please refer to Sect. 2.6 for discussions.

5.6.2 Si Conduction Band

In Si conduction band, if the gate tunneling effect is not considered, the strain energy is typically much smaller compared with the conduction band discontinuity between Si and SiO_2, and thus the potential barrier can be assumed unchanged by strain. The main effect of strain is to shift the Δ_2 and Δ_4 bands differently and thus change the energy splitting between them. When strain shifts the two sets of subbands differently, it also changes the electron occupation of the two types of valleys. This results in a variation of the confining potential profile and consequently affects the confinement splitting. Therefore, for a complete and accurate determination of the subband alignment under both confinement and strain, the strain Hamiltonian and electric confinement need to be considered simultaneously. However, at low inversion charge density, the confinement is primarily due to the depletion, and at high inversion charge density, the confinement splitting is very large, thus typically the strain-induced variation of confining potential profile is small. Calculations show that for an inversion charge density of $10^{13}/\text{cm}^2$, the valley splitting is changed by about 4 meV by 1 GPa [110] uniaxial stress just due to confinement potential shift induced by strain, compared with the valley splitting of about 150 meV for unstrained Si. Therefore, strain effect and spatial quantization can be considered separately for a good approximation.

Then the subband alignment of Si conduction valleys can be qualitatively considered by two steps. First the subband structure is calculated under electric confinement. The subband structure contains two sets of subbands. One set is originated from the Δ_2 valleys, and the other set is originated from the Δ_4 valleys. The second step is to find the strain energy for both sets and shift either set by its corresponding strain energy. The shift of the subbands by strain can result in either increase or decrease of the valley splitting (defined by the splitting of the ground subbands) depending on the form of strain. We may use the uniaxial strain along the [110] direction as an example. Under uniaxial tension, $\varepsilon_{xx} > \varepsilon_{zz}$. Then according to (4.215), Δ_2 valley shifts down, and Δ_4 valley shifts up. The strain splitting is $\Xi_u(\varepsilon_{xx} - \varepsilon_{zz})$. It adds to the electric confinement splitting, and thus the valley splitting becomes larger. Under uniaxial compression, $\varepsilon_{xx} < \varepsilon_{zz}$. The Δ_2 valley shifts up, and Δ_4 valley shifts down. The strain splitting and confinement splitting are offset to each other, and thus the valley splitting becomes smaller. In modern Si CMOS devices, the electric field is very high ($\sim 10^6$ V/cm) at the Si side close to the Si/SiO_2 interface, and thus the confinement splitting is very large. For example, under an effective field of magnitude 1 MV/cm, the splitting between the ground subbands of Δ_2 valley and Δ_4 valley is 121 meV. For a stress of 1 GPa, the strain splitting is about 38 meV, only amounting to about one-third of the confinement splitting. Thus even with a uniaxial compression of 1 GPa, the valley splitting is still about 83 meV with Δ_2 valley being the ground subband. This has a significant result that the electron repopulation between valleys may not be very significant even with relatively large strain.

For consideration of gate tunneling for gated MOS devices or MOSFETs, besides the valley splitting, the uniform shift of the entire conduction band, or the shift of the band weight, shall be also considered, since it alters the potential barrier between Si and SiO_2, and electron tunneling probability depends sensitively on tunneling potential barrier. The uniform shift is given by (4.214), namely

$$\Delta E_{c,av} = (\Xi_d + \frac{1}{3}\Xi_u)(\varepsilon_{xx} + \varepsilon_{xx} + \varepsilon_{xx}). \tag{5.73}$$

The discussion for the Ge conduction band is similar to the Si conduction band. The difference of the conduction valley position in k-space between them only changes the strain energy formulation, as indicated in (4.215)–(4.217).

5.6.3 Valence Band

Strain splits the valence bands in a much more complicated way than in the conduction band, since the valence band is degenerate, and strain strongly warps the energy dispersion. The final subband alignment can only be obtained numerically through diagonalization of the Luttinger–Pikus–Bir Hamiltonian. In MOSFET channels, the strain-induced subband shifts and in-plane band structure warping are the keys for strained-enhanced p-channel MOSFET devices. Qualitatively, in an unstrained confined structure, the ground HH subband is always the ground hole subband. Strain with different symmetry can shift the subband differently. The trend of strain shifts can be understood according to the strain splitting in bulk (see Figs. 4.31 and 4.32). For the two types of technologically important stresses, the biaxial tensile and uniaxial compressive stress, the subband splitting (here only the HH and LH ground subbands are considered) decreases and increases, respectively. The splitting curves for Si valence bands are shown in Fig. 5.30. For an initially split HH and LH ground subbands, decrease of the splitting by biaxial tensile stress eventually makes them cross over each other.

With the relative shift of HH and LH bands by increasing strain, the in-plane subband structures gradually change beginning from around the Γ point. This is more obvious in the uniaxial stress case, and warping in the biaxial stress case is not significant. The 2D in-plane energy contours of uniaxially stressed Si channel are shown in Fig. 5.31 with four stress values. With increasing stress, the warped area enlarges, while beyond the warped area around the Γ point, the energy structure is pretty like the relaxed Si. Warping is the major reason why the hole mobility of Si p-channel MOSFETs

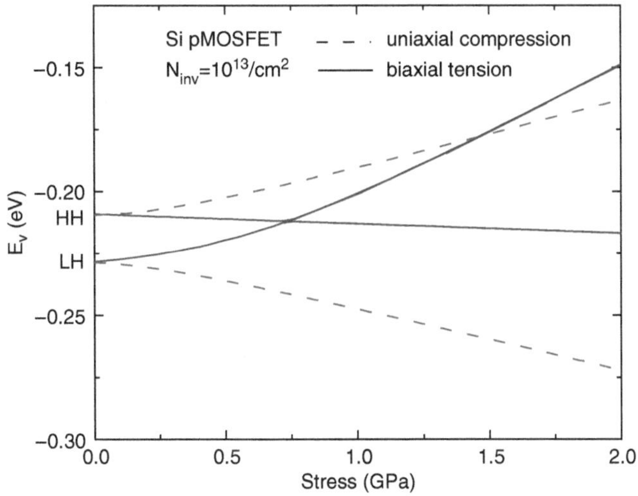

Fig. 5.30. Shift of the ground HH and LH subband for Si 2D hole gas in the Si p-channel MOSFET inversion layer

can be enhanced under uniaxial compression since it reduces the effective masses along the channel direction. larger warping area can hold more portion of conduction holes, and thus the enhancement factor can be raised. In contrast, the negligible in-plane warping effect of biaxial stress brings little change to the hole conductivity mass. The reduction of HH and LH subband splitting is also not favorable to phonon scattering reduction, so the biaxial stress is not a desirable choice for enhancing hole mobility in p-type electron devices.

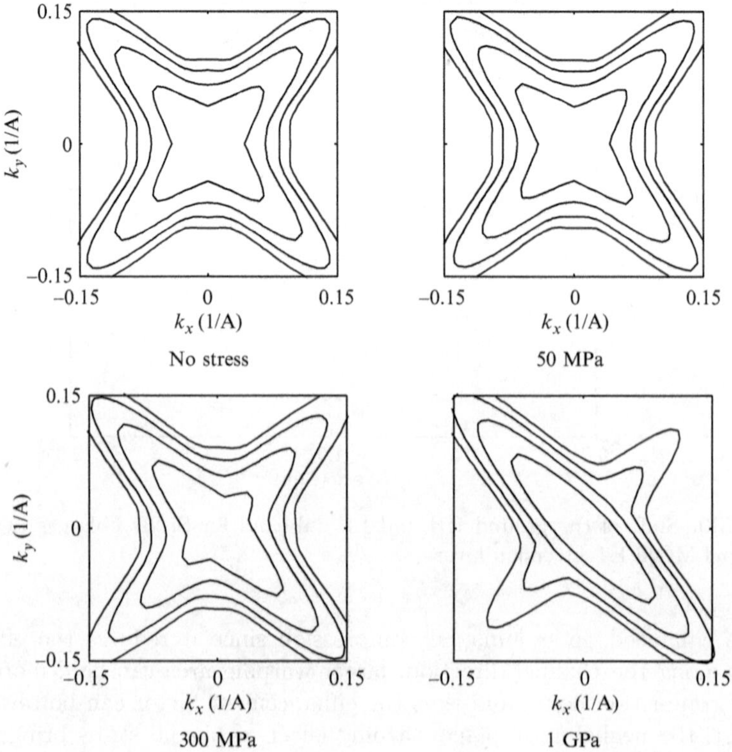

Fig. 5.31. 2D in-plane energy contour of the top valence subband for Si 2D hole gas in the Si p-channel MOSFET inversion layer under [110] uniaxial compressive stress. Four stress situations are presented. With larger stress, a larger area around the Γ point is warped by strain

Part II

Transport Theory of Strained Semiconductors

6

Semiconductor Transport

6.1 INTRODUCTION

Performance of electron devices such as MOSFETs is characterized by parameters such as mobility, transconductance, leakage current, etc. These parameters are some particular aspects or concrete representation of carrier transport properties in semiconductor structures. Device performance enhancement by strain is rooted from strain altered carrier transport properties. Based on the band structures we discussed in earlier chapters, we will discuss some basic transport models in this chapter to understand how strain changes electron/hole transport, to reach theoretical understanding of strain effects on device performance. First in Sect. 6.2 a qualitative overview is given for readers who want just to dabble in transport theories and strain-enhanced mobility in Si MOSFETs. Then in Sects. 6.3 and 6.4 we discuss the various scattering mechanisms in semiconductors and construct a general formulation to treat scattering processes. Boltzmann equation is introduced afterward in Sect. 6.5 to calculate the carrier mobility. In low-dimensional semiconductor structures, scattering processes are significantly different from that of bulk. The new features of scattering in 2D systems are particularly discussed in Sect. 6.6. The strain effects on transport are discussed and summarized in Sect. 6.7.

6.2 CARRIER TRANSPORT: A QUALITATIVE OVERVIEW

In this section, we first introduce the Drude's model, which might be the simplest transport model originally developed for electron transport in metals. However, this very idea can be transferred into semiconductors and is qualitatively satisfactory. Then this simple Drude's model is applied in this section to qualitatively discuss the strain effects on MOSFETs.

Y. Sun et al., *Strain Effect in Semiconductors: Theory and Device Applications*, DOI 10.1007/978-1-4419-0552-9_6,

6.2.1 Drude's Electron Transport Model

Drude's transport model was proposed by Paul Drude in 1900 to account for the transport properties of electrons in materials, especially in metals. It is a crude classic model, which totally neglects the quantum characteristic, and treats the electrons as classic particles which have certain trajectories and acts like pinballs when colliding with lattices. Electrons are fermions, and the electrons at the Fermi surface usually have pretty high energy. The thermal motion of electrons is drastic considering that 1 eV energy corresponds to the thermal energy at over 1.16×10^4 K. The electron velocity of thermal motion then is at the order of magnitude of 10^7 cm/s. Drude's model considered a superimposed drift velocity of the electrons in a constant and uniform electric field, and one electron that moves in the electric field will be scattered repeatedly by a mean time interval τ, which is usually very short and determined by the electron thermal motion, and thus the drift velocity is infinitesimal compared to the thermal velocity. Upon scattering, it loses all its drift velocity and then accelerates in the field until it is scattered again. The scattering rate, number of times for an electron being scattered in a unit time, is then $P = 1/\tau$. Assume an N electron concentration with each electron having momentum mv. Then for a time period of $\mathrm{d}t$, the momentum loss due to scattering is $NPmv\mathrm{d}t$. Thus, the momentum change in an electric field F is

$$\mathrm{d}(NPmv) = (NeF - NPmv)\mathrm{d}t. \tag{6.1}$$

At equilibrium state, the momentum loss rate $\mathrm{d}(mv)/\mathrm{d}t = 0$, then we have

$$v = \frac{eF}{mP} = \frac{e\tau}{m}F. \tag{6.2}$$

From the relation $v = \mu F$, where μ is the electron mobility, we have in Drude's model,

$$\mu = \frac{e\tau}{m}. \tag{6.3}$$

The mean free path, l, is defined as $v\tau$, where v is thermal velocity of the order of 10^7 cm/s, represents the length scale for an electron that travels without a scattering event. Although the Drude's model is very crude, it provides a very good explanation for many experimental observations, and the conceptions such as mean free path, etc. are extensively used also in device physics. Drude's transport model can also serve as an easy qualitative starting point to understand the basic transport properties taking place in semiconductors. It should be noted that the parameter, τ, shall be understood as the momentum relaxation time instead of the mean free time, since the change that matters in scattering for transport is the momentum (or velocity) change in a scattering event. For isotropic scattering, these two parameters are indeed the same, since an electron has the same probability to be scattered to an arbitrary direction. We can see this point from a simple example. Assume that an electron has the momentum $mv(0)$ at time 0, and then we remove the external electric field. From (6.1), the variation of momentum follows

$$\frac{m\mathrm{d}v(t)}{t} = -\frac{mv(t)}{\tau}, \tag{6.4}$$

and then we have

$$mv(t) = mv(0)\exp(-\frac{t}{\tau}). \tag{6.5}$$

This indicates that the mean free time is exactly the momentum relaxation time. For anisotropic scattering, when an electron is scattered into an arbitrary direction, the momentum change, or the velocity change, is different for different directions. In such cases, the momentum relaxation time is different from the mean free time. Proper models have to be developed to account for this anisotropy. Sometimes, only numerical results can be obtained.

6.2.2 Strain Effects on Electron/Hole Transport in MOSFETs

From Drude's model, it is easily understood that strain can affect mobility through two ways, i.e., through changing the conductivity mass m or altering momentum relaxation time τ. Conductivity mass can be changed by carrier repopulation between bands induced by degeneracy lifting, or through band warping by strain. Momentum relaxation time alteration is a little complicated. Momentum relaxation in bulk materials has contributions from various scattering mechanisms among which phonon scattering dominates at room temperature for lightly to moderately doped semiconductors. Strain alters the momentum relaxation time normally by changing the DOS, sometimes by shifting the scattering coupling strength. We still use the most common types of stress/strain in Si MOSFETs: in-plane biaxial stress and the channel direction uniaxial stress as examples. For either stress, strain is created in all three directions. Experimentally, biaxial tensile stress is found to enhance the electron mobility in strained Si and also the hole mobility but only at relatively large stress. However, the uniaxial compressive stress enhances the hole mobility at both low and high stresses and has a higher enhancement rate under the same stress, and the longitudinal (uniaxial) tensile stress has similar enhancement to electron mobility compared to in-plane biaxial tensile stress. Thus, for planar Si MOSFETs, the uniaxial stress has become the preferred method in production of strain and altering of the Si lattice.

Considering a (001) surface-orientated transistor channel experiencing electric confinement along [001] direction, we need to concentrate on three aspects of the strain-altered band structure: (1) the out-of-plane effective mass which determines the magnitude of the energy level shift under the applied gate voltage of the transistor; (2) the effective mass along [110], which is the conductivity mass along the transistor channel, and (3) the energy contours in the k_x-k_y plane, which determine the 2D DOS for a given subband. The

2D DOS is qualitatively understood to be proportional to the area enclosed by the 2D energy contours.

To have general understanding of strain effects on carrier transport, we may first discuss each of these features for biaxial and uniaxial stress for p-channel MOSFETs, just for instance. Note that other types of strain are also extensively utilized in the other structures such as FinFet, channels with other materials such as SiGe. Start with the out-of-plane mass. Because the subband energy under electric confinement is inversely proportional to $m_z^{1/3}$, where m_z is the out-of-plane effective mass, as seen by approximating the confinement with a triangular potential well, then for unstressed Si under electric confinement, the ground subband is HH-like both out-of-plane and in-plane. As we learned from the discussion in the last chapter, along the z-direction, the top valence band is LH-like under uniaxial and biaxial tension, but HH-like under compression for bulk. The properties of the second band are the reverse of the top band for both stress cases. The strained hole subband alignment in the MOSFET channels depends on both the confinement splitting and strain splitting. The ideal situation for the purpose of enhancing the carrier mobility is to induce the LH-like level in the channel direction and the HH-like dispersion in z-direction to the ground subband. Under this condition, the band splitting due to both confinement and strain is additive. This is precisely what longitudinal uniaxial compression brings about. In contrast, biaxial tensile stress induces the HH-like level in channel direction and LH-like level in z-direction to the ground subband, and thus the strain splitting is opposite to the confinement splitting. At small stress region where the splitting-induced interband scattering change is not strong, biaxial stress induces an increased conduction hole mass and a degradation in the hole mobility. So for biaxial stress, mobility enhancement only occurs at very large stress when the effect of interband scattering suppression preponderates over the conductivity mass increase. Although the biaxial compression also gives the same HH/LH arrangement, but the band warping is not as optimal as the uniaxial compression, which generates a very large curvature to the $E-k$ dispersion of the ground valence band.

In the strain altered band structure it is important to populate the subband that has the low in-plane conductivity mass. To accomplish this, a high DOS is needed for that subband. To understand the 2D DOS in the inversion layer of the transistor, the energy contours in the x-y plane are plotted in Fig. 6.1 for unstressed, biaxial tensile, and uniaxial compressive stressed Si. The shapes of the contours match the expectation based on symmetry discussed in Chap. 3. The in-plane energy contours are approximately a circle for biaxial stress while an ellipse for uniaxial stress (for energies less than the strain splitting energy). The in-plane transverse (to the channel direction) effective mass contributes to conduction through its effects on the in-plane DOS. For isotropic bands as in the case of biaxial stress and the conduction bands in many of the direct III-V materials, the conductivity mass and the DOS mass are the same (i.e., a small mass needed for high mobility leads to a low DOS). For anisotropic bands as

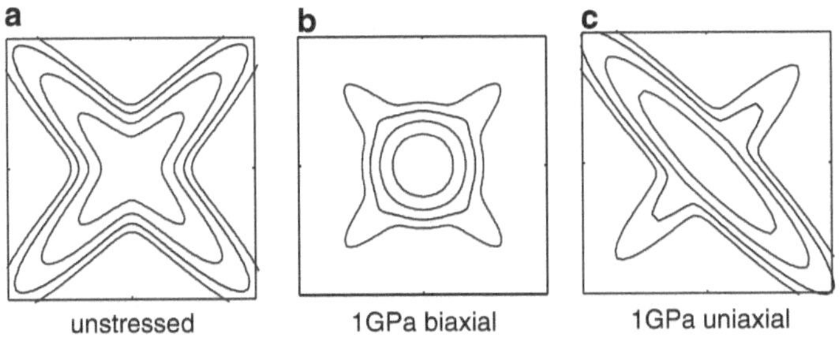

Fig. 6.1. 2D energy contours in the x-y plane for Si valence bands with no stress, 1 GPa biaxial tensile and 1 GPa [110] uniaxial compressive stress. Four contours are respectively correspond to 25, 50, 75, and 100 meV

in the case of uniaxial stress, both a small conductivity mass and a high DOS can be achieved. In the simplest description, the DOS mass is approximated by a single value defined by the mass that gives the same DOS as manifested by the enclosed area of the energy contours. Under uniaxial compressive stress, even though the channel direction effective mass is small, the large in-plane transverse effective mass contributes to a larger in-plane DOS for the (001) device. The combination of both small conductivity mass and large DOS mass due to the uniaxial compressive stress makes it uniquely suited for enhancing the hole mobility in application rather than the biaxial tensile stress.

For multivalley conduction band such as Si, the physics of strain-enhanced electron mobility is similar to what has previously been discussed for holes. First under confinement the Δ valleys already split into the Δ_2 (along z) and Δ_4 valleys (along x and y). In Si Δ valleys, the two valleys along the z direction have smallest effective mass, which equals the transverse mass, m_t, along the channel. The electrons in the other four valleys participate the transport through a conductivity mass $m_c = 2m_l m_t/(m_l + m_t)$, larger than m_t. So the idea to use strain to enhance the electron mobility is to enlarge the valley splitting. This will result into two effects: first the average conductivity is reduced by repopulating the electrons from the Δ_4 valleys into the Δ_2 valleys, and second the intervalley scattering is suppressed. Technologically people use either biaxial or longitudinal uniaxial tensile stress to reach this goal.

Electric confinement-induced splitting is typically around 100 meV for Si n-channel MOSFETs and about 20–30 meV for p-channel MOSFETs at an approximate inversion charge density of $10^{13}/\mathrm{cm}^2$. Stresses that have been implemented in current strained-Si technology are typically around 1GPa. For the conduction band with so large valley splitting, most electrons are already located at the Δ_2 valleys. The strain-induced splitting further increases occupation of the Δ_2 valleys, but the reduction of the conductivity mass by repopulation is limited. For example, at an inversion charge density of $10^{13}/\mathrm{cm}^2$, about 75% electrons are populated at the Δ_2 valleys. At the high stress limit, even if all electrons in the Δ_4 valleys are transferred to the Δ_2

valleys, the reduction of the conductivity mass is only about 14%. The scattering suppression may take a major role for strain-enhanced electron mobility. In contrast, since the strain splitting is typically small for the valence bands compared to the optical phonon energy, the scattering rate change by strain is not significant. However, band warping is very strong in the valence bands. Band warping-induced conductivity mass shift is the predominant factor for strain-altered hole mobility in p-channel MOSFETs.

In the following sections, we give a more thorough, systematic, and theoretical discussion of scattering processes in semiconductors and introduce the method to calculate the carrier mobilities and their strain effects.

6.3 SCATTERING IN SEMICONDUCTORS: GENERAL CONSIDERATION

6.3.1 Scattering Rate

For a perfect periodic crystal, there is no scattering to the carriers. Any factor that induces the change to the perfect periodic crystal potential will introduce a scattering source. Charged and neutral impurities, dislocations, etc., which posses a scattering potential in the length scale comparable to the De Broglie wavelength of the electrons, are effective scattering sources. The thermal vibration of the lattice also causes fluctuation of the periodic potential, thus it becomes an important scattering source. In fact, phonon scattering is the most essential scattering mechanism in crystals because lattice vibration is inevitable even at low temperature.

Under the single electron approximation, an electron transits from one state, which we label as $\psi_{E,\mathbf{k}}$, to the other state, which we label as $\psi_{E',\mathbf{k}'}$ in a scattering process. In this process, the momentum and energy of the scattering system must be conserved. Impurity scattering is an elastic process, since we normally consider that the impurity is static in semiconductors. The projectile of an electron is deflected by the potential of the impurity, and the electron energy before and after scattering are the same. Normally phonon scatter is inelastic. An electron has an energy gain after scattered, and the momentum difference $\mathbf{k}' - \mathbf{k}$ equals the phonon wavelength $\mathbf{q}$.

Let us dwell a little on phonon scattering process in the following text to gain some basic understanding of normal scattering processes. The probability of an electron scattered from $\psi_{E,\mathbf{k}}$ to $\psi_{E',\mathbf{k}'}$ depends on factors such as the scattering coupling strength $C_{\mathbf{q}}$, the state overlap factor $I(\mathbf{k},\mathbf{k}')$, and the availability of the final state, and for phonon scattering, the phonon density $n(\omega_{\mathbf{q},b})$, where $\omega_{\mathbf{q}}$ is the phonon frequency and b labels one phonon branch. The coupling factor between the initial and final states is

$$C_{\mathbf{q},b} = \int \psi^*_{E',\mathbf{k}'} H_{\mathbf{q},b}(\mathbf{r}) \psi_{E,\mathbf{k}} \mathrm{d}\mathbf{r}, \tag{6.6}$$

where $H_{q,b}(\mathbf{r})$ is the phonon–electron interaction. Depending on the specific form of $H_{q,b}(\mathbf{r})$, the integral of $C_{\mathbf{q},b}$ provides selection rules for phonon scattering. Sometimes easy observation can be obtained by just inspecting the symmetry of the system. For example, since electron–nonpolar optical phonon coupling is proportional to atomic displacement and has the odd parity, the transition is forbidden in III–V semiconductor conduction band Γ valley, but the electron–polar optical coupling is a long-range coupling and is almost a constant in the scale of a unit cell, and thus it is allowed in the Γ valley. Both nonpolar and polar optical phonon scattering are allowed in the valence bands.

The state overlap factor is written as

$$I(\mathbf{k},\mathbf{k}') = \int_{\text{unit cell}} \psi^*_{E',\mathbf{k}'}\psi_{E,\mathbf{k}}\mathrm{d}\mathbf{r}. \tag{6.7}$$

The meaning of this term is easy to see. If there is no wave function overlap between the initial and final states, the electron cannot make the transition. This directly excludes the transition between different spin states by phonon scattering. Electrons in alkaline metals all have the s state, so I^2 is unity. States of electrons in semiconductors' valence bands contain admixture of HH, LH and split-off hole composition, and consequently I^2 is not unity. For holes within HH or LH bands (Wiley, 1971),

$$I^2(\mathbf{k},\mathbf{k}') = \frac{1}{4}(1 + 3\cos^2\theta_{\mathbf{k},\mathbf{k}'}), \tag{6.8}$$

and for scattering between the HH and LH bands,

$$I^2(\mathbf{k},\mathbf{k}') = \frac{3}{4}\sin^2\theta_{\mathbf{k},\mathbf{k}'}, \tag{6.9}$$

where $\theta_{\mathbf{k},\mathbf{k}'}$ is the angle between $\mathbf{k}$ and $\mathbf{k}'$.

The availability of the final states is proportional to the density of final electron states $N(E_{\mathbf{k}'})$. It is also proportional to the probability that the final states are not occupied. The DOS of a given energy is already discussed in earlier chapters, and the occupation probability of an electron state depends on the temperature and system Fermi energy.

The probability of absorbing or emitting a phonon is also proportional to the phonon density $n(\omega_{\mathbf{q},b})$ for absorption, and $(1 + n(\mathbf{q}))$ for emission. Phonons are boseons, and $n(\omega_{\mathbf{q},b})$ is given by the Bose–Einstein factor,

$$n(\omega_{\mathbf{q},b}) = \frac{1}{\exp(\hbar\omega_{\mathbf{q},b}/k_B T) - 1}, \tag{6.10}$$

where $\hbar\omega_{\mathbf{q},b}$ is the phonon energy with wave vector $\mathbf{q}$ in the branch b.

Using Fermi's golden rule and combining all the factors we discussed above, we reach an expression for phonon scattering rate of an electron with state

$\psi_{E,\mathbf{k}}$ (FRS, 1999),

$$W(\mathbf{k}) = \frac{V}{8\pi^2 NM'} \int \frac{C_{\mathbf{q},b}^2 I(\mathbf{k},\mathbf{k}')^2}{\omega_{\mathbf{q},b}} \delta_{\mathbf{k}\pm\mathbf{q}-\mathbf{k}',\mathbf{K}} \times (n(\omega_{\mathbf{q},b}) + \frac{1}{2} \mp \frac{1}{2}) \delta(E_{\mathbf{k}'} - E_{\mathbf{k}} \mp \hbar\omega_{\mathbf{q},b}) \mathrm{d}\mathbf{k}'. \tag{6.11}$$

The upper sign in the equation is for phonon absorption and the lower sign is for emission. Here V is the crystal volume, N is the number of unit cells in the crystal, and M' is the reduced mass for unit cell. The capital $\mathbf{K}$ is the reciprocal lattice vector. For transition among the same valley, $\mathbf{K} = 0$. $\mathbf{K} \neq 0$ is called an Umklapp process, which usually happens in a multiple-valley system and the scattering is outside the first Brillouin zone. The Kroniker symbol $\delta_{x,y}$ equals 0 if $x \neq y$, and equals 1 if $x = y$, restricting the momentum conservation. The integration of the δ-function over energy gives the DOS for phonon scattering. The scattering mean free time is

$$\tau(\mathbf{k}) = \frac{1}{W(\mathbf{k})}. \tag{6.12}$$

When there are more than one scattering mechanisms and they have similar strength, the total scattering rate is just the addition of them, i.e.,

$$W(\mathbf{k}) = W_1(\mathbf{k}) + W_2(\mathbf{k}) + W_3(\mathbf{k}) + \cdots, \tag{6.13}$$

and correspondingly, the relaxation time follows

$$\frac{1}{\tau(\mathbf{k})} = \frac{1}{\tau_1(\mathbf{k})} + \frac{1}{\tau_2(\mathbf{k})} + \frac{1}{\tau_3(\mathbf{k})} + \cdots. \tag{6.14}$$

6.3.2 Momentum Relaxation Rate

Momentum relaxation is to randomize the initial electron momentum or to make the electron to lose its initial momentum. When an electron is scattered from $\mathbf{k}$ to $\mathbf{k}'$ as shown in Fig. 6.2, the change of its momentum is $\hbar(\mathbf{k}' - \mathbf{k})$,

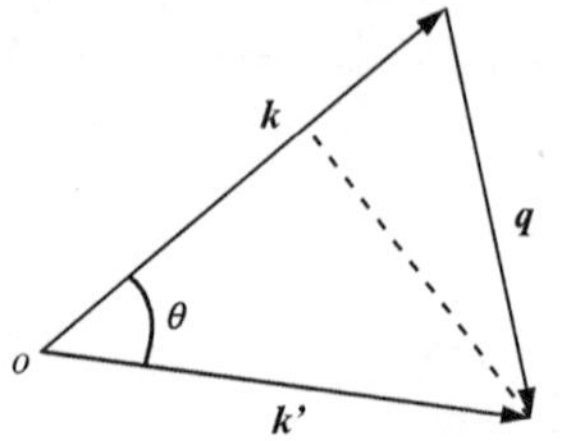

Fig. 6.2. A scattering event. In this event, an electron is scattering from state $\mathbf{k}$ to state $\mathbf{k}'$. The momentum change in this scattering event is $\hbar\mathbf{q} = \hbar(\mathbf{k}' - \mathbf{k})$

and the change at the initial momentum direction is $\hbar(\mathbf{k}' \cdot \mathbf{k}/k - k)$. Then for scattering angle of $\theta_{\mathbf{k},\mathbf{k}'} = 0$, the effect on momentum loss is small. On the contrary, if $\theta_{\mathbf{k},\mathbf{k}'} = 180°$, i.e., the scattering is to the opposite direction, the momentum loss is large. The rate of the momentum loss can be written as

$$\frac{\mathrm{d}\hbar k}{\mathrm{d}t} = \int \hbar(\mathbf{k}' \cdot \mathbf{k}/k - k)W(\mathbf{k},\mathbf{k}')\mathrm{d}\mathbf{k}', \tag{6.15}$$

where $W(\mathbf{k},\mathbf{k}')$ is the scattering rate from state $\mathbf{k}$ to $\mathbf{k}'$, which is also the integrand (6.11). We may still assume that there is a definite time constant, τ, in the system for momentum relaxation, then the momentum relaxation process is described as

$$\frac{\mathrm{d}\hbar k}{\mathrm{d}t} = -\frac{\hbar k}{\tau}. \tag{6.16}$$

For isotropic bands and small phonon energy such as in acoustic scattering case, $k' \approx k$, then $\mathbf{k}' \cdot \mathbf{k}/k - k \approx -k(1 - \cos\theta_{\mathbf{k},\mathbf{k}'})$. Therefore, we have to add a weighting factor $(1 - \cos\theta_{\mathbf{k},\mathbf{k}'})$ to the normal scattering rate to obtain the momentum relaxation rate. Since the integration of the factor $(1 - \cos\theta_{\mathbf{k},\mathbf{k}'})$ is unity over the 2π angle, the momentum relaxation rate equals the scattering rate if the scattering does not depend on angle, i.e., the scattering is isotropic. In realistic cases where k and k' might be very different, then we have to evaluate the momentum loss in each scattering event or adopt some approximations.

6.4 SCATTERING PROCESSES IN SEMICONDUCTORS

In the following, we will briefly introduce two common scattering mechanisms in semiconductors: the lattice scattering and impurity scattering.

6.4.1 Lattice Scattering

Lattice scattering is caused by thermal vibration of the crystal lattice. Lattice vibration causes strain, and strain induces potential shift, which isproportional to strain through

$$\Delta E = \sum_{ij} \Xi_{ij}\varepsilon_{ij}, \tag{6.17}$$

where Ξ are deformation potentials. The deformation potential theory for describing phonon excitation is in fact the same used to treat the homogeneous strain. The difference lies in that phonon-produced strain is a wave in form as shown in Fig. 6.3 while homogeneous strain induces static shift. The varying shift of potential causes coupling between electronic states of the perfect crystal.

For diamond and zinc-blende structure semiconductors, the lattice scattering includes acoustic phonon scattering and optical phonon scattering.

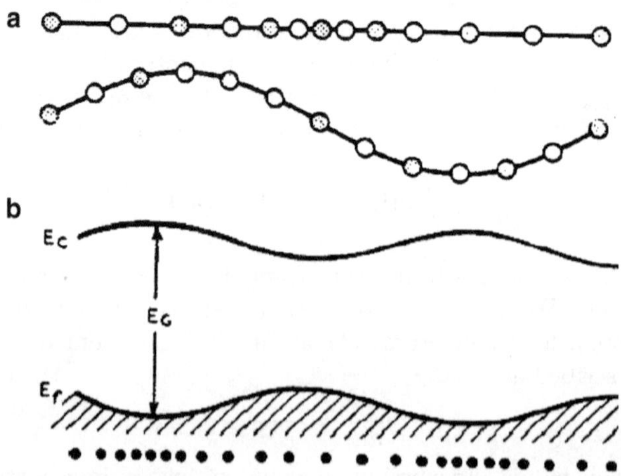

Fig. 6.3. (**a**) Acoustic wave creates atom displacements and thus deformation; (**b**) Deformation induces localized band edge shift, which couples different electronic states

Normal acoustic and optical phonon scattering is due to the deformation-induced potential variation called deformation potentials. Acoustic phonon is caused by the collective vibration of the crystal, and the optical phonon is generated by the relative vibration of multiple atoms in the unit cell. Shown in Fig. 6.4 is the phonon dispersion for Si. A typical characteristic of an acous-

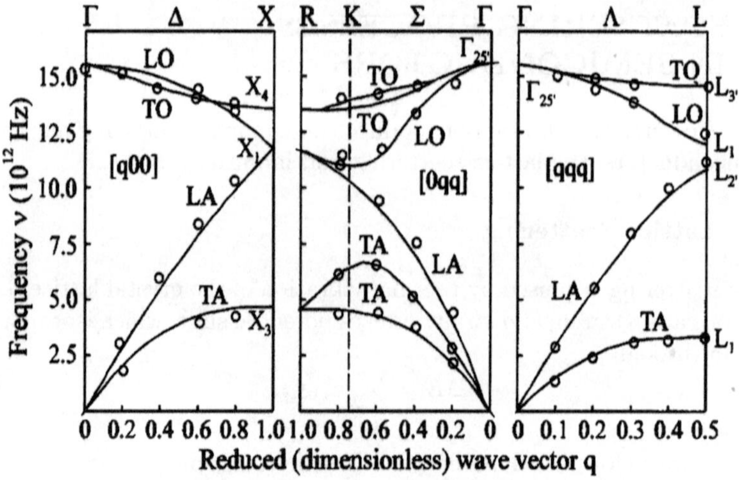

Fig. 6.4. Phonon dispersion of Si

tic mode is that the $\omega - q$ relation in the short wave vector (long wavelength) limit is linear, while for an optical mode, the dispersion is almost flat with a constant larger than zero phonon energy in the short wave vector limit. For polar semiconductors, since they do not have inversion symmetry, acoustic vibration causes long-range electric field, and it can couple with electronic states, so there is piezoelectric scattering with the acoustic vibration. Also the vibrations of oppositely charged atoms also produces long-range electric fields, and thus they possess another scattering mechanism: the polar optical phonon scattering.

6.4.2 Acoustic Phonon Scattering

As mentioned above, acoustic waves can scatter electrons by two mechanisms: deformation potential induced by strain and piezoelectric effect in polar crystal. In this subsection we discuss the first mechanism.

Let us first inspect intravalley scattering. Acoustic phonon energy,

$$\hbar\omega_{\mathbf{q}} = \hbar v_s q, \tag{6.18}$$

where v_s is the sound velocity in bulk crystal, which is small for small q. For typical sound velocity at 3–4 km/s, the acoustic phonon energy is below 2 meV for wave vector as large as 0.1/Å. For nondegenerate semiconductors, the electron energy is typically around k_BT, which is about 26 meV at room temperature. For degenerate semiconductors, the electron energy at the Fermi surface can be much larger than k_BT. Thus, the acoustic scattering has quasi-elastic characteristic. In most cases, for electrons of interest, their electron wave vector is at the order of 0.01/A, and consequently the acoustic phonon wave vector is also at this order, and consequently momentum reservation in scattering processes restricts the acoustic phonon to long wavelength range. Because typical optical phonon energy is around several tens of meV, and this amount of energy only corresponds to a small wave vector change, and thus optical phonon participating the scattering process is also at long wavelength range.

Because of the small acoustic phonon energy, one simple approximation is applicable to acoustic phonon scattering: the equipartition approximation, which assumes the same phonon density for both absorption and emission. Using first-order Taylor expansion, the acoustic phonon density is given by

$$n(\omega_{\mathbf{q}}) = \frac{1}{\exp(\hbar\omega_{\mathbf{q}}/k_BT) - 1} \approx \frac{k_BT}{\hbar\omega_{\mathbf{q}}}, \tag{6.19}$$

for relatively high temperature when $n_{\mathbf{q}} \gg 1$.

In the deformation potential framework, the band edge shift is related to differential displacement (strain) of the crystal by

$$H_{ep}(\mathbf{r}, t) = \Xi_d \nabla \cdot \mathbf{u}(\mathbf{r}, t) \tag{6.20}$$

where $\mathbf{u}(\mathbf{r}, t)$ represents the displacement of the atom site $\mathbf{r}$. The acoustic wave with wave vector $\mathbf{q}$ is expressed as

$$\begin{aligned} \mathbf{u}(\mathbf{r}, t) &= a_q \hat{\mathbf{e}} \sin(\mathbf{q} \cdot \mathbf{r} - \omega_q t + \phi) \\ &= -\frac{i a_q}{2} \hat{\mathbf{e}} \{\exp[i(\mathbf{q} \cdot \mathbf{r} - \omega_q t + \phi)] - \exp[-i(\mathbf{q} \cdot \mathbf{r} - \omega_q t + \phi)]\}, \end{aligned} \tag{6.21}$$

where $\hat{\mathbf{e}}$ is the unit vector along the vibration direction, and a_q is the wave amplitude. Substituting the above equation into (6.20), we have

$$H_{ep}(\mathbf{r},t) = \frac{a_q \Xi_d}{2} \hat{\mathbf{e}} \cdot \mathbf{q} \{\exp[i(\mathbf{q}\cdot\mathbf{r} - \omega_q t + \phi)] + \exp[-i(\mathbf{q}\cdot\mathbf{r} - \omega_q t + \phi)]\}. \quad (6.22)$$

Acoustic phonon has three modes, one longitudinal and two transverse modes. For transverse modes, $\hat{\mathbf{e}} \perp \mathbf{q}$, so $\hat{\mathbf{e}} \cdot \mathbf{q} = 0$, and thus they are not coupled to electronic states. For longitudinal mode, $\hat{\mathbf{e}} \cdot \mathbf{q} = q$. Thus the electron–acoustic phonon coupling strength is

$$C_q^2 = \frac{(\Xi_d a_q q)^2}{4}. \quad (6.23)$$

The acoustic wave with wave vector $\mathbf{q}$ can be considered as an oscillator. At the classic limit $\hbar\omega_q \ll k_B T$, the energy of each degree of freedom is given by $\rho V \omega_q^2 a_q^2 = k_B T$, where ρ and V are the density and volume of the crystal, respectively. Remembering $\omega_q = v_s q$, the coupling strength is

$$C_q^2 = \frac{\Xi_d^2 k_B T}{2\rho V v_s^2} = \frac{\Xi_d^2 k_B T}{2 V c_L}, \quad (6.24)$$

where c_L is the averaged longitudinal elastic modulus, $c_L = \rho v_s^2$. Then the scattering rate from state $\mathbf{k}$ to $\mathbf{k}'$, under the approximation of equipartition, is given by

$$W(\mathbf{k},\mathbf{k}') = \frac{2\pi \Xi_d^2 k_B T}{\hbar V c_L} \delta(E_{\mathbf{k}'} - E_{\mathbf{k}} \mp \hbar\omega_q). \quad (6.25)$$

In elastic approximation, $\hbar\omega_q \sim 0$.

The acoustic scattering as can be seen is not dependent on scattering angle, so it is isotropic. The momentum relaxation rate caused by acoustic scattering equals the scattering rate.

6.4.3 Piezoelectric Scattering

Strain in semiconductor crystals without inversion symmetry creates macroscopic electric fields. This is called piezoelectricity as we introduced in Chap. 2. Acoustic strain also produces varying electric fields, which couples with electronic states, and causes scattering. The piezoelectric scattering is stronger in crystals with less symmetry, so it is stronger in wurtzite than in zinc-blende semiconductors.

The treatment of piezoelectric scattering is complicated, so we will not give further deduction for the electron–piezoelectric field interaction, but just write it down below,

$$C_{\mathbf{q}}^2 = \frac{e^2 K^2 k_B T}{2\epsilon V} \frac{q^2}{(q^2 + q_0^2)^2}, \quad (6.26)$$

where

$$K^2 = \frac{e_{14}^2}{\epsilon}\left(\frac{12}{35c_L} + \frac{16}{35c_T}\right) \tag{6.27}$$

represents the scattering contribution from longitudinal and transverse acoustic waves, and q_0 is the reciprocal screening length. e_{14} is the piezoelectric constant introduced in Chap. 2. We notice that the coupling is phonon wave vector dependent, so it is anisotropic. In zinc-blende semiconductors, the piezoelectric scattering is weak at room temperature. Only at low temperature and in very pure materials, piezoelectric scattering is important.

6.4.4 Optical Phonon Scattering

If we say the acoustic waves create macroscopic strain in crystals, the optical waves create "microscopic" strain by relative displacement of atoms in unit cells. Electrons can be scattered by deformation potentials induced by this microscopic strain. In long-wavelength limit, the cations or anions in polar semiconductors have a uniform shift and thus produce a macroscopic electric field. Polar optical phonon scattering is caused by coupling between electronic states with this field. We first study the deformation potential coupling.

The optical wave-induced deformation potential closely depends on the symmetry of the band structure. In the diamond and zinc-blende structure semiconductors, the relative displacement of the two kinds of atoms in the unit cells does not produce the first order (to the atom displacement) energy shift for simple bands such as the s-band, and thus for direct gap III–V materials the optical phonon scattering in the conduction band is weak and negligible. This is also true for the X valleys. But for scattering in L and degenerate valence bands, optical phonon scattering is important. In Table 6.1 we reproduce the selection rules for acoustic and optical phonon scattering caused by deformation potentials (FRS, 1999).

The intervalley scattering, either the acoustic or optical phonon-induced, is similar to the optical phonon scattering. The phonon energy involved in these processes is usually larger than k_BT, and thus they are inelastic processes.

Table 6.1. The selection rules for acoustic and optical phonon deformation potential scattering

Valley	Phonons allowed
Γ_1	LA
X_1	LA+TA
L_1	LA+TA+LO+TO
Γ_{15}	LA+TA+LO+TO

It is usual to define the deformation potential caused by optical wave to be proportional to the relative displacement instead of the differential displacement. The deformation potential energy then is written as

$$H_{ep} = Du = \frac{Da_q}{2}\{\exp[i(\mathbf{q}\cdot\mathbf{r} - \omega_q t + \phi)] + \exp[-i(\mathbf{q}\cdot\mathbf{r} - \omega_q t + \phi)]\}. \quad (6.28)$$

The optical deformation potential D here has the unit of eV/cm and is at the order of 10^9. For phonon absorption and emission processes, the wave amplitude a_q is replaced by its quantum mechanical counterpart, $(2n_q\hbar/\rho V\omega_q)^{1/2}$ and $[2(n_q+1)\hbar/\rho V\omega_q]^{1/2}$, respectively, since the optical phonon energy is comparable to k_BT. Comparing to the treatment of the acoustic phonon scattering, we can see that here k_BT is replaced by $n_q\hbar\omega_q$ and $(n_q+1)\hbar\omega_q$, respectively, for phonon absorption and emission. For intravalley long-wavelength optical phonons, ω_q can be considered independent of q, and represented by the zone center phonon energy ω_0. Thus the electron–optical phonon interaction is

$$C^2 = \frac{D^2\hbar\left(n_q + \frac{1}{2} \mp \frac{1}{2}\right)}{2\rho V\omega_0}. \quad (6.29)$$

Thus, the scattering rate from $\mathbf{k}$ to $\mathbf{k}'$ can be written as

$$W(\mathbf{k}, \mathbf{k}\pm\mathbf{q}) = \frac{D^2\hbar}{2\rho V\omega_0}\left(n_q + \frac{1}{2} \mp \frac{1}{2}\right)\delta(E_{\mathbf{k}\pm\mathbf{q}} - E_{\mathbf{k}} \mp \hbar\omega_0). \quad (6.30)$$

Although we write the phonon wave vector $\mathbf{q}$ explicitly in the equation, however, the transition probability does not depend on the phonon wave vector. Thus, the optical phonon scattering is actually an isotropic process. The momentum relaxation rate is the scattering rate.

6.4.5 Polar Optical Phonon Scattering

The long-range electric fields in companion with the longitudinal optical wave can scatter electrons, and this mechanism is called polar optical phonon scattering. Transverse optical phonons do not create this type of electric fields. The polar optical phonon scattering exists in polar materials. In fact, it is the most important scattering mechanism in the conduction bands of direct gap semiconductors since the optical phonon deformation potential scattering is prohibited. In valence bands, because the optical deformation potential scattering is allowed, the polar optical phonon scattering is not as important. Considering deformation potential scattering only already gives results close to experiments (Wiley, 1971).

Since only long-wavelength optical wave is important, the long-range electric field can be obtained by considering atom displacement-induced polarization. We assume that each of the ions in the unit cell has effective charge e^*, then the semiconductor polarization is proportional to the relative displacement of the two ions,

$$P = Ne^*u, \tag{6.31}$$

where N is the number of unit cells in the crystal. Since the electric field and polarization are both induced by the optical wave, the electric displacement $D = 0$. Thus, from $D = \epsilon_0 F + P$, we have

$$F = -\frac{P}{\epsilon_0} = -\frac{Ne^*u}{\epsilon_0}. \tag{6.32}$$

The effective charge depends on the polarity of the material and is given here without deduction,

$$e^* = \frac{M'}{N}\omega_L \epsilon_0 \left(\frac{1}{\epsilon_\infty} - \frac{1}{\epsilon}\right), \tag{6.33}$$

where ω_L is the frequency of the longitudinal optical mode, ϵ_∞ is the high-frequency dielectric constant, and M' is the reduce mass in the unit cell. The polar interaction between the electric field and the charge is basically $V = -e\phi$, where $\nabla\phi = -\mathbf{F}$, and the interaction energy is also a wave with the amplitude

$$A(q) = \frac{Nee^*a_q}{2\epsilon_0 q}, \tag{6.34}$$

where a_q is the amplitude of the displacement u. This can be analogized to the optical deformation potential scattering, and we obtain the coupling strength

$$C_q^2 = \left(\frac{Nee^*}{\epsilon_0}\right)^2 \frac{1}{q^2}. \tag{6.35}$$

Considering the screening effect, it should be

$$C_q^2 = \left(\frac{Nee^*}{\epsilon_0}\right)^2 \frac{q^2}{(q^2 + q_0^2)^2}, \tag{6.36}$$

where q_0 is the reciprocal screening length.

The scattering rate from state $\mathbf{k}$ to $\mathbf{k}'$ then is

$$W(\mathbf{k}, \mathbf{k}\pm\mathbf{q}) = \frac{\pi}{\rho V \omega_L}\left(\frac{Nee^*}{\epsilon_0}\right)^2 \frac{q^2}{(q^2 + q_0^2)^2}\left(n_q + \frac{1}{2} \mp \frac{1}{2}\right)\delta(E_{\mathbf{k}\pm\mathbf{q}} - E_{\mathbf{k}} \mp \hbar\omega_0). \tag{6.37}$$

The polar optical phonon scattering is anisotropic. The screening length usually is determined by the electron energy, electron density, and temperature. For low electron density case, the screening effect is not significant. When electrons have higher energy, the phonon wave vector involved is relatively larger, and the screening is also weak. Also, screening for holes is usually much weaker than electrons since holes have much larger mass. Because of the q dependence of the scattering rate, scattering with small angles is stronger than that with large angles. This is disadvantageous for momentum relaxation, and it is the reason why the polar optical phonon may not be so important in the valence bands.

6.4.6 Impurity Scattering

Most impurities in semiconductors are dopants, which are charged impurities. The scattering by charged ions is induced by Coulomb potential of the impurities. Normal Coulomb potential is long-range, and an electron is continuously scattered, and thus the scattering cross-section is infinite. However, in semiconductors, the itinerant charge can effectively screen the Coulomb potential and thus make a scattering event localized around the impurity site. We will introduce here the Brooks-Herring formulation which solves the infinite scattering cross-section problem by considering the impurity screening effect. So first we discuss the screening of the charged impurities.

For a crystal without any potential disturbance, charges are uniformly distributed. The perturbation induced by an impurity results into a deviation of the uniformity of charge distribution. For instance, a positively charged impurity in an n-doped semiconductor can attract electrons resulting into a higher local electron density. Assuming the unperturbed electron density is n_0, then at the impurity site the electron density is

$$n(r) = n_0 \exp\left[\frac{eV(r)}{k_B T}\right], \tag{6.38}$$

where $V(r)$ is the induced potential perturbation by the impurity. Then the additional electron density, the impurity induced, is

$$\rho = -e[n(r) - n_0] = -en_0 \exp\left[\frac{eV(r)}{k_B T} - 1\right] \approx -\frac{e^2 n_0 V(r)}{k_B T}, \tag{6.39}$$

where we assumed the perturbation is small, and thus $|eV| \ll k_B T$, and thus the exponential can be expanded to the first order. The perturbation is obtained by solving the spherically symmetrical Poisson equation

$$\frac{1}{r}\frac{\mathrm{d}^2[rV(r)]}{\mathrm{d}r^2} = -\frac{\rho(r)}{\epsilon}. \tag{6.40}$$

Substituting (6.40) into (6.39), we can obtain

$$\frac{\mathrm{d}^2[rV(r)]}{\mathrm{d}r^2} = \frac{e^2 n_0 [rV(r)]}{\epsilon k_B T} = \frac{rV(r)}{L_{\mathrm{D}}^2}, \tag{6.41}$$

where the Debye's screening length, $L_{\mathrm{D}} = (\epsilon k_B T/e^2 n_0)^{1/2}$. The solution of the potential then is

$$rV(r) = A \exp\left(-\frac{r}{L_D}\right). \tag{6.42}$$

The parameter A can be obtained by considering the boundary condition when $r \to 0$, the screening is negligible, and the potential is in fact the bare Coulomb potential $-Ze/4\pi\epsilon r$. Thus, we have finally the screened Coulomb potential

$$V(r) = -\frac{Ze}{4\pi\epsilon r}\exp\left(-\frac{r}{L_D}\right). \tag{6.43}$$

Thus, the Coulomb potential due to the impurity is reduced by the screening exponentially with distance from the impurity site. The Debye screening length L_D is the parameter to measure the strength of the screening. For an electron density of $1 \times 10^{18}/\mathrm{cm}^3$, $L_D \approx 28\text{Å}$. This distance is approximately one quarter of the average distance of the charged ions (dopants), and thus the impurity scattering can be considered a short-range, two-body collision. Two scattering events are considered independent.

The interaction strength depends on the momentum change of an impurity scattering event. The qth Fourier components, where $\mathbf{q} = \mathbf{k}' - \mathbf{k}$, are given by

$$A(\mathbf{q}) = -\frac{e}{V}\int V(r)\exp(-\mathbf{i}\mathbf{q}\cdot\mathbf{r})\mathrm{d}\mathbf{r}. \tag{6.44}$$

This is a typical Born scattering problem to convert a spherical wave to a series of plane waves. The solution for this integration is

$$A(\mathbf{q}) = -\frac{4\pi e}{V}\int_0^\infty V(r)\frac{r\sin qr}{q}\mathrm{d}r. \tag{6.45}$$

If we assume that there is one impurity in the whole space, and substituting Eq (6.43) into the equation above, we obtain

$$A(\mathbf{q}) = -\frac{Ze^2}{V\epsilon q}\int_0^\infty \exp(-\frac{r}{L_D})\sin qr\mathrm{d}r = -\frac{Ze^2}{V\epsilon q^2}\frac{1}{q^2+q_0^2}, \tag{6.46}$$

where $q_0 = 1/L_D$ is the reciprocal Debye screening length. The impurity scattering coupling strength is

$$C^2_{\mathbf{k},\mathbf{k}'} = A^2(\mathbf{q}) = \frac{Z^2e^4}{V^2\epsilon^2}\frac{1}{q^2+q_0^2}, \tag{6.47}$$

and the scattering rate is

$$W(\mathbf{k}) = \frac{Vmk}{(2\pi)2\hbar^3}\int C^2_{\mathbf{q}}(1-\cos\theta)d\Omega, \tag{6.48}$$

where Ω is solid angle.

Impurity scattering is normally elastic. For isotropic band, $\mathbf{k}' = \mathbf{k}$, we have

$$|\mathbf{k}' - \mathbf{k}|^2 = 2k^2(1-\cos\theta) = 4k^2\sin^2(\theta/2). \tag{6.49}$$

The scattering rate per unit angle θ (the injection direction as the polar direction) then is

$$W(\theta) = \frac{Z^2 e^4 N(E_k)}{2\hbar\epsilon^2 V} \frac{1}{[4k^2 \sin^2(\theta/2) + q_0^2]^2}. \tag{6.50}$$

The impurity scattering is normally not described by state to state probability, but by cross-section. The relation between the scattering rate and cross-section is

$$W(\theta) = \frac{v}{V}\sigma(\theta), \tag{6.51}$$

where v is the group velocity of the electron. The impurity scattering cross-section then is

$$\sigma(\theta) = \frac{Z^2 e^4 N(E_k)}{32\hbar\epsilon^2 v k^4} \frac{1}{[\sin^2(\theta/2) + (q_0/2k)^2]^2}. \tag{6.52}$$

There are also the other formalisms to treat the impurity scattering. Interested readers can refer to "Ridley: Quantum Processes in Semiconductors," Chap. 4 (FRS, 1999). For high doping semiconductors where the screening is strong, the Brooks-Herring result is satisfactory. At such a case, the semiconductor shall be considered degenerate, where degenerate statistics is required.

6.5 BOLTZMANN EQUATION

The Drude's transport model is easy to understand, however, the approximations adopted are intuitive and not well justified. That simple model is based on an exponential distribution in which a unique scattering time (mean free time) τ appears. It assumes that electrons obey classical statistics, and this is not always true in semiconductors. As we can see from the earlier discussion for the various scattering mechanisms, τ is dependent on energy, sometimes even on state vector $\mathbf{k}$. For the current topic of interest, the semiconductor doping is medium to heavy, and the carrier densities are high. Degenerate statistics is required. Thus, we need a more rigorous formalism based on the Boltzmann transport model, which governs the transport from the evolution of electron distribution by external field and scattering.

The electron distribution can be described by a distribution function $f(\mathbf{k}, \mathbf{r})$, which represents the electron occupation probability at position $\mathbf{r}$ and state $\mathbf{k}$. The following factors can affect the distribution function:

Motion of carriers in real space. If the distribution of carriers with velocity $\mathbf{v}$ is not uniform, then the nonuniformity will also move at velocity $\mathbf{v}$ in space, and thus results in the change of distribution function with time.

Motion of carriers in reciprocal space. The $\mathbf{k}$ values of carriers can shift in reciprocal space under external field. This has the similar effect on distribution function as the motion in real space.

Scattering. Scattering tends to randomize the states of carriers, and thus makes the distribution more even.

If we use $(\partial f/\partial t)_m$, $(\partial f/\partial t)_f$, and $(\partial f/\partial t)_c$ to represent the change rate of the distribution function caused by the above factors, respectively, then the total change rate is

$$\frac{\mathrm{d}f}{\mathrm{d}t} = \left(\frac{\partial f}{\partial t}\right)_m + \left(\frac{\partial f}{\partial t}\right)_f + \left(\frac{\partial f}{\partial t}\right)_c . \tag{6.53}$$

Consider the change of number of carriers in a volume element between $\mathbf{k}$ to $\mathbf{k}+\mathrm{d}\mathbf{k}$ and $\mathbf{r}$ to $\mathbf{r}+\mathrm{d}\mathbf{r}$, which is written as $\mathrm{d}^3k\mathrm{d}^3r$. First due to space variation of f, the change of carrier number in this volume in a time period $\mathrm{d}t$ is

$$\begin{aligned} &\frac{1}{(2\pi)^3}[f(\mathbf{k},x,y,z,t) - f(\mathbf{k},x+\mathrm{d}x,y+\mathrm{d}y,z+\mathrm{d}z,t)] \\ &= -\frac{1}{(2\pi)^3}\nabla f \cdot \mathbf{v}\mathrm{d}^3k\mathrm{d}^3r\mathrm{d}t \\ &= \frac{1}{(2\pi)^3}\left(\frac{\partial f}{\partial t}\right)_m \mathrm{d}^3k\mathrm{d}^3r\mathrm{d}t. \end{aligned} \tag{6.54}$$

Thus, we obtain

$$\left(\frac{\partial f}{\partial t}\right)_m = -\nabla_r f \cdot \mathbf{v}. \tag{6.55}$$

Second, similarly, for motion in the $\mathbf{k}$ space under external fields we obtain

$$\left(\frac{\partial f}{\partial t}\right)_f = -\nabla_k f \cdot \frac{\mathrm{d}\mathbf{k}}{\mathrm{d}t}. \tag{6.56}$$

At a stationary state, the carrier distribution does not change with time, and thus we have $\mathrm{d}f/\mathrm{d}t = 0$, then the Boltzmann equation is obtained as

$$\nabla_r f \cdot \mathbf{v} + \nabla_k f \cdot \frac{\mathrm{d}\mathbf{k}}{\mathrm{d}t} = \left(\frac{\partial f}{\partial t}\right)_c . \tag{6.57}$$

If there is no space variation caused by temperature gradient, etc., we can consider the distribution function as a function of solely $\mathbf{k}$. At thermal equilibrium state, $f_0(\mathbf{k}) = f_0(-\mathbf{k})$, thus the current density

$$\mathbf{j} = -\frac{e}{(2\pi)^3}\int \mathbf{v} f_0(\mathbf{k})\mathrm{d}\mathbf{k} = 0, \tag{6.58}$$

where $\mathbf{j}$ is the current density, and $\mathbf{v}$ is the carrier velocity. If there is an external field, then the distribution function will have some deviation from the equilibrium state, and we can write the new distribution function as

$$f(\mathbf{k}) = f_0(\mathbf{k}) + \phi(\mathbf{k}), \tag{6.59}$$

where $\phi(\mathbf{k})$ describes the deviation of the distribution function from the equilibrium state. The current density at such a case is obtained as

$$\mathbf{j} = -\frac{e}{(2\pi)^3}\int \mathbf{v}\phi(\mathbf{k})\mathrm{d}\mathbf{k}. \tag{6.60}$$

Thus, the carrier transport problem is attributed to a problem of finding $\phi(\mathbf{k})$.

Assuming the system exists a relaxation time τ due to scattering, then the system shall obey equation

$$\left[\frac{\partial f}{\partial t}\right]_c = -\frac{\phi}{\tau}, \tag{6.61}$$

then the Boltzmann equation in an external field is

$$\frac{\mathrm{d}\mathbf{k}}{\mathrm{d}t} \cdot \nabla_k f = -\frac{\phi}{\tau}. \tag{6.62}$$

When there is only an electric field, for an electron

$$\frac{\mathrm{d}\mathbf{k}}{\mathrm{d}t} = -\frac{e}{\hbar}\mathbf{F}. \tag{6.63}$$

Substituting (6.63) into (6.62), we obtain

$$\phi = \frac{e\tau}{\hbar}\mathbf{F} \cdot \nabla_k f = \frac{e\tau}{\hbar}\frac{\partial f_0}{\partial E}\nabla_k E \cdot \mathbf{F}, \tag{6.64}$$

where we assume that the deviation is not large and $\nabla_k f \simeq (\partial f_0/\partial E)\nabla_k E$, and use the gradient of E to k to substitute the gradient of f to k. Substituting (6.64) into (6.60), we obtain

$$\mathbf{j} = -\frac{e^2}{(2\pi)^3\hbar}\int \tau\frac{\partial f_0}{\partial E}\mathbf{v}(\nabla_k E \cdot \mathbf{F})\mathrm{d}^3 k. \tag{6.65}$$

Recalling $\mathbf{v} = \nabla_k E/\hbar$, the current density in each direction is

$$j_m = -\sum_n \frac{e^2}{(2\pi)^3\hbar^2}\int \tau\frac{\partial f_0}{\partial E}\frac{\partial E}{\partial k_m}\frac{\partial E}{\partial k_n}\mathrm{d}^3 k F_n = \sum_n \sigma_{mn}F_n. \tag{6.66}$$

Thus, the conductivity is

$$\sigma_{mn} = -\frac{e^2}{(2\pi)^3\hbar^2}\int \tau\frac{\partial f_0}{\partial E}\frac{\partial E}{\partial k_m}\frac{\partial E}{\partial k_n}\mathrm{d}^3 k. \tag{6.67}$$

The conductivity is a second-rank tensor. From the relation $\sigma = ne\mu$, the mobility is also obtained as

$$\mu_{mn} = -\frac{e}{(2\pi)^3 n\hbar^2}\int \tau\frac{\partial f_0}{\partial E}\frac{\partial E}{\partial k_m}\frac{\partial E}{\partial k_n}\mathrm{d}^3 k. \tag{6.68}$$

The mobility depends on band structure through the relation $\partial E/\partial k$ and depends on temperature through $\partial f_0/\partial E$, where $f_0(E)$ is the Fermi-Dirac distribution function. Then we have

$$\frac{\partial f_0}{\partial E} = -\frac{1}{k_B T} f_0(E)(1 - f_0(E)). \tag{6.69}$$

Thus, finally the mobility is

$$\mu_{mn} = \frac{e}{(2\pi)^3 n\hbar^2 k_B T} \int \tau(\mathbf{k}) \frac{\partial E}{\partial k_m} \frac{\partial E}{\partial k_n} f_0(E)(1 - f_0(E)) \mathrm{d}^3 k. \tag{6.70}$$

This result is different from the Drude's model in that here the effective mass is replaced by detailed band structure, which may be complicated and may not have a well-defined curvature. The relaxation time τ is also k-dependent, and the mobility is a tensor. If the Drude's model is appropriate only for a uniform electron gas, the Boltzmann equation has a much larger applicable range including a system with a complicated band structure. But we have to note that if we assume a constant relaxation time, the Boltzmann distribution function instead of the Fermi–Dirac distribution function, and a simple parabolic band, the Drude's transport model is recovered using the Boltzmann equation. In the following, we will use the Si conduction valleys as an example to study the electron conductivity using Boltzmann equation.

6.5.1 Electron Conductivity Mass of Si

Let us define k_1, k_2, and k_3 to be the axes of the crystal principle-axis coordinate system or lab coordinate system, and k'_1, k'_2, and k'_3 to be the axes of ellipsoid in a valley. In the valley coordinate system, the energy is written as

$$E^s = E_0^s + \frac{\hbar^2 k_1^{'2}}{2m_1^*} + \frac{\hbar^2 k_2^{'2}}{2m_2^*} + \frac{\hbar^2 k_3^{'2}}{2m_3^*}, \tag{6.71}$$

where the superscript s labels the sth conduction valley. For Si, $s = 1, 2, 3, ..., 6$. E_0^s is the energy of the valley edge.

By defining the other three new parameters

$$\phi_1 = \frac{\hbar k'_1}{\sqrt{m_1^*}}, \quad \phi_2 = \frac{\hbar k'_2}{\sqrt{m_2^*}}, \quad \phi_3 = \frac{\hbar k'_3}{\sqrt{m_3^*}}, \tag{6.72}$$

the electron energy is rewritten as

$$E^s = E_0^s + \frac{1}{2}(\phi_1^2 + \phi_2^2 + \phi_3^2) = E_0^s + \frac{1}{2}\phi^2. \tag{6.73}$$

In the sth valley, the conductivity in the ellipsoid coordinate system is

$$\sigma_{ij}^{s} = -\frac{2e^2}{(2\pi)^3\hbar^2}\int \tau \frac{\partial f_0}{\partial E^s}\frac{\partial E^s}{\partial k_i'}\frac{\partial E}{\partial k_j'}\mathrm{d}^3k'^s. \tag{6.74}$$

In nondegenerate case, the electron obeys the Boltzmana distribution function, and

$$\frac{\partial f_0}{\partial E^s} = -\frac{1}{k_BT}f_0, \tag{6.75}$$

and

$$\mathrm{d}^3k = \frac{\sqrt{m_1^*m_2^*m_3^*}}{\hbar^3}\mathrm{d}\phi_1\mathrm{d}\phi_2\mathrm{d}\phi_3. \tag{6.76}$$

Thus (6.74) is rewritten as

$$\sigma_{ij}^{s} = \frac{2e^2}{(2\pi)^3k_BT\sqrt{m_i^*m_j^*}}\frac{\sqrt{m_1^*m_2^*m_3^*}}{\hbar^3}\int \tau f_0\phi_i\phi_j\mathrm{d}\phi_1\mathrm{d}\phi_2\mathrm{d}\phi_3. \tag{6.77}$$

Because of the symmetry of the ellipsoid, τ is an even function of $\mathbf{k}'$. Together with the consideration of ellipsoidal symmetry of the valley, $\sigma_{ij} = 0$ for any $i \neq j$. Thus only σ_{11}, σ_{22}, and σ_{33} are not zero.

Let us study σ_{11}. Obviously we have

$$\begin{aligned}\int \tau f_0\phi_1^2\mathrm{d}\phi_1\mathrm{d}\phi_2\mathrm{d}\phi_3 &= \int \tau f_0\phi_2^2\mathrm{d}\phi_1\mathrm{d}\phi_2\mathrm{d}\phi_3 = \int \tau f_0\phi_3^2\mathrm{d}\phi_1\mathrm{d}\phi_2\mathrm{d}\phi_3\\ &= \frac{1}{3}\int \tau f_0\phi^2\mathrm{d}\phi_1\mathrm{d}\phi_2\mathrm{d}\phi_3,\end{aligned} \tag{6.78}$$

then the electron density in a volume $\mathrm{d}k'^3$ can be written as

$$dn^s = \frac{2}{(2\pi)^3}f_0\mathrm{d}k'^3 = \frac{2}{(2\pi)^3}f_0\frac{\sqrt{m_1^*m_2^*m_3^*}}{\hbar^3}\mathrm{d}\phi_1\mathrm{d}\phi_2\mathrm{d}\phi_3, \tag{6.79}$$

and from (6.77), we have

$$\sigma_{11}^{s} = \frac{2e^2}{3k_BT}\frac{1}{m_1^*}\int \tau\frac{\phi^2}{2}\mathrm{d}n^s = \frac{2e^2}{3k_BT}\frac{1}{m_1^*}\langle \tau E^s\rangle, \tag{6.80}$$

where

$$\langle \tau E^s\rangle = \frac{\int \tau E^s\mathrm{d}n^s}{\int \mathrm{d}n^s}. \tag{6.81}$$

We can use $\langle \tau E\rangle$ to replace $\langle \tau E^s\rangle$ since it is the same for any valley. Also from classical statistics, $3k_BT/2 = \langle E\rangle$, so we obtain

$$\sigma_{11}^{s} = \frac{n^se^2}{m_1^*}\frac{\langle \tau E\rangle}{\langle E\rangle} = \frac{ne^2}{6m_1^*}\frac{\langle \tau E\rangle}{\langle E\rangle}, \tag{6.82}$$

where $n = \sum_{s=1}^{6} n^s$ is the total electron density. σ_{22}^s and σ_{33}^s have similar expressions, only with the effective mass changed according to the axes.

The next step is to transform the conductivity measured in the valley ellipsoidal coordinate system to the crystal coordinate system.

The transformation between two coordinate systems can be generally written as

$$\begin{pmatrix} k_1 \\ k_2 \\ k_3 \end{pmatrix} = \begin{pmatrix} c_{11} & c_{12} & c_{13} \\ c_{21} & c_{22} & c_{23} \\ c_{31} & c_{32} & c_{33} \end{pmatrix} \begin{pmatrix} k_1' \\ k_2' \\ k_3' \end{pmatrix}, \tag{6.83}$$

or simply, $K = CK'$. The transformation of a second-rank tensor between the two coordinate system is

$$\sigma = C\sigma^s C^{-1}, \tag{6.84}$$

where for the transformation matrix between two coordinate systems, $C^{-1} = \widetilde{C}$, where $\widetilde{C}$ represents the transpose of matrix C. The left-hand side is the conductivity contribution of the electrons from only one valley. We need to sum over all the valleys to get the electron conductivity of the crystal, so we have

$$\sigma_{ij} = \sum_s \sum_{pq} c_{ip}^s c_{jq}^s \sigma_{pq}^s. \tag{6.85}$$

For Si, the transformation matrices for ellipsoids along the crystals k_x, k_y, and k_z are

$$C^1 = \begin{pmatrix} 0\,0\,1 \\ 1\,0\,0 \\ 0\,1\,0 \end{pmatrix}, \quad C^2 = \begin{pmatrix} 0\,1\,0 \\ 0\,0\,1 \\ 1\,0\,0 \end{pmatrix}, \quad C^3 = \begin{pmatrix} 1\,0\,0 \\ 0\,1\,0 \\ 0\,0\,1 \end{pmatrix}. \tag{6.86}$$

Substituting the above matrix elements into (6.85), remembering that there are two equivalent valleys at one crystal axis, and defining $m_1^* = m_l^*$, $m_2^* = m_3^* = m_t^*$ we can obtain

$$\sigma_{11} = \sigma_{22} = \sigma_{33} = \frac{ne^2}{3}\left(\frac{2}{m_t^*} + \frac{1}{m_l}\right)\frac{\langle \tau E\rangle}{\langle E\rangle}. \tag{6.87}$$

If we just assume $\tau = \langle \tau E\rangle / \langle E\rangle$, and compare the above equation with Drude's model, we have the familiar electron conductivity mass for Si

$$\frac{1}{m_c} = \frac{1}{3}\left(\frac{1}{m_l} + \frac{2}{m_t}\right). \tag{6.88}$$

Although this conclusion can be obtained directly by the cubic symmetry Si has, but the above procedure showed a way of how to calculate the electron conductivity in a general multivalley system.

6.6 NEW FEATURES IN 2D SCATTERING

6.6.1 Broken Symmetry due to Confinement

First thing we have to notice is that in a 2D system, the 3D band structure changes into a series of 2D subbands. Thus, the integration over the 3D Brillouin zone for scattering turns into summation of integration in these 2D subbands. The state density $V/8\pi^3$ in 3D has to be replaced by $V/4\pi^2$ in 2D cases.

Screening effect in 2D system is very complicated. The screening factor differs in intra-subband and inter-subband scattering. Complete treatment needs to consider dynamic screening, which is beyond the scope of this book. But for Si, and the valence bands for normal semiconductors with not too high charge densities, screening effect can be neglected as a good approximation.

For phonon scattering in 2D system, there is significant difference from that in the 3D case since the 2D system is confined in one direction, say, z, and thus the electron momentum in this direction is not well defined. Then for an electron scattered by absorbing or emitting a phonon with wave vector $\mathbf{q}$, the x-y wave vector $\mathbf{k}_\perp$ must change by $\mathbf{q}_\perp$, but there is no restriction for $\mathbf{q}_z$.

An envelope function for a 2D Bloch state can be written as

$$\psi(r) = \frac{F(z)}{\sqrt{A}} \exp(\mathrm{i}\mathbf{k}\cdot\mathbf{R}), \tag{6.89}$$

where $\mathbf{k}$ and $\mathbf{R}$ are the 2D phonon wave vector and space vector in the x-y plane, and A is the area of the x-y plane. Normalization requires

$$\int F^2(z)\mathrm{d}z = \int \rho(z)\mathrm{d}z, \tag{6.90}$$

where $\rho(z)$ is the carrier density function in the z-direction. Similar to the wave function overlap factor in 3D phonon scattering case, the matrix elements for scattering in 2D system are also proportional to the envelope function overlap, but weighted by a phase factor due to breaking of periodic symmetry in the z-direction,

$$I_{ij}(q_z) = \int F_i(z)F_j(z)\exp(\mathrm{i}q_z z)\mathrm{d}z. \tag{6.91}$$

I_{ij} is called the form factor. Since there are no restrictions for q_z, then for I_{ij}^2, we shall integrate over q_z, and obtain

$$\frac{1}{w_{ij}} = \int_{-\infty}^{\infty} |I_{ij}(q_z)|^2 \mathrm{d}q_z, \tag{6.92}$$

where w_{ii} evaluates the effective well width of the ith subband. For acoustic phonon and optical phonon scattering, since the coupling strength is not dependent on phonon wave vector q, (6.92) turns into

$$\frac{1}{w_{ij}} = 2\pi \int |F_i(z)|^2 |F_j(z)|^2 \mathrm{d}z = 2\pi \int \rho_i(z)\rho_j(z)\mathrm{d}z. \tag{6.93}$$

The factor 2π comes from the quantum confinement-induced symmetry breaking. By comparing (6.7), (6.91), and (6.93), and, for example, the bulk Si hole mobility of $\sim$500 cm^2/Vs and the channel mobility of $\sim$70 cm^2/Vs, we can see that the quantum confinement-induced scattering enhancement plays a decisive role affecting the mobility in MOSFET channels.

However, the coupling strength does depend on q for polar optical phonon, piezoelectric and impurity scattering, then we have to also include q in the integration in (6.92). If we neglect the screening, we have

$$\frac{1}{w_{ij}} = \int_{-\infty}^{\infty} \frac{|I_{ij}(q_z)|^2}{q_z^2 + q_\perp^2} \mathrm{d}q_z. \tag{6.94}$$

By the identity

$$\frac{q_\perp}{\pi} \int_{-\infty}^{\infty} \frac{1}{q_z^2 + q_\perp^2} \mathrm{d}q_z = \exp(-q_\perp |z|) \tag{6.95}$$

we have

$$\frac{q_\perp}{\pi} \int_{-\infty}^{\infty} \frac{|I_{ij}(q_z)|^2}{q_z^2 + q_\perp^2} \mathrm{d}q_z = H_{ij}(q_\perp), \tag{6.96}$$

where

$$H_{ij}(q_\perp) = \int\int \mathrm{d}z_1 \mathrm{d}z_2 \rho_i(z_1)\rho_j(z_2) \exp(-q_\perp |z_1 - z_2|). \tag{6.97}$$

Since polar optical phonon scattering is q-dependent, it is very complicated to treat in 2D transport computations. We may first inspect its effect in momentum relaxation compared to other scattering mechanisms such as nonpolar optical phonon scattering. In semiconductor valence bands, the optical phonon scattering is supposed to be the dominant scattering source in high temperature such as the room temperature.

Suppose an electron is scattered from $\mathbf{k_1}$ to $\mathbf{k_2}$, as illustrated in Fig. 6.5. In this event, the relative change of the electron momentum is $q\cos(\theta)/k_1$. Using the cosine theorem, we have

$$\cos\theta = \frac{k_2^2 - k_1^2 - q^2}{2k_1 q}, \tag{6.98}$$

then

$$\frac{q}{k_1}\cos\theta = \frac{k_2^2 - k_1^2 - q^2}{2k_1^2}. \tag{6.99}$$

From (6.96), the transition by polar optical phonon is proportional to $H(q)/q$ (without inducing too much confusion, we use q to represent $q_\perp$ hereafter), then we may combine the $1/q$ factor with (6.99), and have

$$\frac{1}{q}\left(\frac{q}{k_1\cos\theta}\right)=\frac{k_2^2-k_1^2-q^2}{2k_1^2q}=\frac{k_2^2-k_1^2}{2k_1^2q}-\frac{q}{2k_1^2}. \tag{6.100}$$

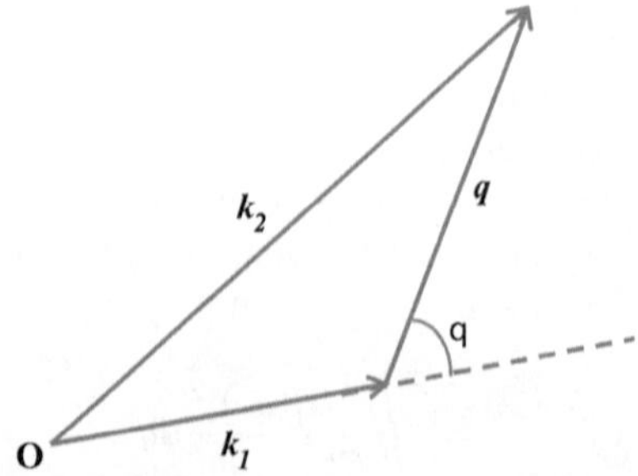

Fig. 6.5. A small angle scattering event. In this case, momentum change is not large, especially when the magnitude of **k** is similar before and after scattering

In the above equation, the second term is proportional to q, and the first term looks proportional to $1/q$, but is in fact definite when $q \to 0$ due to its dependence on k_2. Therefore, the strongest q-dependence is from the exponential relation in $H(q)$. Hence the polar optical phonon scattering is strong for small q scattering, but weak for large q. Thus, polar optical phonon scattering is not a so effective mechanism for momentum relaxation as the nonpolar optical phonon scattering.

6.6.2 Surface Roughness Scattering

For semiconductor heterojunction or semiconductor/oxide electron devices, there is a unique scattering mechanism that does not exist in the bulk: the interface scattering. A good interface in semiconductor devices is extremely important. It affects significantly the device performance and reliability. If there is large density of interface states, e.g., $>10^{12}/\mathrm{cm}^2\cdot\mathrm{eV}$, the fermi level will be pinned relative to the semiconductor band edge, and the MOS or MIS devices cannot even generate an inversion layer. Interface is a historic issue for III-V MIS or MOS devices and it is the reason why Si has been standing out. Given today's Si technology, the interface state density in the Si-SiO_2 interface can be very low, at $\sim 10^8 - 10^9/\mathrm{cm}^2\mathrm{eV}$. But still, the interface can have a pronounced effect in carrier scattering even in Si MOSFET devices.

The semiconductor–oxide interface is not an ideal surface. An interface is as a matter of fact a transition layer. For Si-SiO_2 interface, it consists of the Si-O chemically mismatched zone and SiO_2 strain layer. In the Si-O chemically mismatched zone, silicon oxide exists as SiO_x, where $x = 0.4 - 2$. Measurements reveal that the thickness of this layer can be up to 3 nm. Lattice of Si and SiO_2 can be well matched. Si–O chemically mismatched layer is accommodated by strain. Models for Si–SiO_2 interface states include the Coulomb potential due to the charged sites at the interface, valence bond distortion, and interface defects such as unsaturated bond linkage, unbridged oxygen centers, defect sites, etc. The scatterings due to them are all elastic processes. Except for the interface states which may be interface defects or traps, the most importance scattering mechanism for Si interface may be that caused by interface roughness. A picture of the Si–SiO_2 interface obtained by cross-sectional high-resolution transmission electron microscopy is shown in Fig. 6.6 (Goodnick et al, 1985). When the inversion charge density

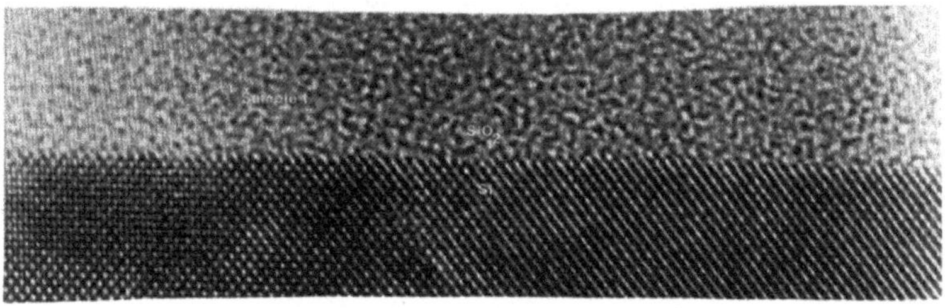

Fig. 6.6. Image of Si/SiO_2 interface. The interface is not ideal. There is a transition layer with width of several atomic layers. From (Goodnick et al, 1985)

is larger than $10^{12}/\text{cm}^2$ in the inversion layer, the carriers are even more strongly confined to the interface; the surface roughness scattering becomes increasingly responsible for the significant decrease in electron mobility in Si inversion layer.

Interface roughness is caused by the nonuniformity of the interface potential, which is a different mechanism from the above-mentioned interface states. In the usual models for the surface roughness scattering, one assumes an abrupt boundary between Si and SiO_2, which randomly varies according to a quasicontinuous function $\Delta(\mathbf{r})$, where $\mathbf{r}$ signifies the 2D position vector in the plane of the interface. This assumption is reasonable given today's advanced technology and was satisfactorily checked by different experimentalists (also see Fig. 6.6). When a gate voltage is applied on the gate, the potential on the semiconductor surface is not even due to the interface roughness. This uneven distribution of electric potential on the surface induces the scattering to the electronic motion. This is exactly analogous to the friction of a rough surface to a moving body. The surface roughness can be simulated by either Gaussian or exponential auto-covariance functions. The surface profiles generated by a Gaussian and exponential auto-covariance function with rms height, Δ_m, of 5 Å and correlation length, L, of 30 Å are shown in Fig. 6.7. The correlation length is a parameter used to describe the potential correlation between surface fluctuations. Shorter correlation length indicates more local-

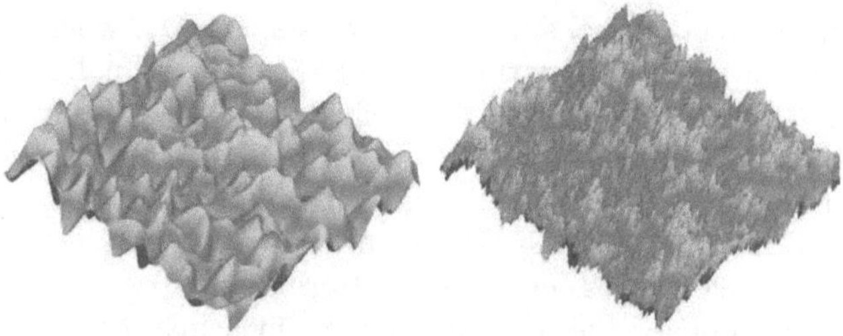

Fig. 6.7. Two simulated Si/SiO_2 interfaces generated by Gaussian (left) and exponential (right) auto-covariance functions

ized and microscopic potentials, and longer correlation length indicate that surface potential at one point may have a larger influence on farther away points. Within the two functions, the exponential auto-covariance function is believed to give more accurate description as adopted by Goodnick (Goodnick et al, 1983, 1985) and Gamiz (Gámiz et al, 1999), et al. Many researchers (Sugano et al, 1980; Krivanek et al, 1978; Yoshinobu et al, 1994) experimentally studied the well-prepared Si-SiO_2 interface and found that the rms height is at between 2 – 5 Å, which is approximately one or two monolayers of the interface width, and the interface roughness correlation length lies between 10 and 25 Å.

The surface potential may be expanded at the interface by

$$V[z+\Delta(\mathbf{r})] = V(z) + \Delta(\mathbf{r})\frac{\partial V(z)}{\partial z}, \tag{6.101}$$

where $V(z)$ is the nonperturbed potential and z is the coordinate perpendicular to the interface, and $\Delta(\mathbf{r})$ is the surface roughness on the x-y plane. The surface roughness scattering rate is obtained by the Fermi's golden rule. First the surface roughness perturbation Hamiltonian is written as

$$H_{\mathrm{SR}} = -e\{V[z+\Delta(\mathbf{r})] - V(z)\}, \tag{6.102}$$

then combine with (6.101), and it gives

$$H_{\mathrm{SR}} = -e\Delta(\mathbf{r})\frac{\partial V(z)}{\partial z} = e\Delta(\mathbf{r})F(z), \tag{6.103}$$

where $F(z)$ is the transverse electric field at the surface. Then the scattering matrix element between state $\mathbf{k}$ in the μth and $\mathbf{k}'$ in the νth band is

$$\begin{aligned} C^2_{\mu\mathbf{k},\nu\mathbf{k}'} &= |\langle \nu, \mathbf{k}'|H_{\mathrm{SR}}|\mu,\mathbf{k}\rangle|^2 \\ &= e^2|\int \psi_\nu(z)E(z)\psi_\mu(z)\mathrm{d}z|^2|\Delta(\mathbf{q})|^2, \end{aligned} \tag{6.104}$$

where $\psi_\mu(z)$ and $\psi_\nu(z)$ are the envelope functions of the $(\mu,\mathbf{k})$ and $(\nu,\mathbf{k}')$ states, respectively, $\mathbf{q}=\mathbf{k}'-\mathbf{k}$, and $\Delta(\mathbf{q})$ is the qth Fourier component of $\Delta(\mathbf{r})$.

Given the surface roughness function, i.e., the Gaussian or the exponential auto-covariance function, the surface roughness scattering rate can be obtained. For exponential auto-covariance surface roughness profile,

$$|\Delta(\mathbf{q})|^2 = \frac{\pi \Delta_m^2 L^2}{[1 + (q^2 L^2/2)]^{3/2}}. \tag{6.105}$$

Surface roughness scattering increases with the transverse electric field at the device interface. At low effective field, because of the weak confinement, phonon scattering dominates the carrier mobility. At relatively high effective field, surface roughness scattering increases rapidly. For n-Si MOSFETs with an effective field around 10^6 V/cm, the contribution of surface roughness scattering to the electron mobility is about the same as the phonon scattering, as can be seen from Fig. 6.8 (Yamakawa et al, 1998) where Monte Carlo simulation results for the mobility of an n-Si MOSFET are shown. For

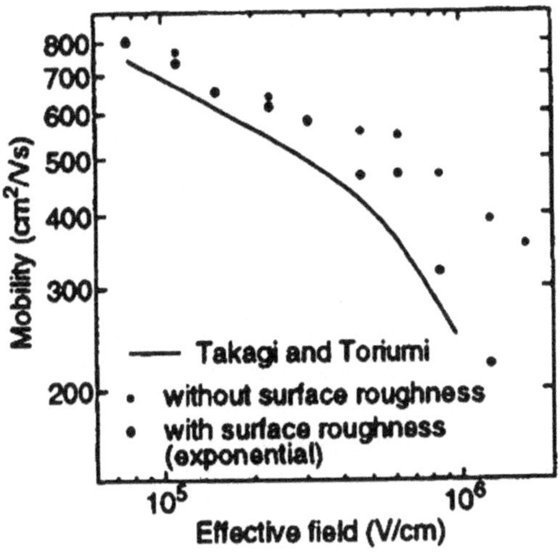

Fig. 6.8. Electron mobility as a function of effective electric field. The *solid line* represents the theoretical results and *large dots* represent simulation results with surface roughness and *smaller dots* for results without consideration of surface roughness scattering. At a field of 10^6 V/cm, surface roughness scattering is equally important as the phonon scattering. From (Yamakawa et al, 1998)

p-Si MOSFETs, surface roughness scattering is not as pronounced as for n-MOSFETs partly because the optical phonon scattering is stronger. Its rate is only about one quarter to half a quarter of the optical phonon scattering rate (Fischetti et al, 2003; Sun et al, 2007). But surface roughness scattering is important at low temperatures, where phonon scattering is strongly suppressed.

6.7 STRAIN EFFECTS ON CARRIER TRANSPORT

Strain effects on carrier transport can be qualitatively understood by Drude's model. From this model, the conductivity change is caused by the mass shift or/and scattering rate variation induced by strain. They can be both ascribed to strain altered band structures discussed in Chaps. 3–5. It is hard to separate the influence of mass from scattering in affecting the mobility, especially when the bands considered are anisotropic. But under certain circumstances such as in Si conduction band or the valence bands, which we will discuss later, one effect is dominant over the other. When resistivity, or conductivity, is concerned, one has to also consider the carrier density variation by strain. This is very important for intrinsic semiconductors, where strain shifts the bandgap and induces significant change of intrinsic carrier density. The carrier density change can cause strong piezoresistive effect. However, for extrinsic semiconductors, bandgap shift does not vary the density of majority carriers. The piezoresistive effect is caused by strain-altered band structures. Correspondingly, the change of resistivity is expected to be smaller. In the following subsections, we are going to discuss the piezoresistive effect for electrons and holes with various electronic structures in extrinsic semiconductors and totally neglect strain-induced carrier density alteration.

6.7.1 Piezoresistance

In Chap. 2, we introduced the piezoresistive effect. In all the factors that affect the resistance, the change of resistivity dominates in semiconductors. Resistivity is inversely proportional to mobility, and thus stress-modified band structure, and subsequently altered mobility is the main cause of piezoresistivity. Here, we rewrite the piezoresistivity equation for convenience

$$\frac{\Delta\rho_i}{\rho} = \sum_{k=1}^{6} \pi_{ik}\tau_k, \tag{6.106}$$

where π_{ik} is a 6×6 matrix. For cubic semiconductors, it has a very simple form as

$$\pi_{ik} = \begin{pmatrix} \pi_{11} & \pi_{12} & \pi_{12} & 0 & 0 & 0 \\ \pi_{12} & \pi_{11} & \pi_{12} & 0 & 0 & 0 \\ \pi_{12} & \pi_{12} & \pi_{11} & 0 & 0 & 0 \\ 0 & 0 & 0 & \pi_{44} & 0 & 0 \\ 0 & 0 & 0 & 0 & \pi_{44} & 0 \\ 0 & 0 & 0 & 0 & 0 & \pi_{44} \end{pmatrix}. \tag{6.107}$$

The piezoresistance coefficients can be obtained for an arbitrary stress by using (6.106). Next, we discuss some typical stress situations. First the hydrostatic stress, the stress vector can be written as $(-P, -P, -P, 0, 0, 0)$. Then we have (by neglecting the geometrical piezoresistive effect) a scalar form piezoresistance

$$\frac{\Delta\rho}{\rho} = -P(\pi_{11} + 2\pi_{12}). \tag{6.108}$$

Under a uniaxial stress along [110], where the stress vector is $(1, 1, 0, 0, 0, 1)T/2$, the piezoresistance $\Delta\rho_i/\rho$ is a six-component vector in the form $(\pi_{11} + \pi_{12}, \pi_{11}+\pi_{12}, 2\pi_{12}, 0, 0, \pi_{44})T/2$. The current direction is not necessarily along the stress direction, and under this situation we can obtain the piezoresistance along the current direction by doing an inverse transformation, $\rho_i \to \rho_{jk}$, so

$$\frac{\Delta\rho}{\rho} = \frac{\Delta\mathbf{E}\cdot\mathbf{j}}{\rho_0 j^2} = \sum_{j,k} \frac{\Delta\rho_{jk}}{\rho_0} l_j l_k, \tag{6.109}$$

where $\Delta\mathbf{E}$ is the change of the electric field by stress and l_x, l_y, l_z are the directional cosines of the current $\mathbf{j}$. The vector component of $\Delta\mathbf{E}$ is simply $\Delta E_i = \sum_j \Delta\rho_{ij} j_j$, which is implicitly used in the above equation. For stress and current are both along the [110] direction, $\frac{\Delta\rho}{\rho} = T(\pi_{11}+\pi_{12}+\pi_{44})/2$. For stress along $[\bar{1}10]$ and current along [110], we have $\frac{\Delta\rho}{\rho} = T(\pi_{11}+\pi_{12}-\pi_{44})/2$. We reproduce the table for piezoresistance under several stress and current situations in Table 6.2.

Table 6.2. Piezoresistance under some stress and current conditions

Stress direction	Current direction	$\Delta\rho/\rho$
[100]	[100]	$T\pi_{11}$
[100]	[010]	$T\pi_{12}$
[110]	[110]	$T(\pi_{11} + \pi_{12} + \pi_{44})/2$
[110]	$[\bar{1}10]$	$T(\pi_{11} + \pi_{12} - \pi_{44})/2$
[111]	[111]	$T(\pi_{11} + 2\pi_{12} + 2\pi_{44})/2$
Hydrostatic		$-P(\pi_{11} + 2\pi_{12})$

Sometimes it is convenient to study the relation between change of resistivity and strain. They are also described by a fourth-rank tensor,

$$\frac{\Delta\rho_{ij}}{\rho} = \sum_{k,l=1}^{3} m_{ijkl}\varepsilon_{kl}, \tag{6.110}$$

where m_{ijkl} is called the elastoresistance coefficients. Similar to the π-coefficient, writing the resistivity also as a six-component array, we can rewrite (6.110) as

$$\frac{\Delta\rho_i}{\rho} = \sum_{j=1}^{6} m_{ij}\varepsilon_j. \tag{6.111}$$

For cubic semiconductors, m_{ij} also has three independent elements, m_{11}, m_{12}, and m_{44}. The relation between m_{ij} and m_{ijkl} is that $m_{1111} = m_{11}$, $m_{1122} = m_{12}$, and $m_{1212} = m_{44}$. Since the stress and strain are related by the compliance matrix,

$$\tau_k = \sum_{j=1}^{6} C_{kj}\varepsilon_j, \tag{6.112}$$

then

$$\frac{\Delta\rho_i}{\rho} = \sum_{j,k=1}^{6} \pi_{ik}C_{kj}\varepsilon_j, \tag{6.113}$$

and thus the elastoresistance coefficients and the π-coefficients are related by equation

$$m_{ij} = \sum_{k=1}^{6} \pi_{ik}C_{kj}. \tag{6.114}$$

From this equation, the three independent elastoresistance coefficients and the π-coefficients have simple relations by

$$\begin{aligned} m_{11} &= C_{11}\pi_{11} + 2C_{12}\pi_{12}, \\ m_{12} &= C_{12}\pi_{11} + (C_{11} + C_{12})\pi_{12}, \\ m_{44} &= C_{44}\pi_{44}. \end{aligned} \tag{6.115}$$

Smith (Smith, 1954) measured the piezoresistance coefficients for bulk Si and Ge for four stress and current configurations, which are schematically shown in Fig. 6.9. From these four configurations, the three independent π-coefficients can be derived. They are listed in Table 6.3.

6.7.2 Electron Transport

Electron piezoresistive effect is only pronounced in semiconductors which have multivalley conduction bands such as Si and Ge. In these semiconductors, stress shifts some valleys with respect to the other valleys and thus electrons repopulate according to the energy shift. This may cause the conductivity mass change and also alter scattering rate due to the change of DOS. One simple example is the Si conduction band. If there is a uniaxial compressive stress along [001], the two Δ-valleys along [001] lower relative to the other four Δ-valleys. If the current is along [001], the resistivity will increase. If the current is along the perpendicular direction, the resistivity will be reduced.

If we assume that stress only shifts the conduction band edges and does not warp the bands, then the conductivity mass change is easiest to treat. In the following, we demonstrate how to derive the elastoresistance coefficients

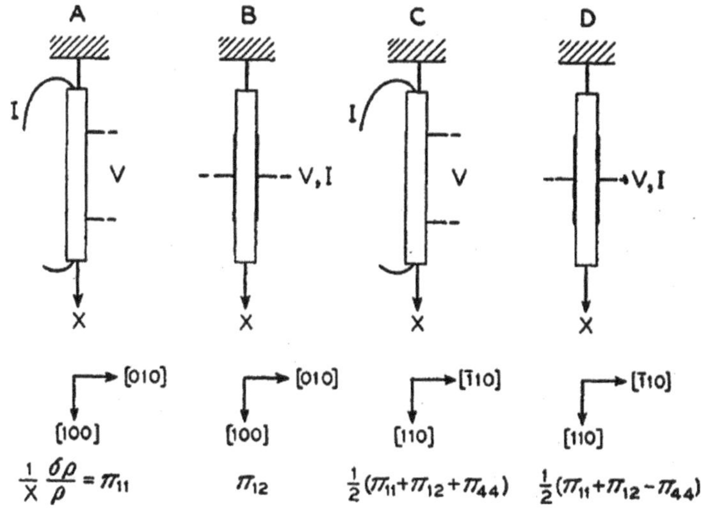

Fig. 6.9. Piezoresistance coefficients and the corresponding stress-current configuration. From (Smith, 1954)

Table 6.3. π-Coefficients for bulk Si and Ge ($10^{-11}\mathrm{Pa}^{-1}$)

	ρ_0 ($\Omega \cdot$ cm)	π_{11}	π_{12}	π_{44}
n-Ge	16.6	−5.2	−5.5	−138.7
p-Ge	15.0	−10.6	5.0	98.6
n-Si	11.7	−102.2	53.7	−13.6
p-Si	7.8	6.6	−1.1	138.1

After Smith (Smith, 1954)

in terms of the deformation potentials and the effective mass ratio for bulk semiconductors by assuming that there is no scattering rate change and the relaxation time is energy-independent.

Assuming there are s conduction valleys, and each valley edge shifts with respect to the Fermi energy by $\Delta(E_c^\nu - E_F)$, then the piezoresistance shall be

$$\frac{\Delta\rho_{ij}}{\rho} = -\frac{\Delta\sigma_{ij}}{\sigma} = -\frac{1}{\sigma_0}\sum_{\nu=1}^{s}\Delta\sigma_{ij}^\nu = -\frac{e}{\sigma_0}\sum_{\nu=1}^{s}\Delta n^\nu \mu_{ij}^\nu. \tag{6.116}$$

Assuming the shift is small, i.e., $\Delta(E_c^\nu - E_F) \ll k_B T$, then from

$$n = N_c \exp\left(-\frac{E_c - E_F}{k_B T}\right) \tag{6.117}$$

we have

$$\Delta n^\nu = \frac{1}{s}\frac{\partial n}{\partial(E - E_F)}\Delta(E_c^\nu - E_F) = -\frac{n}{s}\frac{\Delta(E_c^\nu - E_F)}{k_B T}. \tag{6.118}$$

Then (6.116) becomes

$$\frac{\Delta\rho_{ij}}{\rho} = \frac{e}{\sigma_0}\sum_{\nu=1}^{s}\frac{n}{s}\frac{\Delta(E_c^\nu - E_F)}{k_B T}\mu_{ij}^\nu. \tag{6.119}$$

In the extrinsic case, electron density does not change with stress, so

$$\sum_{\nu=1}^{s}\Delta n^\nu = 0, \tag{6.120}$$

then

$$\sum_{\nu=1}^{s}\Delta(E_c^\nu - E_F) = 0. \tag{6.121}$$

Then the change of Fermi energy is found to be

$$\Delta E_F = \frac{1}{s}\sum_{\nu=1}^{s}\Delta E_c^\nu. \tag{6.122}$$

ΔE_c^ν with strain was discussed in Sect. 4.13.1, and can be expressed as the product of strain and deformation potentials. Then we now have both ΔE_c^ν and ΔE_F, so then piezoresistance can be found.

Let us use Si conduction band as an example. By referring to (4.214) and (4.215), the shift of the mean energy does not cause the electron density repopulation. Only the valley splitting term causes relative band edge shift. Assuming now we have only uniaxial strain component ε_{zz}, let us study the resistance change along the z-direction. The valley shift along [001] is $\delta E_c^{(\|)} = 2\Xi_u^\Delta\varepsilon_{zz}/3$, and the four valleys along [100] and [010] shift by $-\delta E_c^{(\perp)} = \Xi_u^\Delta\varepsilon_{zz}/3$. The electron density shifts by $(2\Xi_u\varepsilon_{zz}/3k_BT \times n/6)$ and $-(\Xi_u\varepsilon_{zz}/3k_BT \times n/6)$, respectively. The piezoresistance is then

$$\begin{aligned}\frac{\Delta\rho_{zz}}{\rho_0} &= -\frac{1}{\sigma_0}\sum_{\nu=1}^{6}\Delta\sigma_{zz}^\nu = \frac{m_c}{n}\left(\frac{2\Delta n^{(\|)}}{m_l} - \frac{4\Delta n^{(\perp)}}{m_t}\right)\\ &= \frac{2}{9}\frac{\Xi_u}{k_BT}\left(\frac{m_c}{m_l} - \frac{m_c}{m_t}\right)\varepsilon_{zz},\end{aligned} \tag{6.123}$$

where m_c is the conductivity mass defined in (6.88). Defining $K = m_l/m_t$, and recalling the relation between $\Delta\rho/\rho$ and m_{ij} defined by (6.110) and (6.111), we obtain

$$m_{11} = -\frac{2}{3}\frac{\Xi_u}{k_BT}\frac{K-1}{2K+1}. \tag{6.124}$$

Similarly,

$$m_{12} = \frac{1}{3}\frac{\Xi_u}{k_BT}\frac{K-1}{2K+1}. \tag{6.125}$$

Note for Si conduction band, $m_{11} = -2m_{12}$. Assume there is no energy shift by ε_{xy}-type shear strain to the $\langle 100 \rangle$ valleys, then it has no piezoresistive effect, and thus

$$m_{44} = \pi_{44} = 0. \tag{6.126}$$

For $\langle 111 \rangle$ conduction valleys that Ge has, the uniaxial stress along $\langle 100 \rangle$ does not split the valleys, then the longitudinal and transverse piezoresistance is zero, i.e.,

$$m_{11} = m_{12} = 0. \tag{6.127}$$

But it can be found that

$$m_{44} = -\frac{1}{3} \frac{\Xi_u}{k_B T} \frac{K-1}{2K+1}. \tag{6.128}$$

From the earlier discussion, we can see that the piezoresistance is strongly dependent on K. If $K = 1$, i.e., the valley band structure is isotropic, then the piezoresistance is zero.

We have noticed that from Table 6.3, π_{11} and π_{12} are much larger than π_{44} in n-Si, and π_{11} and π_{12} are much smaller than π_{44} in n-Ge. This trend is consistent with the analysis above. However, the smaller π-coefficients such as π_{44} for Si are not zero as expected. We need to know what causes this. Still take Si conduction band as an example. Suppose two situations with identical [110] uniaxial stress T but in one situation the current is along [110], and in the other the current is along $[\bar{1}10]$. If there is no band warping-induced effective mass change, we would expect the same piezoresistance, since the electron repopulation-induced mass change is the same and the scattering rate difference is excluded. But in realistic case, $\pi_{44} \neq 0$, the piezoresistance for these two situations differs by $\pi_{44}T$. π_{44} is determined by τ_{xy} and then ε_{xy}. Therefore, we can come to a conclusion that although the ε_{xy}-type shear strain does not shift the band edges, it warps the bands to induce extra mass change. Pseudopotential band structure calculations (Fischetti and Laux, 1996; Uchida et al, 2005) also show that [110] uniaxial stress warps the [001] valleys. Tensile stress tends to reduce the effective mass along the [110] direction, and compressive stress tends to increase it. Hensel, Hasegawa, and Nakayama (Hensel et al, 1965) considered the symmetry reduction by the $\langle 110 \rangle$ uniaxial stress, and came to an expression for the band energy of [001] Si conduction valley with an orthorhombic strain e_{xy},

$$E(\mathbf{k}) = \frac{\hbar^2 (k_z - k_0)^2}{2m_l} + \frac{\hbar^2 (k_x^2 + k_y^2)}{2m_t} + \alpha \hbar^2 e_{xy} k_x k_y, \tag{6.129}$$

where $\alpha = (86.8 \pm 5.0)/m_0$ determined experimentally. With the distortion by the last term, the energy surface is no longer an ellipsoid of revolution, as is schematically shown in the Fig. 6.10 with uniaxial tension along [110].

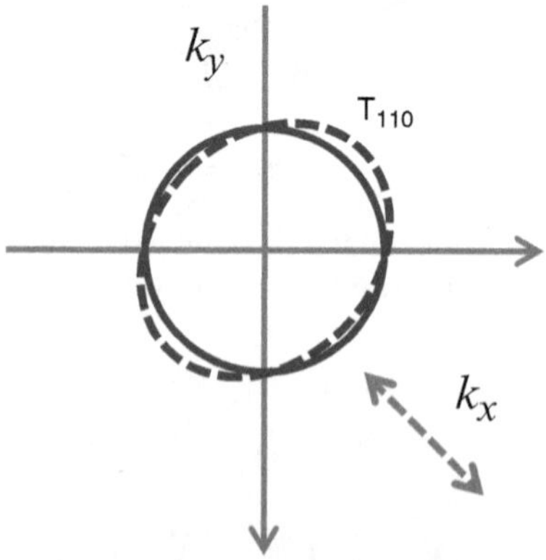

Fig. 6.10. 2D x-y energy contour for the out-of-plane Δ valleys of Si conduction band under [110] uniaxial tensile stress. The in-plane band is warped, and along the stress direction, the effective mass is reduced

With this distortion, the effective mass along [110] follows $m_{110} = m_t/(1 + \alpha e_{xy} m_t)$. The effective mass decreases with the tensile stress as shown in Fig. 6.11. The π_{44} is given by

$$\pi_{44} = \frac{-\alpha m_l s_{44}}{1 + 2K}, \tag{6.130}$$

where s_{44} is the stiffness constant for Si. Using the parameters for Si, π_{44} is evaluated to be $-9.4 \times 10^{-11}\mathrm{Pa}^{-1}$. Because the different origin of π_{11}, π_{12} from π_{44}, their temperature dependence is different. π_{11} and π_{12} are inversely proportional to the absolute temperature, while π_{44} does not have a strong dependence on temperature.

Using the shear deformation potential given in Table 4.8, and $K = 0.92/0.19 = 4.84$, the π-coefficients calculated for n-Si are: $\pi_{11} = -69.1$, $\pi_{12} = 33.0$. Comparing to Table. 6.3, we can see that there is large discrepancy between theory and experiments. This is because we only assume that the resistivity change is caused by the conductivity mass shift only, and the relaxation time is a constant for all valleys. But in fact, the valley splitting caused by stress alters the DOS for intervalley scattering and also alters the intravalley scattering by repopulating electrons to higher energy with respect to the valley edges. If we assume a linear response of strain effect, we may separate the contributions of electron repopulation, valley warping, and scattering reduction to Si conduction band piezoresistance. The result is shown in Fig. 6.12. where the scattering reduction contribution is obtained by subtracting the contribution of repopulation and warping from the total piezoresistance. In bulk case shown in this figure, scattering contribution is not significant; however, it is expected to play an important role for relatively large valley splitting.

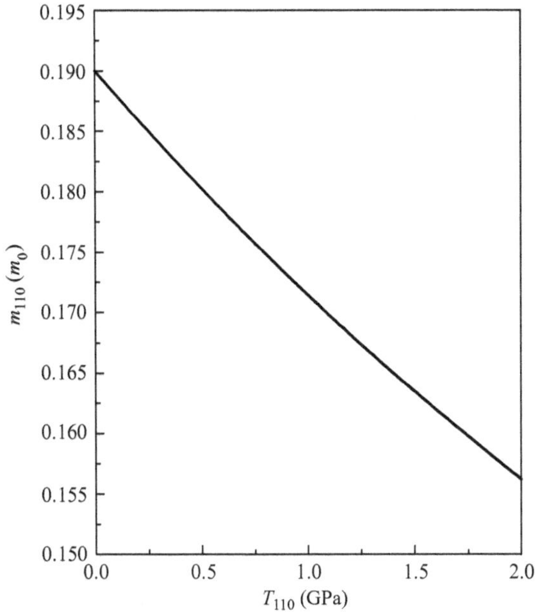

Fig. 6.11. Electron effective mass along [110] of the Δ_2 valley vs. [110] uniaxial tensile stress

6.7.3 Hole Transport

In contrast to the conduction bands where band warping due to strain is small and often neglected, it is nevertheless the predominant factor of piezoresistance in hole transport, especially for semiconductors with relatively large optical phonon energy such as Si. This is because of the degenerate nature of the valence band structures where strain can mix different states and consequently strong warping occurs, as shown in Fig. 6.13. With relatively small stress, strain splitting is much smaller than the optical phonon energy, and strain-induced scattering rate change is very small. We can see this point by looking at the scattering rate change of a hole located at the Γ point of the top valence band. As discussed in Sect. 4.13.3, the split valence bands recover their curvature away from the warped zone around the Γ point, i.e., the top band recovers its HH-like and the second recovers LH-like character. Then we may approximate the DOS of the strained bands using its unstrained DOS masses, with the only difference being that now the HH and LH bands are split by energy ΔE. Because of the proportionality to the DOS, the scattering rate for holes residing at the Γ point at the top valence band can be estimated by

$$1/\tau' \propto \frac{C^2 n(\omega_o)\pi}{\hbar^{3/2}} \left(\frac{1}{\sqrt{\hbar\omega_o}} \left(\frac{2m_D^{hh}}{\hbar} \right)^{3/2} + \frac{\sqrt{\hbar\omega_o - \Delta E}}{\hbar\omega_o} \left(\frac{2m_D^{lh}}{\hbar} \right)^{3/2} \right). \tag{6.131}$$

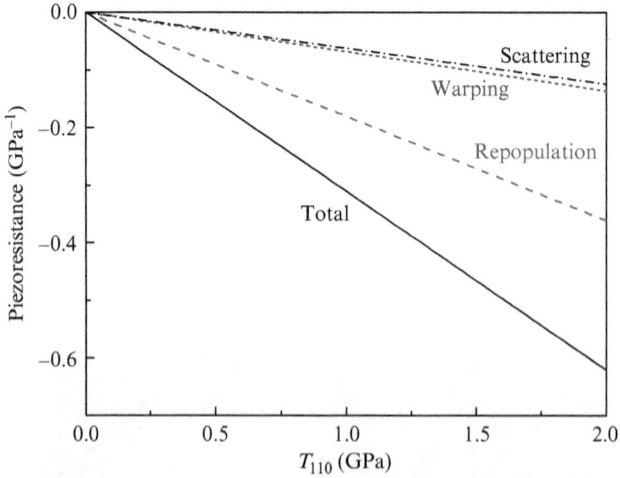

Fig. 6.12. Piezoresistance of Si conduction band, and its components, as functions of [110] uniaxial tensile stress

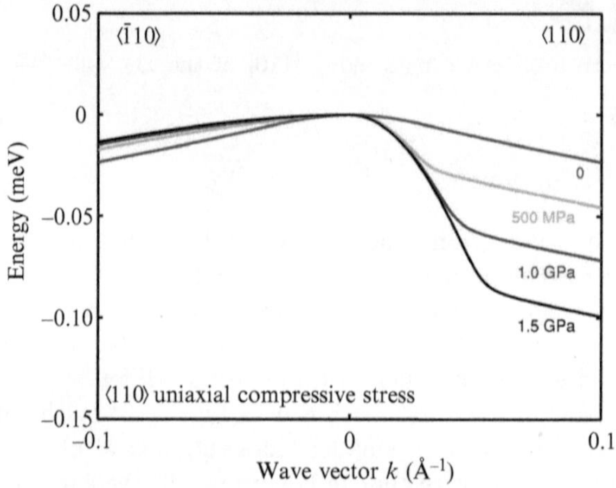

Fig. 6.13. Band structures of Si top valence band under four different ⟨110⟩ uniaxial compressive stress. With increasing stress, the warped region increases

This equation only differs from the unstrained case in that only the scattering contribution from the LH band is reduced due to the strain splitting, which comes into the second term. We can see that when $\hbar\omega_o \gg \Delta E$, this difference is small. For Si, $\hbar\omega_o$~62 meV, and assuming $\Delta E = 20$ meV, which can be induced by ~400 MPa uniaxial stress, $\Delta\tau/\tau \sim 2.5\%$. The scattering-resulted piezoresistance is comparable to the π-coefficients of π_{11} and π_{22} shown in Table 6.3 if we assume a uniaxial stress along ⟨100⟩ where the band warping is negligible. The π_{44} from Table 6.3 is much higher than what would result from the scattering rate change. This indicates that π_{44} can be considered to be solely resulted from the band warping-induced conductivity mass change, which is well illustrated in Fig. 4.33. It is hard to quantify analytically the hole piezoresistance, since the band warping is really complicated by the strong interaction between the degenerate bands, and thus the band parameters such as effective mass and DOS are hard to quantify and need advanced calculation. However, in comparison between the $\langle 110\rangle$ uniaxial stress and the in-plane biaxial stress, it is apparent that the uniaxial stress along $\langle 110\rangle$ can bring much more significant piezoresistive effect. The biaxial stress (or uniaxial stress along along $\langle 100\rangle$) is just too symmetrical to significantly deform the symmetry of the crystal.

6.7.4 Strain on Surface Roughness Scattering

Strain shifts energy levels and changes the bond configuration and thus affects the interface states. However, there is no consensus up to date whether strain affects the surface roughness or not. It seems quite reasonable to assume that strain has little effect on surface roughness, especially strain generated in bulk semiconductors or semiconductor devices is at most several percent. Xie et al. (Xie et al, 1994) experimentally demonstrated this point on SiGe film grown on Si and showed that only for compressive strain at >1.4% did the rms height have noticeable increase. However, experiments on strain-enhanced electron mobility in Si n-channel MOSFETs are found difficult to be explained by electron transfer and phonon scattering reduction only. Extensive strain experiments show that the mobility enhancement in the Si n-channel MOSFETs can reach 80% or above (Currie et al, 2001; Xiang et al, 2003; Takagi et al, 2004; Fiorenza et al, 2004). However, electron transfer effect can only account for ~10% in modern electron devices where the electric field is strong at surface. The effect of phonon scattering suppression is also limited. Takagi (Takagi et al, 1996) had to make an a priori adjustment to the intervalley electron–phonon coupling strength in the 2DEG to reach the consistency between the theory and experiments. To render this puzzling discrepancy, two proposals have been brought forward. One was by Hadjisavvas et al. (Hadjisavvas et al, 2007) from the first-principle calculations (Evans et al, 2005) and tried to interpret the interface roughness from the atomic-level. The other was proposed by Fischetti et al. (Fischetti et al, 2002) who suggested that one has to assume the surface roughness reduced by in-plane tensile stress.

From Hadjisavvas et al., the interface roughness was interpreted as composed of two kinds of defects: the suboxide bond (SB) in the oxide side and the oxygen protrusion (OP) in the Si side, as shown in Fig. 6.14. The SB corresponds to a missing oxygen atom in the Si oxide, and OP is an extra oxygen in the Si. The defects then create a rough interface with potential perturbation. The SB scattering potential is repulsive because it corresponds to a missing atom, whereas the OP scattering potential is attractive as it corresponds to

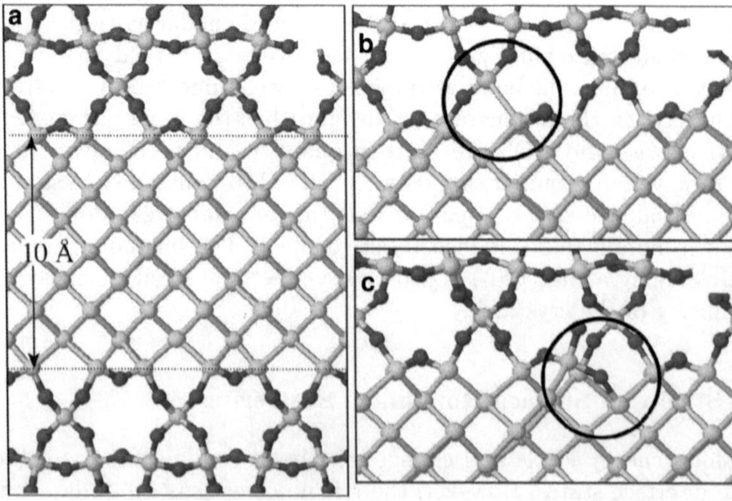

Fig. 6.14. (**a**) Ideal Si/SiO_2 interface; (**b**) Diagram of a suboxide bond defect; (**c**) Diagram of a oxygen protrusion defect

an extra atom. The calculations showed that uniaxial tensile strain led to a reduction of the scattering potential for both defects and has a significant effect on electron mobilities. The results showed good consistency with existing experiments. However, doubts about this theory may arise from two aspects. One is the interpretation of the surface roughness. According to the discussion of surface roughness scattering in Sect. 6.6.2, the surface roughness scattering can exist without interface defects if the interface has waviness in the z direction. The waviness alone can bring about a perturbative potential due to electric differences between the two materials comprising the interface. Thus, surface roughness and interface defects are two different concepts. Second, we have to also notice the significant difference between the common device samples people used in strain-Si experiments, which usually were surface MOSFETs, and the samples the authors used in the calculation, which was a $SiO_2/Si/SiO_2$ structure with the Si layer thickness of only 10 Å. Under this extreme condition, Si film could behave significantly differently from the common surface Si devices such as MOSFETs. If the physics revealed is true, we shall see continuous mobility enhancement with strain at large effective field and low temperature, where the electrons are all brought to the Δ_2 valley by large confinement splitting and sharp Fermi level. However, existing strain experiments are not consistent in low-temperature mobility enhancement. Some show enhancement saturation with decreasing temperature, while some show continuous enhancement. Some of the strain-enhanced electron mobility experimental examples will be shown in the next chapter.

To confirm the assumption by Fischetti et al., Bonno et al. (Bonno et al, 2008) examined the biaxial strained Si/SiO_2 interface by means of atomic force microscopy (AFM). In their experiments, plain Si wafers and Si film grown on SiGe substrate were individually investigated. Results showed that the interface roughness of the 0.8% biaxial strained sample had significantly reduced roughness amplitude. Calculations for mobility with the reduced surface roughness gave good consistency with the earlier experiments of electron mobility enhancement by biaxial strain. Although this result was encouraging, a flaw existed in that in the experiment, different samples were compared so that they may have different surface conditions even before applying strain. Recently, Thompson et al. investigated the interface change on a single Si wafer by controlling the stress applied on the wafer externally and thus a precise comparison of the interface roughness can be made with and without strain. The AFM power spectral density (PSD) curves they obtained for uniaxial tensile and compressive stress are shown in Fig. 6.15a and b, respectively. The low-frequency part of the PSD curve corresponds to macroscopic waviness of the surface, while the high-frequency part is depicting the microscopic roughness of the interface, which is responsible for the surface roughness scattering. It is obvious that the both types of stress do not change the surface in the macroscopic scale, which is consistent to our common sense, while the tensile stress reduces the microscopic roughness amplitude and compressive stress does not. The comparison of the converted roughness amplitude under zero stress and 200-MPa uniaxial tensile stress is made in Fig. 6.16 where it is more clearly seen that the roughness amplitude is greatly reduced by the uniaxial tensile stress. Therefore, in strain-enhanced electron mobility in Si n-channel MOSFETs by tensile strain, contribution from the surface roughness reduction plays a substantial role.

Other than the change to the surface roughness, strain also affects the surface roughness scattering by altering the energy band DOS and shifting the carrier distribution along the z-axis. The uniaxial tensile and compressive stress enhance the electron and hole mobility by electron transfer and band warping, respectively, but at the same time, carriers are brought closer to the oxide interface, thus they experience stronger surface potential perturbation and are more strongly scattered. However, compared to the surface roughness change induced by strain, these effects are not substantial.

6.7.5 Transport in High Effective Field

Strictly, mobility is only a quantity to describe the electronic transport in low electric field regime. At high fields, carriers experience velocity saturation due to enhanced phonon scattering and nonparabolicity of the bands at relatively high energy. At this section we discuss the strain effects on electronic transport at high electric fields, especially for MOSFETs. There is no doubt that strain-enhanced mobility can enhance the drain current in the linear region in an $I_d - V_d$ chart where the lateral field is low, since in this region, the linear relation between the carrier drift velocity v_d and the (lateral) electric field F holds: $v_d = \mu F$. Under high lateral fields, velocity saturation occurs, due

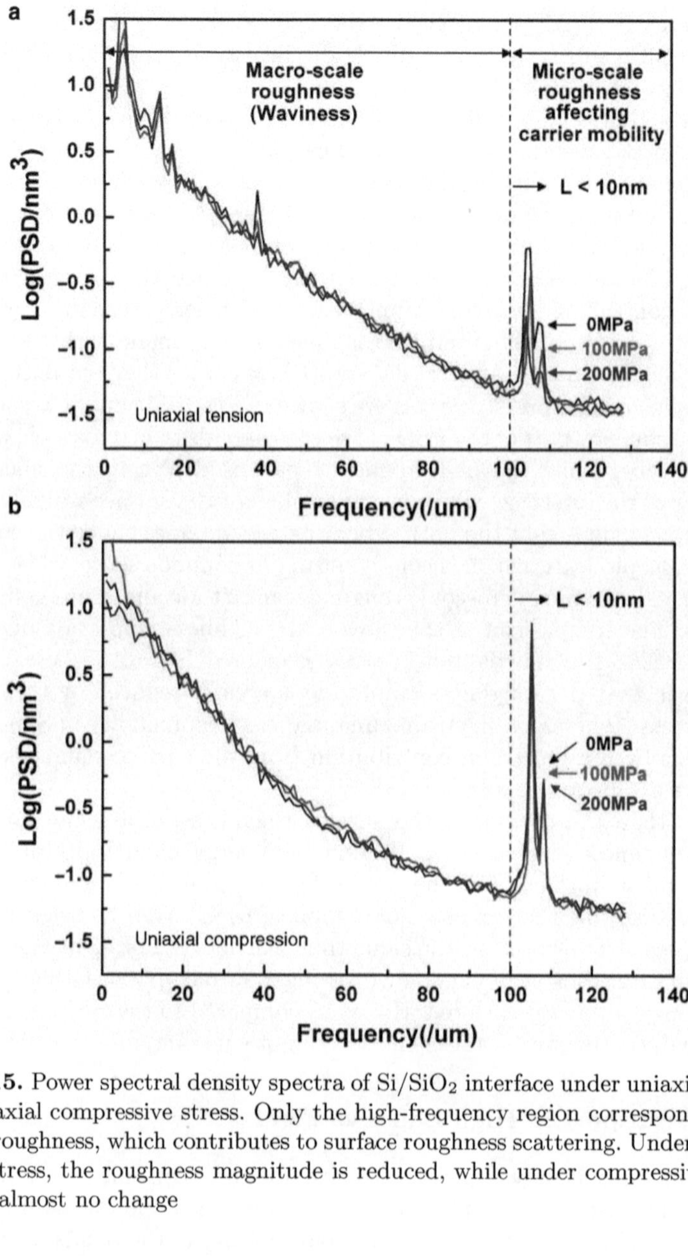

Fig. 6.15. Power spectral density spectra of Si/SiO_2 interface under uniaxial tensile and uniaxial compressive stress. Only the high-frequency region corresponds to the surface roughness, which contributes to surface roughness scattering. Under uniaxial tensile stress, the roughness magnitude is reduced, while under compressive stress, there is almost no change

to the reason we mentioned. The mobility at such a situation is no longer a constant, but a function of the lateral field, and may not be a good quantity to describe the carrier transport.

The saturation velocity can be simply estimated as follows. First we assume that in the lattice scattering processes, optical phonon dominates. When

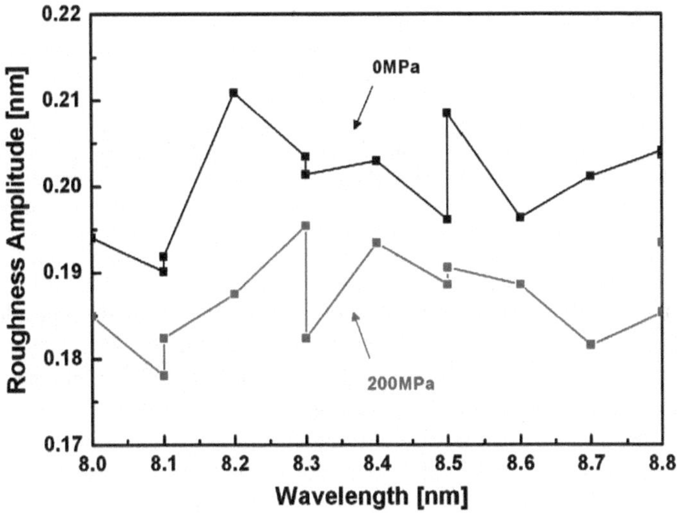

Fig. 6.16. Roughness amplitude as a function of wavelength. Under 200 MPa uniaxial tensile stress, the roughness amplitude is significantly reduced

the electric field increases, the average energy of electrons increases, and a large amount of hot electrons emit optical phonons and experience energy loss, which eventually leads to velocity saturation. Suppose an electron with high enough energy emits a phonon in a period of time τ_o, which is the relaxation time of interaction between electrons and optical phonons, then we have

$$\left(\frac{\mathrm{d}E}{\mathrm{d}t}\right)_{\mathrm{c}} = -\frac{\hbar\omega_o}{\tau_o}, \tag{6.132}$$

where the subscript c indicates collision. On the other hand, the electron keeps gaining energy from the fields. The rate of energy gain is

$$\left(\frac{\mathrm{d}E}{\mathrm{d}t}\right)_{F} = ev_d F = e\mu F^2, \tag{6.133}$$

where μ is the electron mobility (Here we still assume the relation $v_d = \mu F$, but μ is no longer a constant, and may vary with electric field F). At a steady state, the energy gain and loss are equal, and hence we have

$$\frac{\hbar\omega_o}{\tau_o} = e\mu F^2 = \frac{e^2\tau_o}{m^*}F^2. \tag{6.134}$$

Hence the dependence of relaxation time τ_o on the field F is

$$\tau_o = \frac{(m^*\hbar\omega_o)^{1/2}}{eF}. \tag{6.135}$$

The mobility then is

$$\mu = \frac{e\tau_o}{m^*} = \left(\frac{\hbar\omega_o}{m^*}\right)^{1/2}\frac{1}{F}. \qquad (6.136)$$

We can see that the mobility is inversely proportional to electric field at high electric field limit. The drift velocity of the hot electrons hence can be obtained as

$$v_d = \mu F = \left(\frac{\hbar\omega_o}{m^*}\right)^{1/2}. \qquad (6.137)$$

Note that the drift velocity at high field now is a constant, independent of the electric field. The typical lateral field where the velocity saturation occurs is $F > 10^4$ V/cm. For short-channel MOSFETs, this condition is easily satisfied in the channel, especially at the drain side. For example for a MOSFET with channel length of ~100 nm, the average channel lateral field is 0.1 MV/cm with 1 V drain bias. Particularly, the lateral field is not constant along the channel. The field at the drain side is much larger than at the source side (Taur and Ning, 1998).

When velocity saturation occurs, does strain continue to be effective to boost the saturation current I_{dsat} as they does to the linear current I_{dlin}? By inspecting (6.137), we can see that the saturation velocity can be enhanced by reducing effective mass m^*, but how about the Si nMOSFETs where a major contribution of mobility enhancement comes from the scattering reduction? Does the low-field mobility have no significant meaning at all when velocity saturation occurs? Or can we find a relation between the saturation current and the low-field mobility?

The answer to these questions can be sought from the scattering theory in MOSFETs proposed by Lundstrom (Lundstrom, 1997; Ren and Lundstrom, 2000; Lundstrom, 2001), who found that the saturation current is ultimately determined by the injection velocity and mobility at the source side where the electric field is low, rather than the high electric field region, for instance, the drain side. We briefly introduce the theory in the following under a non-degenerate semiclassical limit.

We can divide a MOSFET transistor into three regions, the source, the channel, and the drain, separated by barriers, as shown in Fig. 6.17. Suppose

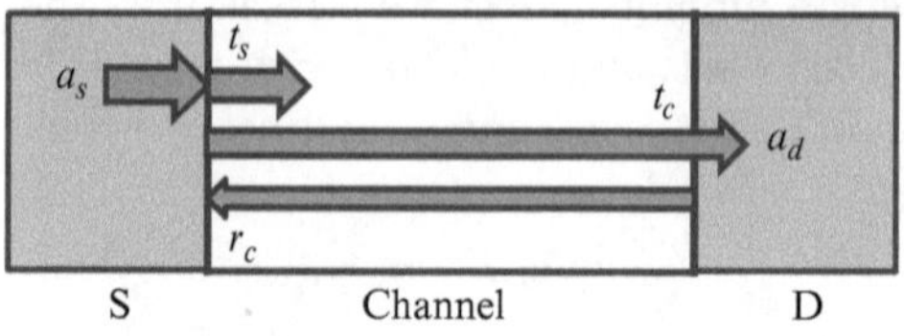

Fig. 6.17. A three-section diagram for a transistor. A flux of electrons, $a_s t_s$, gets injected from the source into the channel. A fraction t_c of the flux get absorbed by the drain, and a fraction r_c of the flux gets scattered back and reenters the source

a flux of electrons $a_s t_s$ get injected from the source into the channel where t_s is the transmission rate of the electrons at the source/channel barrier. Among the injected electron flux, a fraction t_c transmits across the channel and collected by the drain, and the rest fraction $r_c = 1 - t_c$ back scatters and reenters the source. The steady-state flux that enters the drain is

$$a_d = a_s t_s t_c, \tag{6.138}$$

if no backscattering from the drain is assumed for saturation condition. Normally when the drain bias is not high enough, thermal velocity is much higher than the drift velocity, and electrons from the drain can also be injected into the channel and enter the source. The approximation that no backscattering from the drain applies only when the drain bias is high enough and the drain current is in the saturation region. The electron density at the source side of the channel is

$$n(0,y) = \left(\frac{a_s t_s + a_s t_s r_c}{v_{inj}}\right) = \frac{a_s t_s (1 + r_c)}{v_{inj}}, \tag{6.139}$$

where v_{inj} is the injection velocity, $v_{inj} = \sqrt{2k_BT/\pi m^*}$ (please note the difference between the injection velocity, which is unidirectional, and the thermal velocity, $v_{th} = \sqrt{3k_BT/m^*}$, and Richardson velocity, $v_R = \sqrt{k_BT/2\pi m^*}$, which is related to thermionic emission and related to the injection velocity by $v_{inj} = 2v_R$), and y is channel depth, and $n(0,y)$ is related to the inversion charge density at the beginning of the channel N_s by

$$N_s = \int_0^{y_{max}} n(0,y)\mathrm{d}y = \frac{C_{\text{eff}}}{e}(V_g - V_t), \tag{6.140}$$

where C_{eff} is the effective gate capacitance, V_g is the gate voltage, and V_t is the threshold voltage. By combining (6.138) and (6.139), we have

$$a_d = n(0,y) v_{\text{inj}} \frac{t_c}{1 + r_c}. \tag{6.141}$$

By integrating the above equation and using the relation in (6.140), we obtain

$$I_{\text{dsat}} = eW \int_0^{y_{max}} a_d \mathrm{d}y = C_{\text{eff}} W v_{\text{inj}} (V_g - V_t) \left(\frac{1 - r_c}{1 + r_c}\right), \tag{6.142}$$

where W is the device width. This is the key result obtained by Lundstrom who linked the drain current to the backscattering rate, r_c. The merit of this equation is that it does not contain the mobility, which may not have significance at saturation situation. The backscattering rate r_c in (6.142) plays an essential role in determining the drive current. At the ballistic limit, $r_c = 0$, and (6.142) gives the upper limit of the drain current for a given gate voltage.

Next, there exists an observation that if a carrier travels a distance down a potential drop along the channel, it is unlikely to reemerge even if it backscatters. Hence the backscattering rate is largely determined by the transport properties at the source side where the lateral field is weak. For a channel without lateral field, the backscattering rate is

$$r_{c0} = \frac{L}{L+\lambda}, \tag{6.143}$$

where L is the channel length and λ is the electron mean free path. When the lateral field is present in the channel, the channel length is replaced by a backscattering critical length l, over which the potential drops by k_BT/q, and then $l = k_BT/qF(0^+)$, where $F(0^+)$ is the lateral field at the source side of the channel. The critical length, $l = k_BT/qF$, is called the "k_BT" layer, defined by Berz (Berz, 1985), indicating a length scale above which the backscattering vanishes. Typically for a transistor under high drain bias, l is a small portion of the channel length. The mean free path, λ, is related to the low-field mobility, μ_0, by (Tanaka and Lundstrom, 1994)

$$\lambda = \frac{2\mu_0 k_B T}{e v_{\mathrm{inj}}}. \tag{6.144}$$

Hence the backscattering rate is given by

$$r_c = \frac{1}{1 + 2\mu_0 F(0^+)/v_{\mathrm{inj}}}. \tag{6.145}$$

Let us define a parameter, ballistic efficiency B_{sat}, and

$$B_{\mathrm{sat}} = \frac{1-r_c}{1+r_c} = \frac{\lambda/2l}{1+\lambda/2l}. \tag{6.146}$$

When $r_c = 0$, $B_{\mathrm{sat}} = 1$, representing a fully ballistic transport. For MOSFET devices, the quantity B_{sat} is a ratio of drive current to the fully ballistic current, which is the upper limit of the drive current for a given gate voltage. The mobility dependence of the saturation current is obtained by take the derivative of I_{dsat} to μ. Using Eqs. (6.142) and (6.144), the result is obtained as (Lundstrom, 2001)

$$\frac{1}{I_{\mathrm{dsat}}}\frac{\partial I_{\mathrm{dsat}}}{\partial \mu} = \frac{1}{\mu}(1 - B_{\mathrm{sat}}), \tag{6.147}$$

Hence

$$\frac{\partial I_{\mathrm{dsat}}}{I_{\mathrm{dsat}}} = \frac{\partial \mu}{\mu}(1 - B_{\mathrm{sat}}). \tag{6.148}$$

For current technology, $B_{\mathrm{sat}} \sim 0.5$ (Assad et al, 1999),(Lochtefeld and Antoniadis, 2001), so that the percentage change of the saturation current is about half of that of mobility.

This theory can be applied to strain-enhancement device performance and explains the saturation current enhancement by strain. However, we need to notice that in the above derivation, the injection velocity v_{inj} is assumed unchanged, and I_{dsat} enhancement is due to the mean free path increase by enhanced mobility. Whereas the injection velocity does change with strain if strain alters the effective mass. For strain-enhancement I_{dsat} in Si pMOSFETs, increased injection velocity by reduced effective hole mass also contributes, which will be discussed next.

6.7.6 Strain Effects in Ballistic Transport Regime

At the drain side of the channel, carriers are more easily to gain high energy at high electric fields, and statistically, there is always an amount of carriers that exit the drain without sufficient heat exchange with the lattice, and have velocity higher than the saturation velocity. This phenomenon is called velocity overshoot. If velocity overshoot takes place only in the high field drain side of the channel, according to the discussion in the lase subsection, it does not substantially affect the drive current, because the current in the drain is determined by the injection velocity and the low field mobility at the source side. For short channel devices, situations are a little different.

Let us inspect the (6.146). It can be rewritten as

$$B_{\text{sat}} = \frac{I_{\text{dsat}}}{I_{\text{dsat,ballistic}}} \tag{6.149}$$

where $I_{\text{dsat,ballistic}}$ is the upper limit of the drive current. There are two ways to increase the ballistic efficiency and thus enhance the drive current. One is by increasing the mean free path λ, which is helped by strain-enhanced mobility; the other is to reduce the critical length l. For short-channel transistors, the electric field at the source end of the channel is enhanced, which reduced l, and thus ballistic efficiency increases. The ballistic efficiency increase in short-channel devices may also be understood as follows. For short-channel devices, the time for a carrier staying in the channel is greatly reduced and the probability being scattered also decreases. The ballistic transport is an extreme limit of velocity overshoot where carriers do not experience scattering at all in the channel. For ballistic transport, $B_{\text{sat}} = 1$ and the drive current does not depend on mobility and field. It only depends on the source side injection velocity.

Interpreting from the scattering theory presented in the last section, we can see that strain can have twofold effects: (1) improve the ballistic efficiency by enhancing the low field mobility and (2) increase the injection velocity by reducing the effective mass. With the trend of current device scaling, MOSFET device transport will eventually be dominated by the ballistic transport. How far shall we go into device scaling to have ballistic transport? For room temperature operations, simulations show that the carrier transport will become near ballistic if the size is scaled to around 30 nm (Frank et al, 1992). Although for nonballistic transport, mobility enhancement can effectively improve the device performance, but for ballistic carriers, the low-field mobility has no significant meaning. This is obvious from (6.148), that when $B_{\text{sat}} = 1$, $\delta I_{\text{dsat}}/I_{\text{dsat}} = 0$ regardless of mobility. Let aside the device geometries, the ballistic current is only determined by the injection velocity. If we say that the ballistic current is the upper limit of the drive current, strain may be used to raise this limit.

Under ballistic limit, (6.142) is nothing but $I_{\text{dsat}} = Q_s v_{\text{inj}} = eN_s v_{\text{inj}}$. A general form for the injection velocity assuming only one subband is occupied is (Lundstrom and Ren, 2002)

$$v_{inj} = \sqrt{\frac{2k_BT}{\pi m^*}} \left\{ \frac{\mathcal{F}(\eta)}{\ln[1 + \exp(\eta)]} \right\}, \tag{6.150}$$

where $\eta = (E_F - E_1)/k_BT$, E_1 is the subband edge energy, and $\mathcal{F}(x)$ is the Fermi–Dirac function. Under nondegenerate semiclassical limit, the term in the bracket in (6.150) approaches 1, and v_{inj} recovers the expressions we used in the last subsection. At degenerate limit ($\eta \longrightarrow \infty$),

$$v_{\text{inj}} = \frac{4}{3\pi} v_F = \frac{4}{3\pi} \sqrt{\frac{4N_s}{m^* D_{2D}}}, \tag{6.151}$$

where v_{F} is the Fermi velocity and D_{2D} is the 2D DOS of the subband. Under a parabolic band approximation, we have

$$v_{\text{inj}} = \frac{4}{3\pi} \sqrt{\frac{4N_s \pi \hbar^2}{m_x m_D}}, \tag{6.152}$$

where m_x is the conductivity mass along the channel direction, and m_D is the 2D DOS mass. Therefore, for the nondegenerate case, the injection velocity is inversely proportional to the square root of the effective mass in the injection direction, and for degenerate case, the injection velocity not only depends on the effective mass along the injection direction, but also depends on the DOS of the subband. In the case of a strained Si nMOSFET, which is the easiest to discuss, the injection velocity is increased both by reduced conductivity mass and reduced DOS by further splitting of the Δ_2 and Δ_4 valleys by the advantageous strain, e.g., the uniaxial tensile strain. Currently the experimental data on ballistic Si MOSFETs are limited, but many simulations (Takagi, 2003; Bufler and Fichtner, 2003; Ferrier et al, 2006; Fortuna et al, 2006; Huet et al, 2007) show greatly enhanced drive current in ultrashort Si MOSFETs by strain.

Part III

Strain in Semiconductor Devices

7

Strain in Electron Devices

Today's strain-Si technology has been based on the physics of strain effects on bulk and low-dimensional band structures and carrier transport. Although the strain effects on bulk semiconductors have been studied since the 1950s, and the effects on Si MOSFETs have been discovered and studied since 1980s, strain has never really been adopted in mainstream Si CMOS technology until 2002s by Intel, because before then, the speed of the VLSI/ULSI chips could be boosted just by geometrical scaling, i.e., increasing the transistor density, which now numbers typically at $\sim 10^9$ in a single chip, by shrinking the transistor size. However, this comes with price. Short channel effects significantly degrade the device performance when the transistor size is too small, and the leakage current results in large power dissipation, and the traditional scaling reaches a bottleneck. Strain then as a "performance adder" is eventually adopted. Today, the chips with strained-Si devices (see Fig. 7.1) serve the three dominant markets of computer, communication and consumer electronics and highlight the wide success of this technology in not only high performance but also low cost markets.

7.1 STRAIN-SI TECHNOLOGY

Among the two approaches that induce strain to device channels, i.e., the wafer-based global biaxial strain and process-induced uniaxial strain, only the latter is applied in real production up to date. The global strain is much more complex in process and more expensive in cost, while the process-induced strain is easy to adopt. So in this section, we will focus on the technology of the process-induced uniaxial strain. The techniques in production to induce strain include high-stress tensile and compressive SiN capping layers and selective epitaxial SiGe deposited in recessed/raised source and drain (S/D). These techniques induce uniaxial tensile stress to the nMOSFETs and compressive stress to the pMOSFETs. Future generations will bring the SiGe closer to the channel and increase the Ge concentration and increase the stress

Y. Sun et al., *Strain Effect in Semiconductors: Theory and Device Applications*, DOI 10.1007/978-1-4419-0552-9_7,

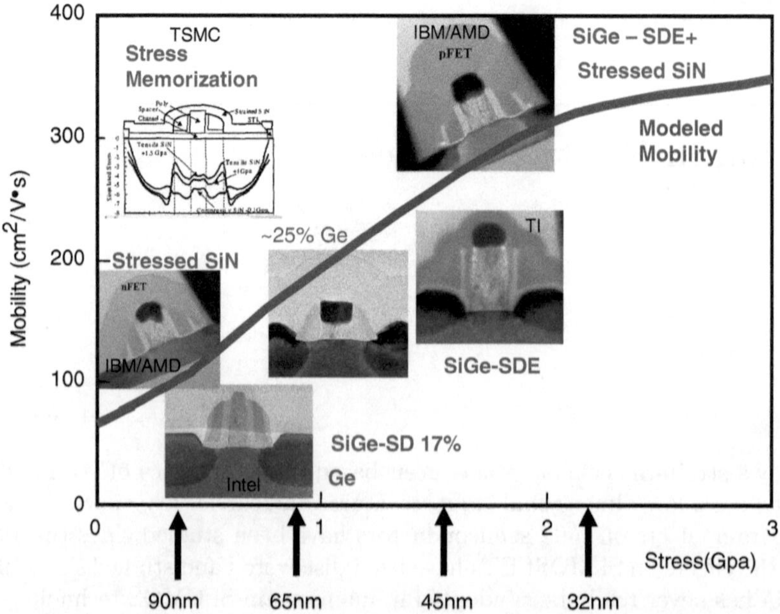

Fig. 7.1. Strain technology for pMOSFETs in different Si technology nodes. The line in the figure is the modeled hole mobility as a function of longitudinal channel strain

in the capping layers, which is approaching 3 GPa. Other possible future techniques for process stress or mobility enhancement may include tensile shallow trench oxide, embedded SiC for nMOSFETs, and hybrid oriented (110) wafers. Together, these techniques are expected to lead to channel stresses of 1–2 GPa approaching the stress level of wafer-based biaxial stress.

Strained-Si technology is completely compatible with the traditional Si CMOS technology. Only slight modifications to a standard CMOS logic technology process flow are needed to insert the longitudinal compressive and tensile stress into the p- and n-MOSFETs, respectively, as shown in Fig. 7.2, where the left panel is a typical 130 to 45-nm front-end CMOS process flow. For strained Si pMOSFETs, a Si recess etch is inserted post spacer formation and followed by selective epitaxial chemical vapor deposition (CVD) of SiGe in the S/D region. The uniaxial compressive stress is induced into the channel by the lattice mismatch between Si and SiGe. The Si lattice constant is 5.43 Å, and that for $Si_{1-x}Ge_x$ is $(5.43 + 0.20x + 0.027x^2)$ Å. Germanium concentration of $17 - 35\%$ is generally used, which causes the smaller lattice constant Si channel to be under compressive stress, whose magnitude also depends on the spacing of the SiGe epitaxy to the channel in addition to the Ge concentration. A 30-nm thick recessed SiGe at the S/D location results in a 250-MPa stress in the middle of the channel, while the same SiGe layer when present at

• Typical CMOS Process Flow	Strained-Si
Gate stack	
Gate patterning	
Offset spacers (optional)	
n-type LDD + halo/pocket implants	
p-type LDD + halo/pocket implants	
Spacer deposition and etch (STI)	
←	Recess and SiGe deposition (pMOSFETs)
n-type HDD implants	
p-type HDD implants	
RTA	
Silicide protection deposition and etch	
Silicidation	
←	High stress capping layer (nMOSFETs)

Fig. 7.2. Schematic of strained-Si process flow, as compared to a standard CMOS process flow

the S/D extension location stresses the channel to 900 MPa. As an example, a 3D finite element analysis shown in Fig. 7.3 demonstrates the cross-sectional stress distribution for a 45-nm gate length transistor after the SiGe deposition.

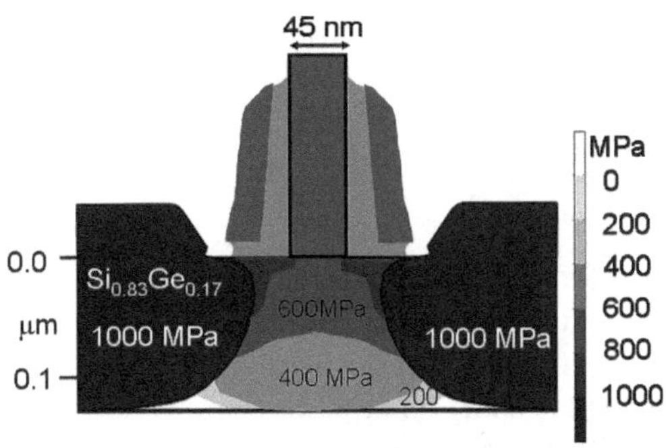

Fig. 7.3. Stress distribution of a pMOSFET with embedded $Si_{0.83}Ge_{0.17}$ S/D

Strain can also be engineered at other process steps like shallow trench isolation and silicide. Integrating the SiGe layer in the S/D extension location also has a key advantage to reduce the effect of the SiGe deposition thermal budget on S/D. Furthermore, by confining the SiGe to the S/D region and introducing it late in the process flow, integration challenges, such as misfit dislocations, yield, and increased self-heating due to the low thermal conductivity of SiGe, are reduced. Longitudinal uniaxial tensile stress is introduced into the nMOSFETs by engineering the stress and thickness of the Si nitride-capping layer. There are several techniques to nearly completely neutralize the capping layer strain on the p-type MOSFETs; one is the use of a Ge implant and masking layer. Another technique to relax the strain is to selectively remove the capping from the p-type transistors. Wafer-based strained-Si, e.g., Si channel grown on virtual SiGe substrate, also attracted a lot of attention in the past decade. However, compared to process-induced strain, the strain induced in channel is biaxial tensile strain, which has been proven not as efficient as the longitudinal tensile strain for nMOSFETs and even degrades the performance of the pMOSFETs; second, thermal annealing can cause strain relaxation in the strained-Si layer. In contrast, the unique advantage of the uniaxial stress Si process flow is that on the same wafer compressive stress is introduced into the p-type and tensile stress in the nMOSFETs to improve both the electron and hole mobility. Since the nitride capping layer is already present to support unlanded contacts, only a few new process steps are introduced at less than a 2% wafer cost increase.

With the maturity and progress of the strained-Si technology, it is more promising in new generation technology nodes. In Table 7.1, the performance enhancement by strain in Intel's 90 and 65-nm technology nodes is shown for comparison. The mobility, saturation current (I_{dsat}), and linear current

Table 7.1. CMOS device performance enhancement by strain in 90 and 65-nm technology nodes

	90 nm		65 nm	
	nMOS	pMOS	nMOS	pMOS
Mobility	20%	55%	35%	90%
I_{dsat}	10%	30%	18%	50%
I_{dlin}	10%	55%	18%	80%

(I_{dlin}) enhancement for the 65-nm technology node are significantly larger than those for the 90-nm technology node, for both nMOS and pMOS. It is reasonable to expect higher performance enhancement for shorter scale technology node, because short channel devices have a better chance to have higher and more uniform channel stress as can be imagined by inspecting Fig. 7.3. However, for short channel devices, the S/D series resistance may be comparable to the channel resistance, and thus the resistance change induced by strain for the whole device might not be as significant as is in the long channel devices, and this raises the issue of its limitation in aggressively scaled logic devices. We will come back to this point later.

Although the potential of wafer-based biaxial strain was realized in MOS devices since the early 90's in the last century, the implementation of it to the real product has been very late. But the research of the biaxial strained CMOS devices has never halted. In companies such as IBM, AMD and Freescale, wafer-based biaxial strain has been combined with the Si-on-insulator technology, and great advances have been obtained. In late 2008, IBM introduced 45-nm SOI foundry services, and claimed that the strained

SOI devices offered up to 30% higher performance and 40% lower power consumption, compared to the traditional bulk CMOS devices.

The wafer-based biaxial strain is introduced by growing Si channel on SiGe "virtual" substrate, i.e., the SiGe substrate is as a matter of fact also a layer epitaxially grown on a Si wafer rather than a single crystal, because single crystal SiGe is difficult to grow with an arbitrary Ge/Si ratio. Another practical reason is to accommodate the traditional Si technology where the process is very mature and the Si wafer diameter can be very large and then cost is reduced. As we will introduce later, one drawback of the global wafer-based strain is the strain relaxation-induced defects in the channel. In the early attempts, thick uniform SiGe buffer layers were deposited directly on the Si wafer, followed by the strained-Si channel epitaxy. The SiGe buffer layers underneath the strain-Si epitaxy were strongly strained by the Si substrate, and it often created threading dislocations or stacking faults, which would most likely penetrate into the channel. To solve this problem, fully relaxed SiGe buffer layer was introduced later. The techniques were to gradually increase the Ge concentration with layer thickness. To minimize the dislocation density, SiGe graded buffer layers were typically grown under temperature of 750°C or greater to nucleate the dislocations and enhance their glide velocity. Strain in the buffer layer was actually accommodated by the nucleation and glide of the threading dislocations. After the glide of the dislocations, the dislocations were literally annihilated. But very often, especially in high Ge content (e.g., >50%), the glide of the dislocation nucleation was impeded by the dislocation pileups. Once the dislocations were trapped by the dislocation pileups, they no longer contributed to the strain relaxation and hence new dislocations had to be generated. In 1998, Currie et al. (Currie et al, 1998) improved the old process to minimize surface roughness and dislocation pileup formation for high Ge content graded SiGe buffer layer growth. He introduced a chemical-mechanical polishing process when the Ge content reached 50% to remove the crosshatch surface roughness and to reduce the dislocation pileup formation. Then, compositional grading was grown on the polished surfaces. By applying this process, pileups and total dislocation densities were found to decrease. By these techniques, threading dislocation density at the order or lower than $10^6/\text{cm}^2$ can be readily obtained (Leitz et al, 2001) at almost any Ge content.

7.2 STRAINED ELECTRON DEVICES

7.2.1 Strained Planar MOSFETs

At first, strained-Si MOSFETs grown in laboratories were all based on SiGe virtual substrates. The first strained-Si n-channel MOSFET grown on SiGe virtual substrate to improve electron mobility was demonstrated by Welser et al. (Welser et al, 1992) in 1992 in IEDM conference. In their experiment, strained-Si channel was grown on relaxed $Si_{0.71}Ge_{0.29}$ layers, which were on top of the graded buffer layer. The mobility of the strained and unstrained n-MOSFETs is reproduced in Fig. 7.4, where the peak mobility at room temperature was enhanced by about 2.2 times. The first wafer-based strained-Si

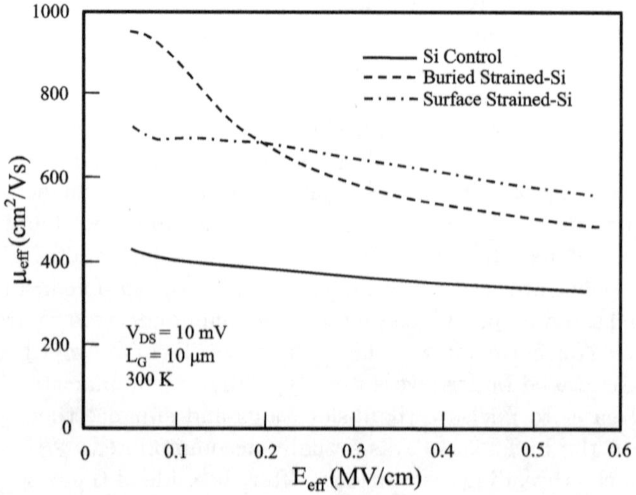

Fig. 7.4. Comparison of effective electron mobilities between the surface channel, buried-channel strained-Si nMOSFETs, and Si control nMOSFETs. From Welser (Welser et al, 1992)

p-channel MOSFET was first reported by Nayak et al. (Nayak et al, 1993) in 1993. The 130-nm strained-Si p-channel was grown on relaxed $Si_{0.75}Ge_{0.25}$ virtual substrate. He obtained both enhanced transconductance and hole mobility, as shown in Fig. 7.5, where strain enhanced the hole mobility by 1.5 times. The idea of using longitudinal uniaxial stress for improving device performance in MOSFETs was activated by the investigations by Ito et al. (Ito et al, 2000) and Shimizu et al. (Shimizu et al, 2001) in the late 1990's by introducing high-stress capping layers deposited on MOSFETs to induce channel stress, and by Gannavaram et al. (Gannavaram et al, 2000) who proposed SiGe in the source and drain area for higher boron activation and reduced external resistance, which furnished the technical means to employ uniaxial channel stress. In the 2002's IEDM conference, Thompson et al. (Thompson et al, 2002) from Intel demonstrated the first process-induced uniaxial stressed 90-nm Si p-channel MOSFETs, which showed prominent advantages over the wafer-based strained p-MOSFETs and heralded the extensive industrial application of the stained-Si technology. The strain in the channel was induced by SiGe deposition in the S/D region, which is compatible with the traditional planar CMOS technology. The mobility gain they obtained was much higher than that of wafer-based strained Si p-MOSFET, as shown in Fig. 7.6.

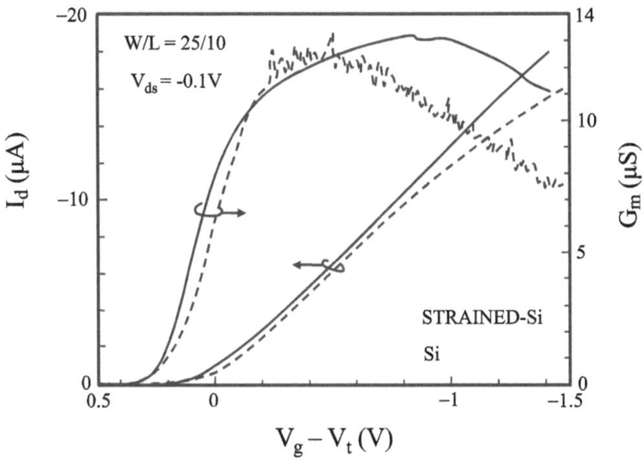

Fig. 7.5. Comparison of mobility and transconductance between strained and unstrained Si pMOSFETs. From Nayak (Nayak et al, 1993)

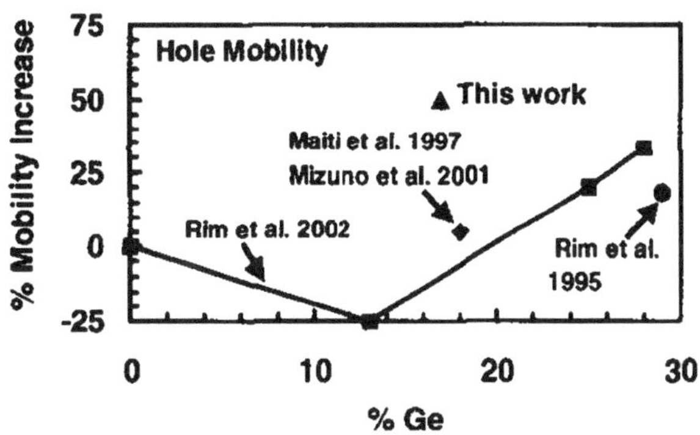

Fig. 7.6. Strain-induced hole mobility enhancement. Except the work labeled as "This work" where the Ge content is referred to that in the SiGe epitaxy, all the % Ge values in other cited works in this figures are the Ge content in the SiGe virtual substrates

7.2.2 Strained Si-on-Insulator (SOI)/SiGe-on-Insulator (SGOI) Devices

Physical limitations such as leakage current and power density when MOS devices come to a deep submicron scale impede the pace of performance enhancement. In search for innovations that can extend CMOS device scaling and performance trends, new device architectures such as Silicon-on-insulator (SOI) are also emerging and promising. IBM has been utilizing SOI in CMOS technology since 1998 (Lammers, 1998; Assaderaghi and Shahidi, 2000) (Lammers, 2001). SOI MOSFETs provide many advantages over bulk surface MOSFET devices. The substrate leakage is eliminated, and the smaller junction capacitance at the S/D reduces the capacitive load in CMOS circuits which speeds up the circuit speed and can also lower power consumption. In late 2008, IBM introduced 45-nm SOI foundry services and claimed that the SOI devices offered up to 30% higher performance and 40% lower power consumption, compared to the traditional bulk CMOS devices.

By merging strained Si with SOI, it may be possible to combine the benefits of strained Si with those of SOI. A typical strained-Si channel SOI structure is shown in Fig. 7.7 (Takagi et al, 2004). Strained SOI/(SGOI) wafers can be

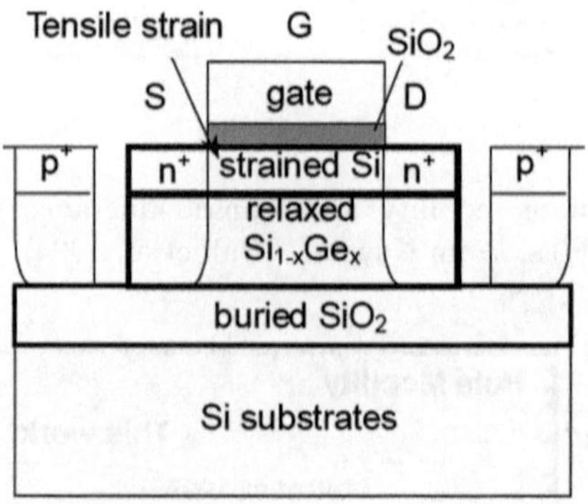

Fig. 7.7. A typical SOI MOSFET structure that combines strained-Si channel. From Takagi (Takagi et al, 2004)

formed using methods like wafer bonding (Taraschi et al, 2004). This combination brings many advantages to the device performance over the traditional bulk CMOS devices: (1) high mobility of strained Si combined with lower impurity scattering and lower gate field, (2) low junction capacitance and low junction leakage current, (3) suppression of short channel effects, (4) reduction in statistical variation of threshold voltage, V_t, (5) suppression of floating body effects due to hole current flow through the SiGe pn-junction in the source region, (6) no need of thick graded-SiGe buffer layers, (7) low dislocation density in relaxed SiGe due to relaxation through slip at the interface between SiGe and buried oxide. The benefits of strained Si and SOI structures are not expected to interfere with each other, since the enhancements occur in two different regions of a MOSFET, under the active region and in the channel region itself. Therefore, strained Si and SOI are potentially complementary feature enhancements (Takagi et al, 2001; Lee et al, 2002; Mizuno et al, 2003).

Structures employing strained-Si channel on SGOI have also been reported by IBM and others (Huang et al, 2002; Mishima et al, 2005). First, SiGe is epitaxially grown which is followed by layer transfer or thermal treatment of the layers to create SiGe-on-insulator structures. Then, a strained Si layer is epitaxially grown on the SGOI before CMOS device fabrication to fabricate strained-Si channel on SGOI. Shown in Fig. 7.8 is the drive current enhancement of the strained-Si on SGOI structures. Significant drive current enhancement is observed.

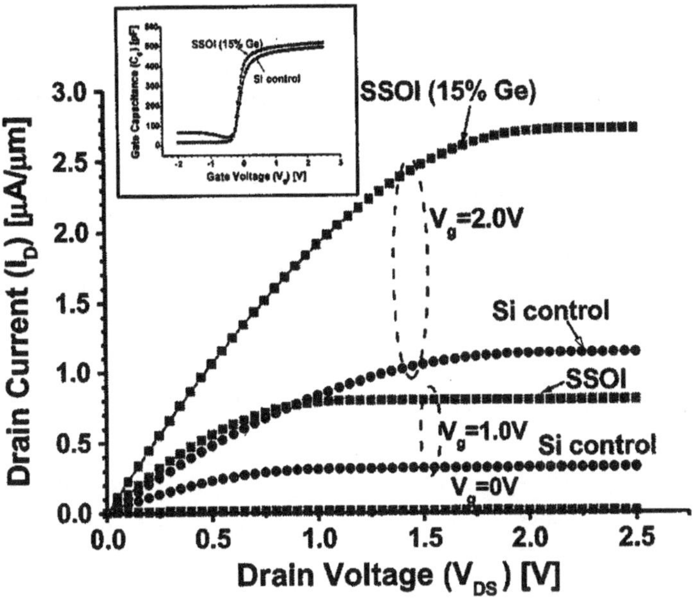

Fig. 7.8. Drive current enhancement of SSOI compared to Si control. From Huang (Huang et al, 2002)

IBM's initial strained SOI researches had been mainly concentrated on wafer-based strain, unlike the process-induced strain in the traditional bulk devices produced in companies such as Intel. Wafer-based strain is a global strain built into the entire active area of the device. Mobility enhancement is primarily for the n-channel SOI MOSFETs. Later, uniaxial stress was achieved in strained SOI through process modification, and thus both n and p-channel MOSFETs can be benefited. However, due to process issues, the biaxial strain methods may be the most scalable in SOI devices for its simplicity.

For the past few decades, geometrical scaling coupled with a series of material and structural enhancements necessitated by the scaling have enabled the exponential density and performance growth characterized by Moore's law. These enhancements have been as a rule complementary. As direct scaling benefits diminish, additive feature enhancements such as proper strain, high-k dielectric, metal gate, and SOI are expected to continue to drive performance gains.

7.2.3 Strain in Other Electron Devices

Strain has also found use in many other electron device applications, such as MOSFET memories and power MOSFETs. MOSFET memory is an important application of MOSFETs. The most important norm of memory technology is the switching time, which is proportional to the ratio of gate capacitance to drive current. Strain-enhanced drive current reduces the switching time and thus improves the memory performance. Power MOSFET, as shown in Fig. 7.9a, is a specific type of MOSFET designed to handle large power

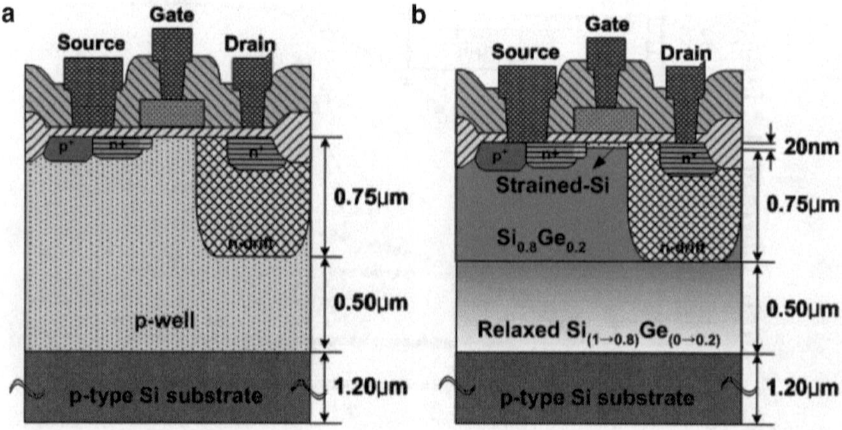

Fig. 7.9. Comparison of power MOSFET structures (**a**) without and (**b**) with process strain. From Cho (Cho et al, 2005)

including switching and gradually control the signal flow. Compared to the other power semiconductor devices such as IGBT and thyrister, the main advantages of power MOSFETs are high commutation speed and good efficiency at low voltages. The conventional Si channel high-voltage MOSFETs have many drwabacks such as a large electric field, a high on-resistance, a low transconductance, and a low current drivability. Strained-Si channel power MOSFET, as shown in Fig. 7.9b, has been reported to have great improvement of the drive current (20%) (see Fig. 7.10) and reduction of on-resistance (25%) (Cho et al, 2005). The transconductance has been improved by 28% and 52% at linear and saturation regions. In fact, because proper type of strain enhances the carrier mobility, any kind of electron devices that rely on carrier speed will be benefited from strain-Si engineering.

7.3 STRAIN ENHANCED MOBILITY

For wafer-based strained-Si n-MOSFETs, as shown in Fig. 7.11, the electron mobility enhancement increases monotonically as a function of biaxial strain in the Si up to 0.8% (20% Ge content in SiGe substrate) where the enhancement

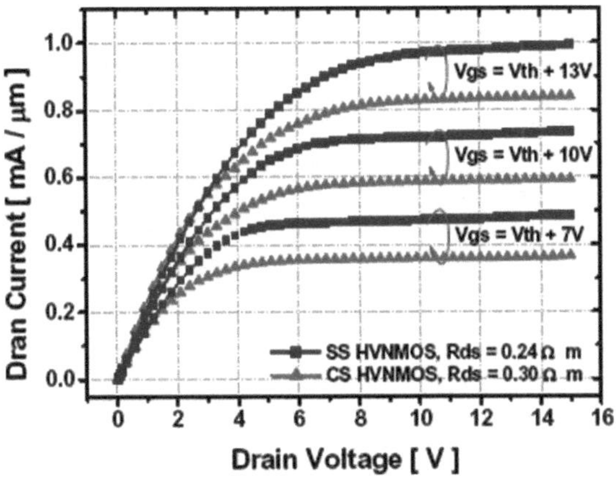

Fig. 7.10. The drive current gain of strained power MOSFET compared to the Si control device. From Cho (Cho et al, 2005)

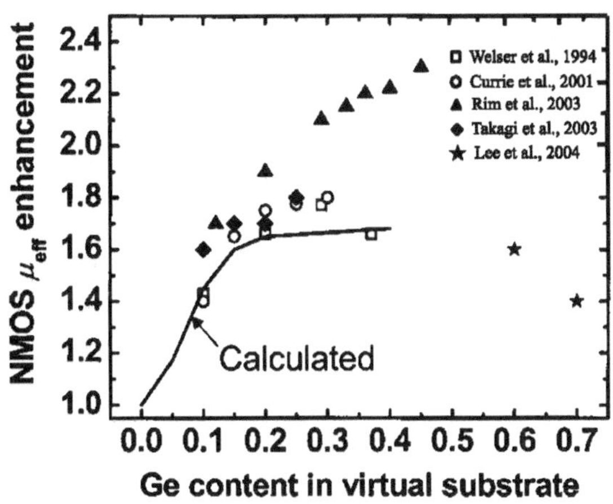

Fig. 7.11. Electron mobility gain with Ge fraction in the SiGe virtual substrate in the wafer-based global strain case. From Lee (Lee et al, 2005)

factor saturates at about 1.8 (Lee et al, 2005). At such large strain, the Si conduction valley splitting is so large to completely populate the electrons in the lower Δ_2 valley. Little enhancement in electron mobility can be achieved with further straining. However, this statement sometimes conflicts with some experiments. What produces this discrepancy is still controversial. This might be related to the strain-altered surface roughness as discussed in last chapter.

For splitting the Si conduction valleys, uniaxial and wafer-based biaxial tensile stress are both very effective, and thus the electron mobility enhancement by both types of strain is similar. The advantage of the uniaxial stress to warp the out-of-plane Δ_2 valleys and directly reduce the conductivity effective mass brings extra 10% enhancement. The drive current enhancement as a function of process-induced uniaxial stress is shown in Fig. 7.12. At the

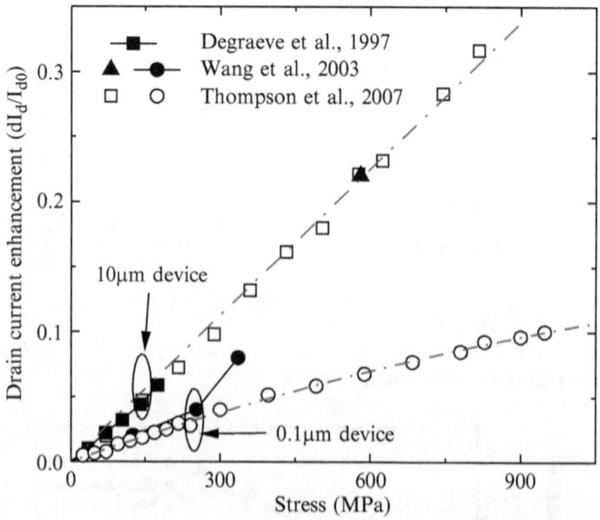

Fig. 7.12. Drive current enhancement in uniaxial stressed nMOSFETs. Short-channel devices show much smaller enhancement than long-channel devices

stress range shown in the figure, the drive current enhancement is linear to the uniaxial stress. However, a conspicuous trend is noticeable. The enhancement is much higher for long channel devices than for short channel devices. The same situation takes place for the saturation current as is evident in Fig. 7.13

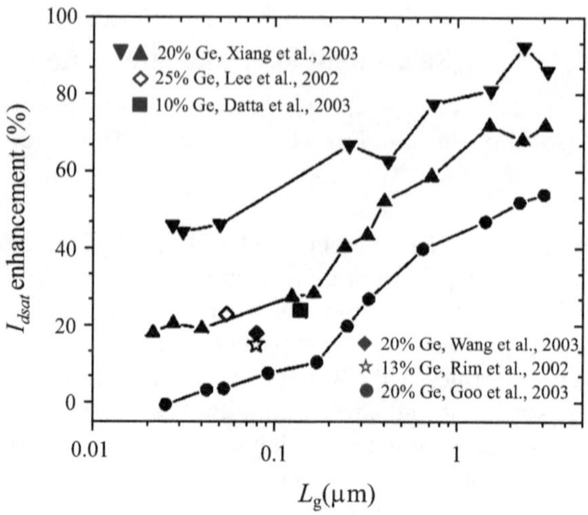

Fig. 7.13. Saturation current gain as a function of gate length for various Ge contents in SiGe virtual substrates in biaxial strain nMOSFETs

where I_{dsat} enhancement as a function of gate length is shown for various Ge content in the SiGe substrate for wafer-based global strain case. However, this does not indicate the degradation of the electron mobility enhancement with shrinking of the channel length, but is due to the S/D series resistance, which is not negligible compared to channel resistance and not reduced by strain. If we exclude the S/D resistance from the measurement, the electron mobility enhancement is almost identical for both long- and short-channel devices for the same magnitude of stress, as shown in Fig. 7.14.

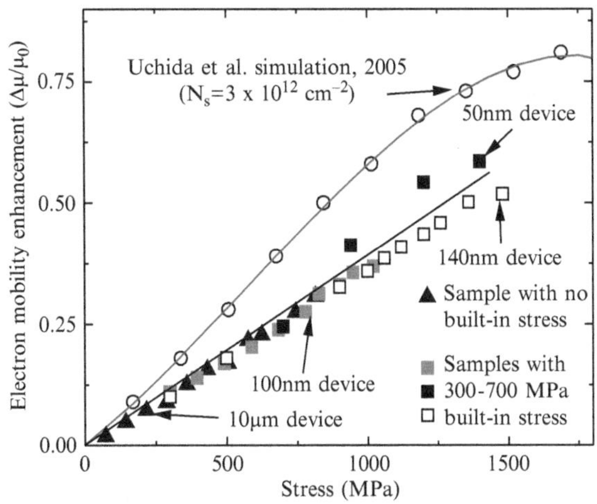

Fig. 7.14. Electron mobility enhancement as a function of uniaxial stress. In some samples, the high stress is achieved by addition of the externally applied mechanical stress and the built-in longitudinal stress. The solid lines are for the guide of eyes

The π-coefficients for Si nMOSFET, which can be directly measured, are shown in Table 7.2, along with an experimental example in Fig. 7.15 showing the longitudinal π-coefficient for uniaxial stress. We can compare the π-coefficients to Smith's bulk data. Little difference is found for ⟨110⟩ channel MOSFET and bulk Si with current along ⟨110⟩. The basic physical mechanisms of the electron mobility enhancement in MOSFETs are the same as for the bulk Si. However, there is some difference in detail. For bulk

Table 7.2. π-Coefficients in Si (001) surface MOSFET ($10^{-11}Pa^{-1}$)

	π_l	π_t
n ⟨100⟩ channel	−63	−20
n ⟨110⟩ channel	−32	−15
p ⟨100⟩ channel	−15	9
p ⟨110⟩ channel	71	−32

Si, the six Δ valleys are degenerate, whereas in MOSFETs, they are already split by electric confinement, as discussed in Chap. 5. At an inversion

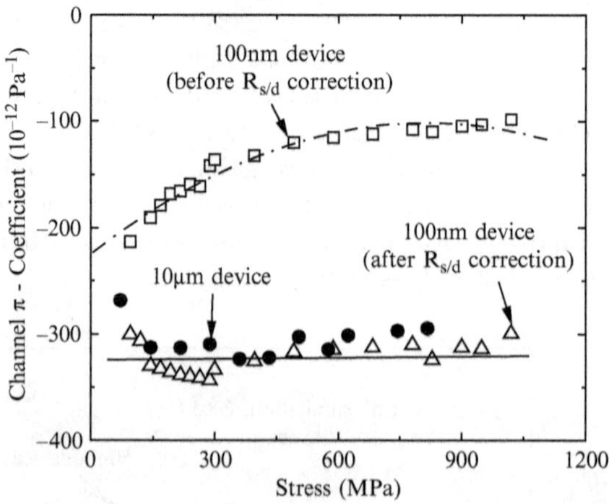

Fig. 7.15. Channel piezoresistance coefficient vs. uniaxial stress. After the reduction of the contribution of the S/D series resistance, the piezoresistance coefficients of both long- and short-channel Si nMOSFETs are consistent

carrier concentration of $10^{13}/\text{cm}^2$, the splitting between Δ_2 and Δ_4 valleys is about 95 meV. This energy is large enough to place about 75% of all electrons in the Δ_2 ground subband. The maximum mobility enhancement due to repopulation alone is only about 14%, much lower than the observed data. There are three other factors that can cause further increase of the mobility. The first is the suppression of the intervalley scattering, the second is the band warping of the Δ_2 valley, and the third is the assumption we discussed in last chapter: stress-smoothed surface. The band warping is independent of the band splitting, thus can provide a continuous source of mobility enhancement. For Si n-MOSFETs, the large electron population in the Δ_2 valley results in higher efficiency of the band warping than in bulk. This is the reason why $|\pi_{44}|=|\pi_l - \pi_t|\sim 17$ is larger in MOSFET than in bulk, which is ~ 13. For uniaxial tensile stress along the $\langle 110\rangle$ Si channel, the band-warping-induced piezoresistance contributes about 26% of the total piezoresistance. The piezoresistance due to repopulation-induced mass change is expected to be smaller. Intervalley scattering could be dominant because due to the high energy of the electrons in the upper Δ_4 subbands, the scattering from Δ_4 valley to Δ_2 valley is a phonon emission process, which is more than ten times stronger than the phonon absorption process. However, when adopting the bulk intervalley phonon energy and neglecting band warping, the mobility enhancement in Si n-MOSFET obtained theoretically was remarkably smaller than that experimentally observed. This could result from three uncertain factors: 1) enhanced intervalley electron–phonon interaction in Si inversion layer; 2) smoothed surface by stress; and 3) enhanced band warping. The factor 3) is only applicable to uniaxial stress along $\langle 110\rangle$. Takagi (Takagi et al, 1996) considered the first factor for wafer-based strained-Si n-MOSFET and obtained satisfactory theoretical results by assuming that the deformation potentials in 2D inversion layer were multiplied by a factor of 2.4. In his calculation, using the bulk intervalley deformation potentials gave him ∼40% saturated enhancement factor at large strain, and after the multiplication, the enhancement factor went up to ∼68%, which coincided with some experiments. However, this is doubtable because there lacks solid support for enhanced intervalley interaction in low-dimensional systems. Whether this is true or not can be examined by low-temperature piezoresistance measurements, since at sufficiently low temperature, the intervalley scattering can be totally suppressed, and it does not contribute to piezoresistance at all, and the enhancement goes to zero. But the existing low-temperature measurements are vastly discrepant. Almost vanishing enhancement (e.g., Ref. (Lime et al, 2005)) as shown in Fig. 7.16, and almost constant enhancement (e.g., Ref. (Sugii and Washio, 2003)) as shown in Fig. 7.17 have been reported.

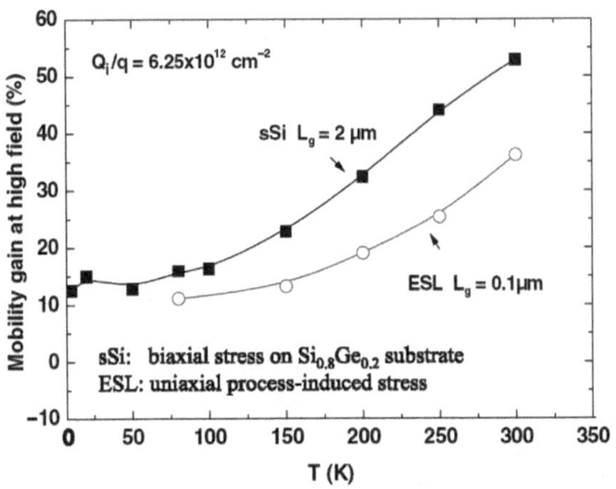

Fig. 7.16. Electron mobility gain at high electric field in Si nMOSFETs vs. temperature. At low temperature, mobility gain saturation is observed. From Lime (Lime et al, 2005)

This experimental discrepancy makes it difficult to determine which mechanism dominates the piezoresistive effect in Si n-channel MOSFETs.

Unlike strain-enhanced electron mobility, very significant differences have been observed for strain-enhanced hole mobility between the global biaxial stress and process-induced longitudinal uniaxial stress. Measured mobility enhancement factor for wafer-based strained-Si p-channel MOSFETs is shown in Fig. 7.18. Because strain splitting actually offsets the confinement splitting in the inversion layer, the mobility enhancement is very small, even negative for small Ge content in the SiGe substrate. Only with significant stress where the subband alignment is reversed and split further, does the mobility begins to shoot up. It is technologically and economically impractical to employ wafer-based strained p-MOSFETs, since high enhancement requires

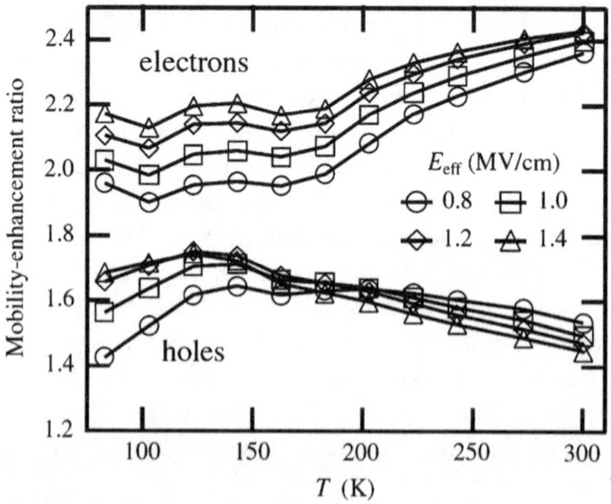

Fig. 7.17. Electron and hole mobility gain in Si n- and pMOSFETs vs. temperature. No significant enhancement degradation is observed with temperature lowering both in n- and pMOSFETs. From Sugii (Sugii and Washio, 2003)

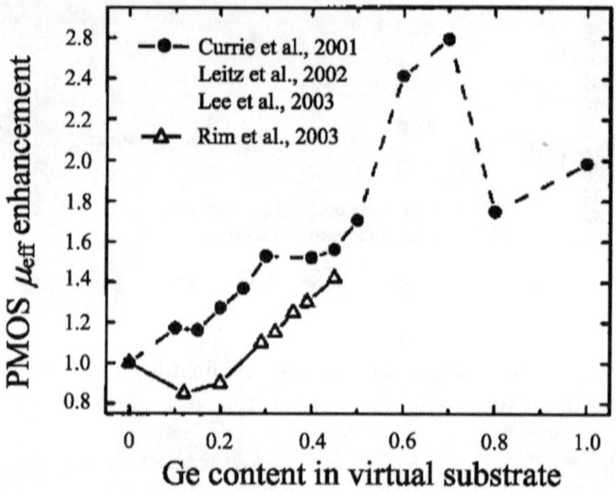

Fig. 7.18. Hole mobility enhancement in Si pMOSFETs under wafer-based biaxial stress. From Lee (Lee et al, 2005)

very large biaxial strain, whereas such large strain between the Si-channel and SiGe substrate can cause strain relaxation and defects, and thus impact the device reliability and yield. The other consideration is that the process of the wafer-based strained-Si is very different from the conventional Si CMOS process. Wafer-based Si p-MOSFET has never been adopted in semiconductor industry, unlike the popular process-induced longitudinal uniaxial compressive stress, which enhances the hole mobility monotonically, up to 2.8 reported by Washington et al. (Washington et al, 2006). Some measurements of the hole mobility enhancement under process-induced longitudinal uniaxial strain are shown in Fig. 7.19. The dominant factor for piezoresistive effect

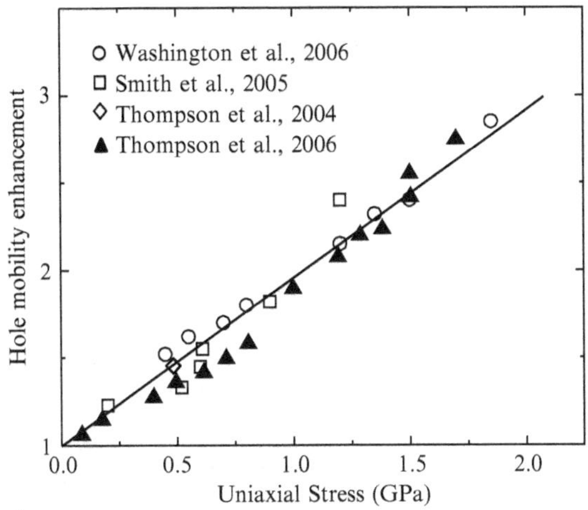

Fig. 7.19. Hole mobility enhancement in Si pMOSFETs under longitudinal uniaxial stress. The solid line is for the guide of eyes

is still band warping. π_{44} extracted from Table. 7.2 is $103 \times 10^{-11}\text{Pa}^{-1}$, and $\pi_{11} + \pi_{12} \sim 38 \times 10^{-11}\text{Pa}^{-1}$. Repopulation and scattering-induced effect is larger than in bulk Si, but far from dominant. This larger effect comes from the stronger interband scattering change due to split subbands under electric confinement. For GaAs and Ge, which have remarkably larger difference between HH and LH effective masses than Si, the subband splitting due to confinement alone is large enough to populate most holes in the ground subband in the normal MOSFET operation gate field, and thus the peizoresistive effect is almost solely due to band warping. Thus, the piezoresistance for Si n- and p-MOSFETs is from two distinctive sources.

7.4 SIGE DEVICES

Apart from being applied as the virtual substrate in strained-Si technology, SiGe itself can also serve as the MOSFET channel. Si and Ge have the same crystal structure, and they can be alloyed with arbitrary content. $\text{Si}_{1-x}\text{Ge}_x$ has new properties. Its lattice constant is larger than that of Si and smaller than that of Ge, and follows (Dismukes et al, 1964):

$$a(x) = 5.431 + 0.01992x + 0.027x^2(\text{Å}), \tag{7.1}$$

where 5.431 Å is the lattice constant of Si crystal at room temperature. Thus, $\text{Si}_{1-x}\text{Ge}_x$ grown on Si substrate is under biaxial compressive stress. The bandgap of SiGe can be tuned by modifying the Ge content. There is a gradual decrease of the bandgap with increasing Ge content, with all conduction valleys lowered. But the L-valleys decrease more rapid than the Δ-valleys, and at $x \sim 0.85$, the L-valleys become the lowest conduction valleys, and the SiGe changes from Si-like to Ge-like. Under biaxial compressive stress, the six Δ-valleys split into the lower in-plane Δ_4 valleys and upper out-of-plane Δ_2 valleys. The valence band degeneracy is lifted with HH band constituting the valence band edge. The band splitting of SiGe strained layer is shown in Fig. 7.20. A thin SiGe layer and the Si substrate form a heterostructure. The consensus for the band alignment is that the valence band offset is much greater than the conduction band offset. The computed valence band offset, the energy distance between the HH hole bands, by $\mathbf{k} \cdot \mathbf{p}$ theory, is about $\Delta E_v = 0.73x$. The small conduction band offset is controversial, and sometimes we may just assume it is zero. The valence band alignment favors holes occupying the SiGe layer, which has a higher mobility than the Si substrate.

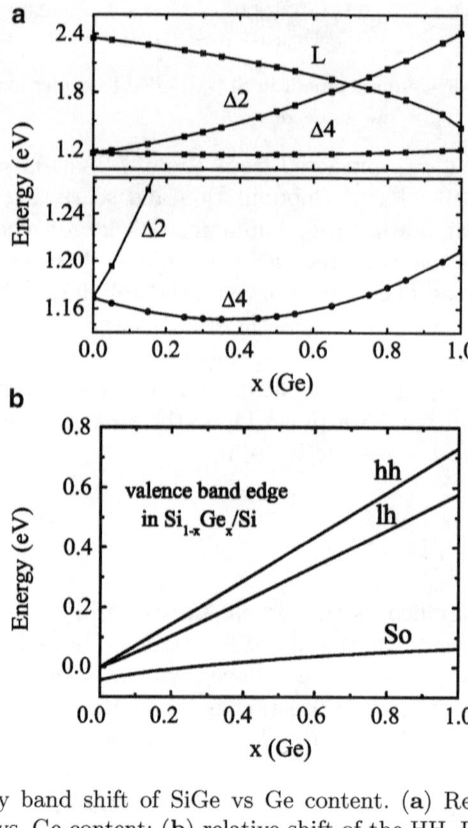

Fig. 7.20. Energy band shift of SiGe vs Ge content. (**a**) Relative shift of SiGe conduction valley vs. Ge content; (**b**) relative shift of the HH, LH and split-off hole band of SiGe vs. Ge content. From Dismukes (Dismukes et al, 1964)

SiGe is considered as a promising material for deep submicron CMOS channel engineering. Relaxed SiGe with Ge fraction less than 80% does not show mobility advantage over Si (Gaworzewski et al, 1998). This is mostly because of the alloy scattering. However, many studies carried out over the last decade have proven that compressively strained SiGe channels could provide an important mobility gain for long-channel pMOSFETs (Alieu et al, 1998; Höck et al, 2000; Lindgren et al, 2002; Loo et al, 2004). This gain is attributed both to the reduction of the hole effective mass and to the splitting between heavy and light hole subbands. Pure Ge channel is rare since the lattice constant between Si and Ge is very large and the direct growth of Ge on Si induces too much strain, which may incur potential reliability issues. There is also a leakage issue related with narrower band gap of Ge.

SiGe devices include strained SiGe surface channel and Si/SiGe dual channel devices. The typical band alignment for a Si/SiGe/Si dual channel pMOSFET is shown in Fig. 7.21. If the Si layer underneath the SiO_2 is

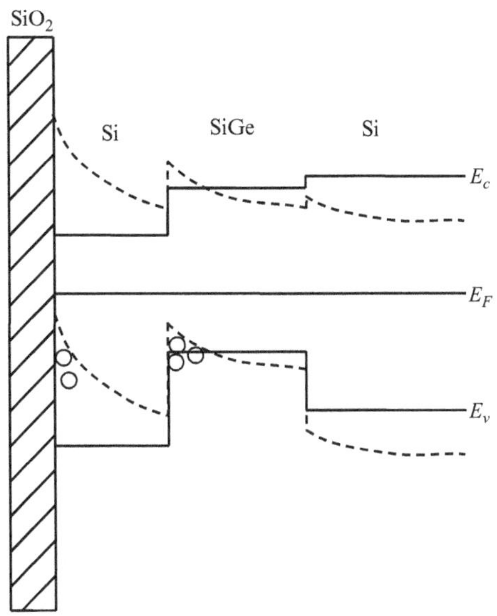

Fig. 7.21. Schematic of the built-in potential (solid line) and band bending profile (dashed line) of a Si/SiGe/Si dual channel pMOSFET

removed or becomes very thin, it turns into a SiGe channel MOSFET. Compressive stress significantly enhances the hole mobility of SiGe MOSFETs. Shown in Fig. 7.22 is the hole mobility comparison between 10-nm thick $Si_{0.8}Ge_{0.2}$ surface channel grown on Si and Si control MOSFET (Shima, 2002). The hole mobility in the 0.1-μm-long ⟨110⟩ channel SiGe MOSFET exhibits 20% enhancement over Si pMOSFET, and the ⟨100⟩ channel SiGe MOSFET shows 25% further hole mobility enhancement over the ⟨110⟩ channel SiGe

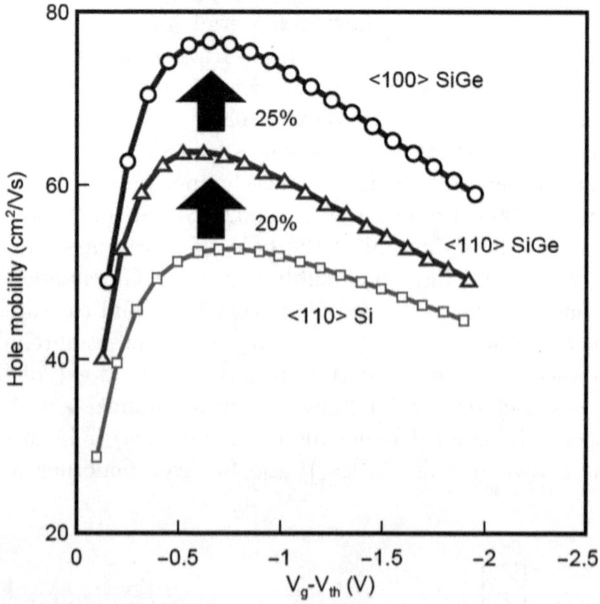

Fig. 7.22. Comparison of hole mobilities for pMOSFETs with ⟨110⟩ $Si_{0.8}Ge_{0.2}$ channel, ⟨100⟩ $Si_{0.8}Ge_{0.2}$ channel, and ⟨110⟩ Si channel. From Shima (Shima, 2002)

MOSFET. Reports from Alieu et al. (Alieu et al, 1998) demonstrated 150% drain current (I_d) enhancement in long-channel ($\geq 0.3\ \mu$m) SiGe pMOSFETs and 40% I_d enhancement in relatively short-channel (0.15 μm) pMOSFETs.

The distinction between SiGe channel and Si/SiGe dual-channel MOSFETs is sometimes vague, because for SiGe channel MOSFETs, there is usually a Si cap layer on top of the SiGe channel. Formation of channel in the SiGe layer or in both Si cap layer and SiGe layer depends on the Si cap layer thickness and gate bias. The electric confinement in Si/SiGe heterostructures is from the combination of both the built-in potential and band bending induced by gate bias. If the Si cap payer is thin, the confinement is mainly from the built-in potential and the quantum well is formed in the SiGe layer, which confines the holes. If the effective field is strong, the band bending at the Si/SiO_2 surface will also form a triangular potential well, and the device becomes a dual-channel device. Depending on the goal of enhancing the mobilities of different carriers, the thickness of Si cap layers can be optimized for nMOSFETs and pMOSFETs. Shown in Fig. 7.23 (Andrieu et al, 2003) is the hole mobility dependence on the Si cap layer thickness for pMOSFETs with $Si_{0.85}Ge_{0.15}$ channels with different cap layer thicknesses. Too thin Si layer increases the interface scattering and too thick Si layer impedes the formation of channel in the SiGe. The thickness of 2-nm shown in the figure gives the best hole mobility enhancement factor. For SiGe channel layer grown on a SiGe virtual substrate with less Ge fraction, it is under compressive

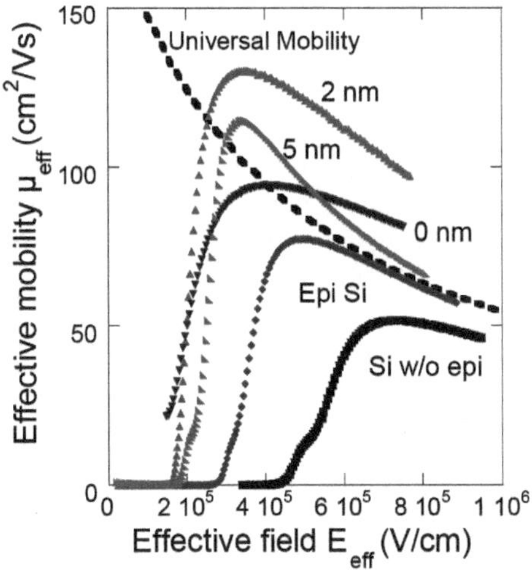

Fig. 7.23. Hole mobility vs. effective electric field for pMOSFETs with $Si_{0.85}Ge_{0.15}$ channels with different cap layer thicknesses. From Andrieu (Andrieu et al, 2003)

strain. The Si cap layer grown on top of the SiGe channel layer is under tensile strain. Because of the distinctive response of the electron and hole mobility to different types of strain, it is advantageous to confine the electrons in the Si layer for nMOSFETs and confine the holes in the SiGe layer for pMOSFETs. Thus, for a similar gate bias operation for both n and pMOSFETs, optimization of Si cap layer leads to thicker Si layer thickness for nMOSFETs and thinner Si layer thickness for pMOSFETs. Wang et al. (Wand et al, 2005) reported a 16% and 12% drive current enhancement for n and pMOSFETs, respectively, by using the Si cap layer thickness of 15 and 5-nm, with channel lengths at 0.13 μm. Similar trend was also confirmed by Mheen et al. (Mheen et al, 2005).

7.5 LEAKAGE AND RELIABILITY OF STRAINED-SI

With strain being introduced to the CMOS process, one serious concern is whether it will induce leakage degradation and reliability issues. In this section, strain effects on CMOS device leakage and its impact on device reliability are discussed.

7.5.1 Strain on Threshold Voltage

Strain-induced threshold voltage change is important to account for in performance benchmarking in strained MOSFETs. The traditional definition of the threshold voltage is (here an n-channel CMOS device is considered for an example),

$$V_t = V_{fb} + (2m - 1)2\psi_B, \tag{7.2}$$

where V_{fb} is the flat-band voltage, m is the body-effect coefficient, and $e\psi_B = |E_f - E_i|$ is the energy difference between the extrinsic and intrinsic Fermi levels. If we assume the body-effect coefficient is not changed by strain, then strain shifts V_t by shifting the flat-band voltage and ψ_B. If we only consider strain on the channel and neglect any effects on the gate, then the threshold voltage change is from two factors: shift of the Fermi level, and change of the bandgap. The change of the bandgap apparently changes the intrinsic Fermi level. The shift of the extrinsic Fermi level and change of the bandgap are related, but the shift of the Fermi level is not fully dependent on the change of bandgap. It is determined by the net shift of both the conduction and valence bands, and their relative shift, which causes the change of the bandgap. The extrinsic fermi level shift is also related to the strain altered DOS in the conduction and valence bands. For example, let us assume that strain does not change the electron density. Then if the conduction band DOS is reduced by strain, the Fermi level will shift up to increase the occupation probability of the states at the band edge. The relative shift between the extrinsic and intrinsic Fermi levels changes ψ_B, and the shift of the extrinsic Fermi level will also shift the flat-band voltage.

The threshold voltage shifts for Si nMOSFETs with the process-induced uniaxial tensile stress and wafer-based biaxial tensile-stress have been investigated by Lim et al. (Lim et al, 2004). In process-induced uniaxial stress case, both the gate and the channel are stressed simultaneously by SiN tensile capping layers, while in wafer-based biaxial stress case, only the channel is stressed. The threshold voltage shifts are given by

$$eV_t(\sigma) = \begin{cases} (m-1)\left[\Delta E_g(\sigma) + k_B T \log \frac{N_v(0)}{N_v(\sigma)}\right], & \text{uniaxial} \\ \Delta E_c(\sigma) + (m-1)\left[\Delta E_g(\sigma) + k_B T \log \frac{N_v(0)}{N_v(\sigma)}\right], & \text{biaxial} \end{cases} \tag{7.3}$$

where ΔE_c is the conduction band shift by strain, and N_v is the effective DOS of the valence bands. Warping of the conduction band is neglected, and the conduction band DOS is assumed unaltered. The extra term in the biaxial stress case obviously comes from the strain-induced flat-band shift, which is absent in the uniaxial stress case since where both the poly-Si gate and the Si channel are stressed under the same condition.

The values of ΔE_c and ΔE_g can be obtained by employing the deformation potential theory. The formulation is already discussed in earlier chapters. The alteration of the valence band effective DOS can be expressed as $N_v(0)/N_v(\sigma) = [m_p(0)/m_p(\sigma)]^{3/2}$. The expression for the hole effective mass m_p is obtained by Thompson et al. (Thompson et al, 2002) as

$$m_p(\sigma) = \begin{cases} \left[m_{lh}^{3/2} H(\sigma)^{3/2} + m_{hh}^{3/2} \right]^{2/3}, & \text{uniaxial} \\ \left[m_{lh}^{3/2} + m_{hh}^{3/2} H(\sigma)^{3/2} \right]^{2/3}, & \text{biaxial} \end{cases} \tag{7.4}$$

where

$$H(\sigma) = \exp\left[-\frac{|\Delta E_{lh}(\sigma)| + |\Delta E_{hh}(\sigma)|}{k_B T} \right]. \tag{7.5}$$

With this information, they obtained the V_t shift induced by biaxial and uniaxial strain, and the figures are reproduced as in Fig. 7.24. The V_t shift is

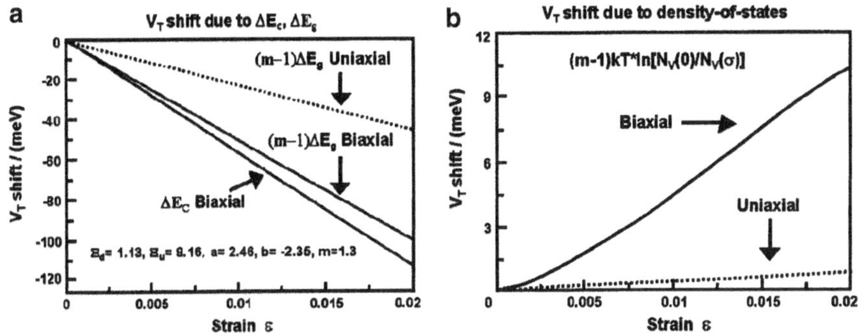

Fig. 7.24. Threshold voltage shift for Si nMOSFETs under biaxial and uniaxial strain. (a) V_t shift due to the shift of conduction band edges and bandgap; (b)V_t shift due to the change of DOS in the valence bands. From Lim (Lim et al, 2004)

much more significant in the biaxial stress case than in the uniaxial stress case not only because of the flat-band voltage shift due to the work function shift in the Si channel by stress. We notice that the bandgap narrowing for biaxial stress is also much larger than for uniaxial stress. Compared to the effects of working function shift and bandgap narrowing, band warping-induced V_t shift is one order of magnitude smaller, even though once again the magnitude of V_t shift by biaxial stress is much larger than that by uniaxial stress.

7.5.2 Leakage Current in Strained-Si Devices

Many researches show that advantageous strain not only enhances channel mobility, but also greatly benefits gate leakage (Yang et al, 2006; Yan et al, 2008). This can be understood from the shift of subbands under the strain introduced in the strained-Si process. For Si nMOSFETs, the advantageous strain type is tensile strain, including both the biaxial tensile and longitudinal uniaxial tensile strains. First with tensile strain, bandgap shrinks. This is a combined effect of both raised valence band and lowered conduction band edges. Then in addition with the lowering of the average conduction band energy, Δ_2 and Δ_4 valleys split further, and more electrons repopulate into the Δ_2 valleys, which are lower than without strain. The total effect is that electrons in the inversion layer are now facing a higher tunneling potential barrier. In addition, due to the electron repopulation, the average electron effective mass in the tunneling direction also increases. The tunneling probability is a strong function of the barrier height and electron effective mass. With both increased barrier height and effective mass, the gate tunneling current in strained Si nMOSFETs is significantly reduced. The mechanism is schematically shown in Fig. 7.25a. The advantageous strain type for pMOSFET

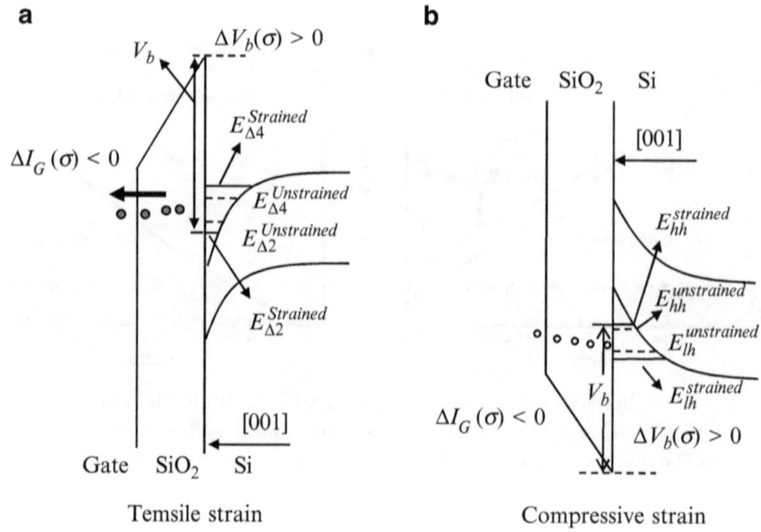

Fig. 7.25. Band shift and gate direct tunneling current change under strain for (**a**) tensile-strained Si nMOSFET and (**b**) compressive-strained Si pMOSFET

is compressive strain, especially the longitudinal uniaxial compressive strain. We can see from Fig. 5.30 that under uniaxial compressive stress, the HH subband shifts up, even though the LH subband shifts down, as shown in Fig. 7.25b. Similar to the conduction band, the further splitting between HH and LH subbands repopulates more holes into the HH bands, which now are energetically further away from the SiO_2 valence band edge, and thus the hole tunneling barrier height is increased, resulting in a reduced hole gate leakage current. Yang et al. (Yang et al, 2006) studied the relation of gate leakage current to various strain conditions, and their experimental and model results are shown in Fig. 7.26b, which are consistent with the above observations. Fig. 7.26a demonstrates the gate leakage dependence on uniaxial strain for Si nMOSFETs. Recently, up to 90% gate leakage improvement has been observed (Yan et al, 2008) for biaxial strain compared to bulk Si nMOSFETs, due to the very large biaxial tensile strain induced by lattice mismatch.

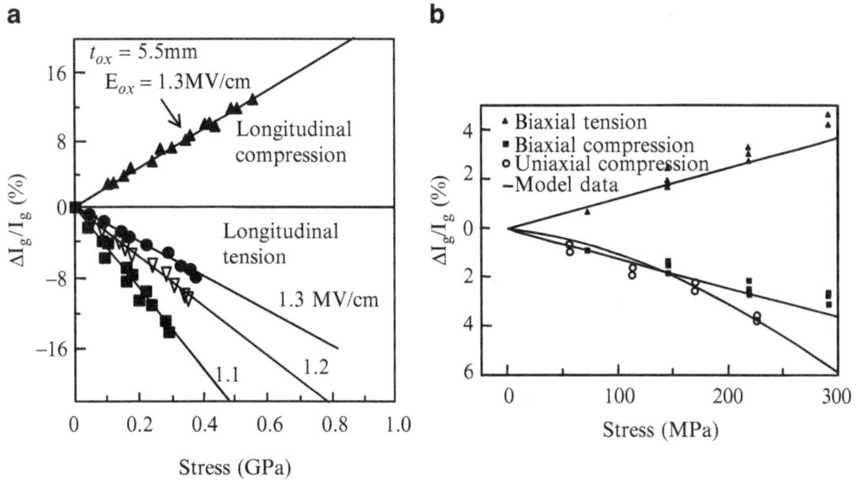

Fig. 7.26. Measured gate tunneling leakage current vs. stress for (**a**) Si nMOSFETs and (**b**) Si pMOSFETs. In (**a**), two types of stress, i.e., longitudinal tensile and longitudinal compressive stress are used in experiments, and the gate leakage current is measured under three different effective oxide fields. Solid lines are for the guide of eyes. In (**b**), three types of stress, i.e., biaxial tensile, biaxial compressive and uniaxial compressive stress are explored for their effect on gate leakage current. Solid lines are modeled results. (**b**) is from Yang (Yang et al, 2006)

However, increased junction leakage and source to drain leakage may arise with the SiGe virtual substrate in wafer-based strain during MOSFET operation. Increased junction leakage is partly related to the interface between SiGe and STI, and both the junction leakage and source to drain leakage suffer from the reduced bandgap due to tensile strain. Fortunately, the junction leakage is much smaller than the gate leakage under normal MOSFET operation condition.

Abnormally large leakage may also arise due to defects induced by strain process. Process of growing strained-Si grown on SiGe virtual substrate normally involves high temperature. Under high temperature, yield stress in Si thin film is reduced and thus strain relaxation can occur at relatively lower strain. If the threading dislocations caused by strain relaxation penetrate from the source to the drain or across a junction, this device will have an abnormally large leakage current. Defect-related issues will be discussed in more detail in the next section.

7.5.3 Reliability of Strained-Si

Device self-heating is the dominant factor that compromises the performance gains in short-channel wafer-based strained-Si devices, where the channels are separated from the Si substrate by low thermal conducting SiGe layers, whose thermal conductivity is at leat 15 times lower than Si in general (Dismukes et al, 1964). Also, the Si/SiGe interface also acts as a phonon boundary which blocks phonon transmission, which is the primary means for heat transfer. The degradation of the device performance comes from the reduced mobility therefore the drive current at high temperature. Self-heating presents a severe obstacle to the use of strained-Si devices in analog applications where sustained power is required. Although self-heating of strained-Si is less severe in digital applications, where significant power dissipation is only during switching, the heating issue is already serious in VLSI circuits operated at high frequency. Therefore, reducing self-heating in wafer-based strained-Si devices is equally important even in digital applications.

Since the self-heating is due to the SiGe virtual substrate, the obvious approach for tackling this issue is to reduce the thickness of the virtual substrate. Novel ultrathin virtual substrates (e.g., 200-nm, as compared to 1–4 μm in thick SiGe virtual substrates) have been attempted in researches (Hackbarth et al, 2003; Olsen et al, 2006; Yan et al, 2008), which showed great potential in reducing the self-heating, and at the same time, improving the gate oxide integrity and channel strain integrity by reducing the surface roughness. Threading dislocation in thin SiGe substrate may pose a concern. However, the experiments showed that the strain relaxation in the ultrathin virtual substrate technology was through point defects rather than the threading dislocation, and thus do not impact the channel. Other attempts or approaches have been also proposed to address this issue (Nicholas et al, 2005).

In strained-Si devices, the enhanced mobility leads to more energetic carriers. Also, the reduced bandgap in the tensile stress Si nMOSFETs makes the impact ionization easier. This favors more severe hot carrier effect in strained-Si devices. However, injection into oxide also depends on the oxide barrier height. Increased barrier height not only reduces the gate leakage, but also reduces the injection probability of hot carriers into the oxide. Researches show improved hot carrier effect in strained-Si devices over the bulk Si devices (Onsongo et al, 2004).

Time-dependent dielectric breakdown (TDDB) in strained-Si devices is also improved by proper type of strain introduced in the devices, i.e., tensile strain for nMOS and compressive strain for pMOS devices. Irisawa et al. (Irisawa et al, 2007) found that TDDB reliability of nMOS was significantly improved while that of pMOS is slightly degraded for biaxial tensile stress. This can be understood from gate leakage and hot carrier injection change induced by strain. The gate leakage and hot electron injection into oxide are suppressed by the tensile strain. The strain-induced leakage current reduction also leads to major improvement in stress-induced leakage current. Thus, the improvement of TDDB reliability for nMOS devices is resulted. Yan et al. (Yan et al, 2008) reported one order of magnitude improvement in time to oxide hard breakdown in biaxial strained nMOSFETs compared to the bulk Si control devices following high-field stressing at 17 MV/cm. Compared to nMOS devices, the gate leakage current is enhanced for pMOS devices under biaxial stress. Hot hole injection can be reasonably assumed to have a positive relation with the gate leakage current, and thus it is safe to assume that the hot hole injection is also enhanced for pMOS devices. Therefore, a degradation of the TDDB reliability for pMOS devices is understandable. When the same physics applied to pMOS devices with process-induced uniaxial compressive strain, we expect that TDDB reliability be improved rather than deteriorated by strain. This has been confirmed by industrial practice, where they found better TDDB behavior for uniaxial strained pMOSFETs.

Another very important reliability issue pertains to the defects in strained-Si devices. Strain relaxation is a major cause of the misfit dislocations, which is detrimental to the channel integrity. The strain field around the dislocations captures metal impurity atoms and increases the dopant diffusion by several orders of magnitude (Braga et al, 1994). Dislocations that are contained entirely within the neutral space charge region of the source and drain junctions do not impact transistor function, while those extending from source to drain cause high leakage current and nonfunctional transistors. Defects will have a more deleterious impact on the yield as the channel length of the MOSFET is scaled. At shorter channel length, the dislocations are more likely to penetrate through the source to the drain and cause a nonfunctional device. Because of the deleterious nature of these defects, study of these defects is an important subject for improving the quality of strained-Si in high-volume manufacturing.

7.6 DEFECTS IN STRAINED-SI

The misfit dislocations in strained-Si devices are mainly induced by film strain relaxation. For an epitaxial layer grown on a substrate, there exists a critical thickness h_c, below which the lattice mismatch can be entirely accommodated by strain, and beyond which, part of the mismatch must be accommodated by dislocations. In process-induced strain cases, for thick SiGe epitaxy grown in the S/D region, the S/D annealing in high temperatures such as 900–1000°C can cause strain relaxation and create dislocations. For epitaxial layers with thickness less than the critical thickness, annealing at high temperature such as 900°C for even 30 min does not cause observable difference (Houghton et al, 1990). A typical type of dislocation in Si crystal is the perfect $\frac{a_0}{2}\langle 110\rangle$ 60° dislocation, which is along the $\langle 110\rangle$ directions on the $\langle 111\rangle$ glide surfaces, where a_0 is the Si crystal lattice constant. The other dislocation type is the $\frac{a_0}{6} < 112 >$ dislocation. The $\frac{a_0}{6} < 112 >$ dislocations are partial dislocations, which are accompanied by stacking faults. A perfect $\frac{a_0}{2}\langle 110\rangle$ 60° dislocation can dissociate into two $\frac{a}{6}\langle 112\rangle$ partial dislocations, i.e., a 90° and a 30° partial dislocation. Of necessity, a stacking fault between the two partial dislocations must also be generated. This dissociation depends critically on strain and lattice temperature. For (001) oriented films under tension such as Si/SiGe, stacking faults may extend through the strained-Si film thickness, whereas most studies consider that the partials are pushed together at the interface and the dislocations are perfect 60° dislocations for compressively strained films such as SiGe/Si.

A dislocation cannot end inside a crystal of a film. It must end at the surface or many of the dislocations glide together to form a dislocation pile-up. Some dislocations in industrial strained-Si product chips are shown in Fig. 7.27, where the dislocations are contained in the charge neutral region in both cases. If strain is relaxed in the Si epitaxy on SiGe virtual substrate or

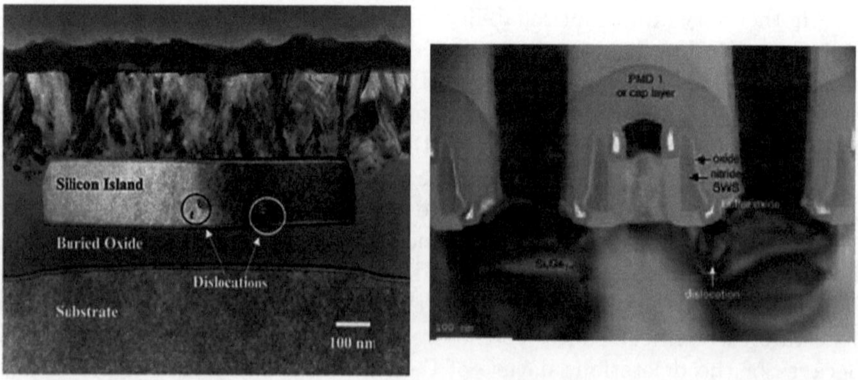

Fig. 7.27. Demonstration of threading dislocations in strained-Si MOSFETs

in the SiGe layer in the global wafer-based strain case, or in the SiGe epitaxy in the S/D or raised S/D region and the dislocations are generated at the S/D-channel edge, they probably penetrate into the channel and cause non-functional devices. For controlling the defects, the thickness of these epilayers must be limited under the critical thickness, or, very importantly, treated with special thermal cycles. The latter is what has been done in Si industry.

The equilibrium critical thickness h_c was studied by Ball and van der Merwe (Ball and van der Merwe, 1983) and by Matthews and Blakeslee (Matthews and Blakeslee, 1974), who found that for $Si_{1-x}Ge_x$ films grown on Si, the relation between h_c (nm) and the Ge atom fraction x_c can be written as

$$x_c = \frac{0.55}{h_c} \ln \left(\frac{4h_c}{b} \right), \tag{7.6}$$

where b is the Burgers vector. For 60° glide dislocations, which are typical in tetrahedral crystals, the Burgers vector $b = \frac{a_0}{2}\langle 110 \rangle \sim 0.4$ nm for Si, where a_0 is the Si crystal lattice constant. Obviously, the critical thickness for SiGe film decreases with Ge content. For SiGe film containing 20% Ge, the critical thickness is about 14-nm. On the other hand, the critical thickness for the Si cap layer on the SiGe virtual substrate is shown in Fig. 7.28 (Samavedam et al, 1999), where we can see that Si cap critical thickness decreases quickly with increasing Ge fraction in the virtual substrate.

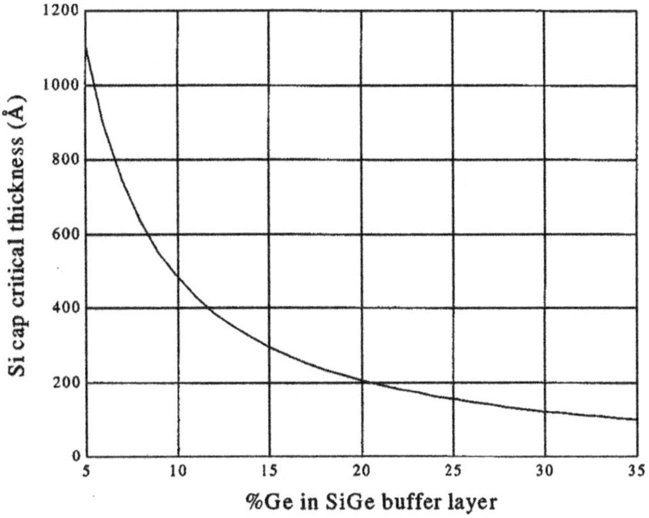

Fig. 7.28. Si cap layer critical thickness as a function of Ge fraction in the SiGe virtual substrate. From Samavedam (Samavedam et al, 1999)

To induce and sustain high stress or for the purpose of application, the thickness of SiGe epitaxy in the S/D in process-induced stress case, or of the Si channel layers grown on the SiGe virtual substrate in the global strain case is usually much higher (e.g., ~100-nm for SiGe epitaxy in S/D). So how to control strain relaxation? In the design of process flow, special procedures are adopted, which are based on one observation that has been confirmed by both theory and experiments. That is, the thickness for low-temperature MBE grown layers can be much larger than the critical thickness without inducing any dislocations (Houghton et al, 1990; Timbrell et al, 1990). This is because the epitaxy and the substrate are not in a thermal equilibrium state, which needs higher temperature to reach. These layers are in the so-called metastable state. For film with thickness larger than the critical thickness, strain relief needs more stress, which decreases with temperature.

To minimize defect density, the process-induced strain, unlike wafer-based strain, is introduced later in the process flow. Epitaxial SiGe is introduced post isolation and gate formation by etching the S/D and growing selective epitaxial SiGe (Chidambaram et al, 2004; Thompson et al, 2004a; Lee et al, 2005). The high stress capping layers are introduced even later post-silicide formation. In some cases when the high stress capping layers for stress memorization are introduced before S/D anneal, they can cause drain shorting defects at S/D anneals at elevated temperatures. In global stress case, because the high-temperature S/D anneal and gate formation have to be after the Si layer growth on the SiGe virtual substrate, the thickness of Si cap layers have to be controlled under the critical thickness that can be looked up from Fig. 7.28.

Because the dislocations must end at the surface, apart from keeping the strained-Si layer thickness under the critical thickness for the wafer-based global strain case, the growth of the SiGe substrate with the lowest defect density is also crucial, because these dislocations can grow into the Si layers on top of the SiGe substrate. However, to have a fully relaxed SiGe layer, dislocations have to be generated to accommodate the strain. Fortunately, the generated dislocations can glide into regions of maximum stress under the combined influence of stress and high-temperature processing. As discussed by Lee (Lee et al, 2005), the general strategy for growth of a low-defect, fully-relaxed SiGe layer is to minimize dislocation nucleation rate and simultaneously maximize the dislocation glide velocity. The former may be attained by increasing or grading the Ge fraction relatively slowly over the film thickness such as through a grading rate of 10% Ge per μm. Since the dislocation glide is thermally activated, the latter may be achieved by growth at high temperature, greater than 750 C. Threading dislocation densities lower than 10^6 cm^{-2} have been reported for relaxed $Si_{1-x}Ge_x$ with $x = 0-0.5$ (Lee et al, 2005).

7.7 SCALABILITY OF STRAIN

Strain-Si technology substantially improves the performance of CMOS devices and pushed back the ending date of the traditional geometrical scaling, so that people can enjoy higher computational speed that strained-Si technology delivers. Strain was first introduced into the 90-nm CMOS technology, and now it has also been used in the latest 45 and 32-nm CMOS technology nodes. One safe prediction is that the geometrical scaling will continue at least to ~20 nm technology node, with some process improvements. What will be the successor of the current Si technology after that is still hard to foresee. One question that naturally arises for strained-Si technology before scaling comes to an end is then: Can strained-Si be scaled?

As discussed earlier in Sect. 7.3, strain-induced performance enhancement decreases with channel length reduction. This trend has also been confirmed by other researchers. Goo et al. (Goo et al, 2003) found that both the linear current and saturation current decrease almost linearly with respect to the logarithmic gate length, as shown in Fig. 7.29. The efficiency fall of strain in improving device performance with gate length shrinking comes from various sources. The most important one of them is the S/D series resistance, whose role has been discussed already in Sect. 7.3. Besides this, STI-induced stress, which is compressive along the channel direction, compensates the tensile strain in the channel for nMOS, which prevents the ideal drive current enhancement in especially short-channel devices where the S/D length is also scaled and thus the STI stress becomes more influential to the channel. The scalability of strain-Si technology is also intertwined with some other considerations with reduced gate length, the most important of which is reducing the

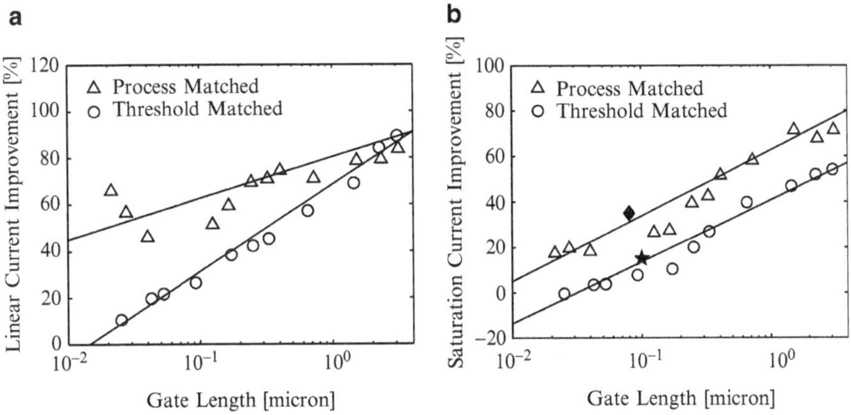

Fig. 7.29. Device performance improvements by strain vs. device gate length. (**a**) shows the linear drive current enhancement and (**b**) shows the saturation drive current enhancement vs. device gate lengths. From Goo (Goo et al, 2003)

short channel effect. To minimize the short channel effects, doping level in the S/D is increased and halo implant becomes heavier to prevent punch-through with decrease of the channel length. They have adverse effects on the effectiveness of strain to improve device performance, which is self-evident from Fig. 7.29 where the threshold-matched samples for which the halo implant is heavier almost show no improvement at all for saturation current at gate length ~25 nm, compared to the process-matched samples.

Nevertheless, with proper and sophisticated process optimized for strained-Si, strain is proven to be still very effective. First and most intuitive approach is to increase the stress level in the channel. For process-induced strain, this may be achieved by bringing the SiGe epitaxy closer to the channel and simultaneously increasing the Ge fraction in SiGe for SiGe S/D, and introduce new strain memorization methods and improving the stress level in the stress capping layers or stress liners. Strain can also be introduced to the channel by STI engineering (Kahng et al, 2007). As we mentioned earlier, the compressive STI stress may compensate the tensile channel stress in nMOS devices, but it enhances the compressive channel stress in pMOS devices. The stress characteristic can also be altered by specially designed STI engineering approaches (Kahng et al, 2007), which can intentionally increase (decrease) the STI stress to improve the pMOS (nMOS) device performance. Other than tensile stress capping layer and global tensile strain, tensile strained devices with embedded $Si_{1-x}C_x$ S/D has been reported in 25-nm gate length strained nMOSFETs (Ang et al, 2007). The lattice of SiC has smaller lattice constant than Si, and their lattice mismatch thus induces tensile strain into the channel. Strain-induced mobility enhancement leads to a significant drive-current I_{dsat} enhancement of 52% over the control transistor. The performance enhancement was achieved for the devices with similar subthreshold swing and drain-induced barrier lowering. Furthermore, the SiC embedded S/D can be combined with the tensile stress SiN capping layer to further improve the channel stress.

To counter the threshold roll-off in short channel devices, metal gate instead of poly-Si gate is more desirable in strained-Si, since the metal gate with proper work functions can be chosen to compensate the threshold roll-off without excessive doping, and metal gate also eliminates the gate depletion. Goo et al. (Goo et al, 2003) reported a 25% further I_{dsat} improvement by employing a SiNi metal gate MOSFET with a channel length of 35 nm over the poly-Si gate control which was incorporated with advanced channel engineering and demonstrated 20% I_{dsat} enhancement already.

Recently, high-κ dielectrics are under study to replace SiO_2 and achieve lower leakage currents at a comparable effective oxide thickness. HfO_2 is the most promising oxide among the high-κ materials. With HfO_2 replacing thermally grown SiO_2, gate leakage is greatly reduced, and short-channel effects are almost eliminated. However, among various materials and process integration challenges, one drawback with HfO_2 is that it causes channel mobility degradation. Compared to SiO_2/Si pMOSFET, a HfO_2/Si pMOSFET exhibits a peak mobility degradation of about 25%. In this regard, strained-Si technology complements high-k technology since the degraded mobility can be recovered by the strain enhancement of mobility. Strained-Si n-MOSFETs with HfO_2 exhibited 35%–60% higher mobility than bulk Si with HfO_2 and a 15%–30% enhancement compared to bulk Si with SiO_2 (Rim et al, 2002; Datta et al, 2003). Strained Si pMOSFETs with high-κ dielectrics show similar mobility gain over the unstrained MOSFETs. This holds the promise for the trade-off between mobility and gate leakage reduction, which is especially attractive for low power, high-performance CMOS technology.

8

Piezoresistive Strain Sensors

8.1 INTRODUCTION

In contrast to the fixed strain incorporated in logic devices for a fixed or constant improvement in device performance, piezoresistive strain sensors respond to variable strain through a modulation in the device resistance. The underlying physics of performance improvement in logic devices and strain transduction in piezoresistive strain sensors is the same: symmetry-breaking strain of the semiconductor crystal lattice warps the energy bands, splits the energy levels, and changes the carrier scattering rates, which changes the carrier mobility and the device resistance. While improvement of logic device performance requires an increase in mobility, which dictates the "sign" of the fixed strain, strain sensors respond to both negative (compressive) and positive (tensile) strains. Since the strain is fixed in logic devices, the linearity of mobility increase with strain is not an issue since the strain is theoretically frozen into the device by stressors incorporated into the device structure during the manufacturing process. In contrast, piezoresistive strain sensors are designed to transduce or detect varying strains by producing a proportional change in resistance. Hence, linear resistance change with strain is important to sense/transduce strains of varying amplitudes into an electrical signal without introducing distortion. For a transducer, the measured resistance vs. strain curve can be used to calculate the input strain from the strain sensitivity or calibration slope of the sensor. For a piezoresistive strain sensor, the upper limit of the measurable strain is usually defined as the maximum strain above which nonlinear deformation occurs. In contrast, there is no maximum allowable stress in strain-enhanced logic devices as long as there is performance enhancement, provided that the stress is manufacturable and the device is reliable.

A resistive transducer whose resistance changes in response to a stimulus is attractive for several reasons including a lower impedance than other transduction mechanisms such as capacitive or piezoelectric mechanisms and a simpler interface circuit to convert the response into an analog voltage or

Y. Sun et al., *Strain Effect in Semiconductors: Theory and Device Applications*, DOI 10.1007/978-1-4419-0552-9_8,

current output. A lower transducer impedance reduces signal attenuation between the transducer and a preamplifier and lessens the capacitive coupling from nearby potentials or crosstalk. Interface circuits for measuring resistance employ constant voltage or current bias and record the resulting current or voltage change. However, resistive transducers are inherently dissipative and hence are not energy conserving. Furthermore, the output and inputs cannot be interchanged to obtain the same transfer impedance. As a result, resistive transducers operate only as a sensor and not as an actuator. Overall, the relative simplicity and manufacturing process compatibility with integrated circuit technology makes piezoresistive transducers attractive for many strain and stress-sensing applications.

Akin to discrete and integrated components for integrated circuits, piezoresistive strain transducers may be subdivided into two categories, discrete and integrated.

8.2 RESISTOR AS DISCRETE STRAIN TRANSDUCER

A discrete piezoresistive strain transducer may be used for a local or "point" strain measurement if certain conditions are met. In this case, it must be assumed that the strain is constant over the dimensions of the strain transducer. Hence, discrete strain transducers are designed for application to structures larger than the transducer itself where the constant strain approximation is valid.

Consider a discrete resistor as shown in Fig. 8.1. As described in Sect. 2.5, the normalized change in resistance of an isotropic conductive solid with a scalar resistivity, ρ, and dimensions, $L \times W \times H$, is given by,

$$\frac{\Delta R}{R} = \frac{\Delta L}{L} - \frac{\Delta W}{W} - \frac{\Delta H}{H} + \frac{\Delta \rho}{\rho}. \tag{8.1}$$

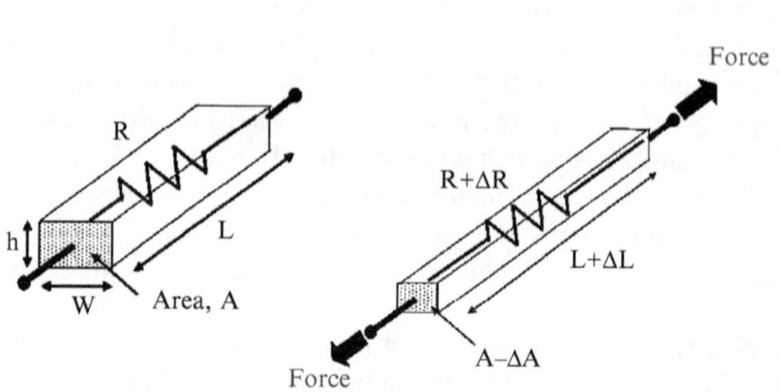

Fig. 8.1. Illustration of the dimensional change in a conductor with an applied force

By substituting the axial strain, $\varepsilon_{\text{axial}} = \Delta L/L$, the lateral strain, $\varepsilon_{\text{lateral}} = \Delta W/W = \Delta H/H$, and the relation between the two known as the Poisson ratio, $\nu = -\varepsilon_{\text{lateral}}/\varepsilon_{\text{axial}}$ into Eq. (8.1),

$$\frac{\Delta R}{R} = (1 + 2\nu)\varepsilon_{\text{axial}} + \frac{\Delta\rho}{\rho}. \tag{8.2}$$

It is seen that the normalized resistance change contains a geometric component, $(1 + 2\nu)\varepsilon_{\text{axial}}$, and an electronic component, $\frac{\Delta\rho}{\rho}$.

8.2.1 Gauge Factor

Since piezoresistive strain sensors produce a resistance change in response to a mechanical strain, a useful figure of merit is the gauge factor (GF) defined as the ratio of the normalized resistance change with axial strain,

$$\text{GF} = \frac{\Delta R/R}{\varepsilon_{\text{axial}}} = (1 + 2\nu) + \frac{\Delta\rho/\rho}{\varepsilon_{\text{axial}}}. \tag{8.3}$$

Metals

In metal strain sensors, the geometric component dominates the GF, and the electronic contribution to the GF caused by the strain-induced change in carrier resistivity or conductivity ($\sigma = 1/\rho$) is negligible. Hence, the second term in Eq. (8.3) is small, and the GF for metals becomes

$$\text{GF}|_{\text{metals}} = \frac{\Delta R/R}{\varepsilon_{\text{axial}}} \simeq (1 + 2\nu). \tag{8.4}$$

Since the Poisson ratio for metals is typically $\sim 0.3 - 0.4$, the GF is approximately 2. Note that $\nu = 0.5$ in an incompressible solid. From the measured change in resistance and a specified GF, the axial strain may be computed. Furthermore, for a homogeneous isotropic elastic material, from the product of the axial strain and the scalar modulus of elasticity, E, the axial stress, τ_{axial}, may be computed (Boresi and Schmidt, 2003),

$$\tau_{\text{axial}} = E\varepsilon_{\text{axial}}. \tag{8.5}$$

Semiconductors

In contrast, for a crystalline cubic semiconductor, the pivotal role of the diamond lattice on carrier transport makes it very sensitive to symmetry-breaking strain effects as detailed in the last five chapters in Parts I and II. As a result, the electronic component of the GF, arising from the change in resistivity with strain, is dominant and results in a GF that is up to two orders of magnitude larger for semiconductors compared to metals,

$$GF|_{\text{semiconductors}} = \frac{\Delta R/R}{\varepsilon_{\text{axial}}} = \frac{\Delta\rho/\rho}{\varepsilon_{\text{axial}}}. \tag{8.6}$$

As indicated in the previous chapters, the effect of strain on the electronic properties of a crystalline semiconductor can be drastic, essentially modifying the fundamental equilibrium lattice constant or the crystalline structure. Combined with the lattice system and the atomic potential, the lattice constant establishes the equilibrium energy band structure and through it, the equilibrium carrier concentration, and strongly affects nonequilibrium carrier transport. To understand the GF of a crystalline semiconductor, which is dominated by the strain dependence of the electronic resistivity, we revisit the concept of the resistivity or its inverse, the conductivity.

8.2.2 Piezoresistance

It is important to emphasize that the observable in a piezoresistor, its resistance or resistivity, requires the presence of an applied current. In an isotropic conductor at constant temperature, the current density is proportional to the electric field where the proportionality factor is defined as the conductivity, $\sigma = 1/\rho$ (Reitz et al, 1979),

$$\mathbf{J} = \sigma \mathbf{E} = \frac{1}{\rho}\mathbf{E}, \tag{8.7}$$

or in terms of each component,

$$E_i = \rho j_i. \tag{8.8}$$

Hence, another equivalent observable is the electric field or voltage. Equation (8.8) reduces to the familiar scalar form, $V = IR$, for a one-dimensional isotropic conductor where V is voltage and I is current.

In a general anisotropic conductor, the resistivity is a second-rank tensor, ρ_{ij}. ρ_{ij} is represented by a 3×3 matrix, which describes the relation between the three components of the electric field vector and the three components of the current density vector. Thus, the electric field components may be computed by matrix multiplication,

$$E_i = \sum_j \rho_{ij} j_j, \tag{8.9}$$

where the summation is over x, y, and z. By symmetry, since the off-diagonal terms are equivalent with interchange of the i and j indices, only six of the nine resistivity components are unique, ρ_{xx}, ρ_{yy}, ρ_{zz},ρ_{yz}, ρ_{xz}, and ρ_{xy} (Nye, 1984). For crystalline semiconductors whose space lattice is cubic, the x, y, and z directions are aligned with the equivalent $\langle 100 \rangle$ directions. In its unstrained state, cubic semiconductors such as Si and Ge are isotropic conductors. Therefore, the general resistivity tensor must reduce to a scalar. As noted by (Mason and Thurston, 1957), the tensor elements for an isotropic conductor simplify to

$$\rho_{ij} = \rho \delta_{ij}, \tag{8.10}$$

which is equivalent to the scalar resistivity, ρ, in (8.8). For notation simplification in the subsequent discussion, let the six unique resistivity coefficients, ρ_{xx}, ρ_{yy}, ρ_{zz},ρ_{yz}, ρ_{xz}, and ρ_{xy} be numbered by a single index, ρ_i, $j = 1, 2, ..., 6$.

Because strain of a crystalline lattice changes the electronic properties of a crystalline semiconductor, the occurrence of strain changes the resistivity tensor, $\rho_i(\varepsilon_m) = \rho_i(0) + \Delta\rho_i(\varepsilon_m)$. Moreover since stress, τ_k, and strain, e_m, are related by the stiffness, C_{km}, and compliance, S_{mk}, coefficients, respectively,

$$\tau_k = \sum_m C_{km} e_m, \tag{8.11}$$

and

$$e_m = \sum_k S_{mk} \tau_k, \tag{8.12}$$

the resistivity is equivalently a function of stress, $\rho_{\tau_k} = \rho_i(0) + \Delta\rho_i(\tau_k)$. Under applied stress, the change in the ith component of the electric field (or voltage) observable is given by

$$\Delta E_i(\tau_k) = \sum_j \Delta\rho_{ij}(\tau_k) j_j, \tag{8.13}$$

where j_j are the current density components flowing through the sample. In matrix form, (8.13) can be written as

$$\begin{bmatrix} \Delta E_1 \\ \Delta E_2 \\ \Delta E_3 \end{bmatrix} = \rho \begin{bmatrix} \frac{\Delta\rho_1}{\rho} & \frac{\Delta\rho_6}{\rho} & \frac{\Delta\rho_5}{\rho} \\ \frac{\Delta\rho_6}{\rho} & \frac{\Delta\rho_2}{\rho} & \frac{\Delta\rho_4}{\rho} \\ \frac{\Delta\rho_5}{\rho} & \frac{\Delta\rho_4}{\rho} & \frac{\Delta\rho_3}{\rho} \end{bmatrix} \begin{bmatrix} J_1 \\ J_2 \\ J_3 \end{bmatrix}. \tag{8.14}$$

The change in resistivity is related to the applied stress by coefficients known as the piezoresistance or π coefficients,

$$\frac{\Delta\rho_i}{\rho} = \sum_{k=1}^{6} \pi_{ik} \tau_k, \tag{8.15}$$

(also given in Chap. 6 as (6.106)). Specifically, the change of the six resistivity coefficients, $\Delta\rho_i$, $i = 1, 2, ..., 6$, normalized to the unstressed resistivity, ρ, is related to the six elements of the stress tensor, τ_k, via a 6×6 piezoresistance matrix, π_{ik}, where the piezoresistance matrix for cubic semiconductors has the form,

$$\pi_{ik} = \begin{pmatrix} \pi_{11} & \pi_{12} & \pi_{12} & 0 & 0 & 0 \\ \pi_{12} & \pi_{11} & \pi_{12} & 0 & 0 & 0 \\ \pi_{12} & \pi_{12} & \pi_{11} & 0 & 0 & 0 \\ 0 & 0 & 0 & \pi_{44} & 0 & 0 \\ 0 & 0 & 0 & 0 & \pi_{44} & 0 \\ 0 & 0 & 0 & 0 & 0 & \pi_{44} \end{pmatrix}. \tag{8.16}$$

By substituting (8.15) and (8.16) into (8.14), the change in the ith component of the electric field (or voltage) observable may be directly related to the applied stress and the current density (Mason and Thurston, 1957),

$$\begin{aligned}
\frac{\Delta E_1}{\rho} &= \{\pi_{11}\tau_1 + \pi_{12}(\tau_2 + \tau_3)\}J_1 + (\pi_{44}\tau_6)J_2 + (\pi_{44}\tau_5)J_3, \\
\frac{\Delta E_2}{\rho} &= \{\pi_{11}\tau_2 + \pi_{12}(\tau_3 + \tau_1)\}J_2 + (\pi_{44}\tau_6)J_1 + (\pi_{44}\tau_4)J_3, \qquad (8.17) \\
\frac{\Delta E_3}{\rho} &= \{\pi_{11}\tau_3 + \pi_{12}(\tau_1 + \tau_2)\}J_3 + (\pi_{44}\tau_5)J_1 + (\pi_{44}\tau_4)J_2.
\end{aligned}$$

To illustrate the role of the different terms on the observable electric field, consider a special case as illustrated in Fig. 8.2 where electrodes are placed such that only the electric field or voltage drop in the "1" direction is measured when only a current density in the same "1" direction flows through the piezoresistor. Then, (8.17) simplifies to

$$\frac{\Delta E_1}{\rho} = (\pi_{11}\tau_1 + \pi_{12}\tau_2)J_1. \qquad (8.18)$$

Since the stress τ_1 is longitudinal with respect to the current density, J_1, π_{11} may be considered as the longitudinal coefficient, π_l. Similarly, the stress τ_2 is transverse with respect to the current density, J_1, so π_{12} may be considered as the transverse coefficient, π_t. Hence, the normalized resistivity change in the "1" direction can be written as,

$$\frac{\Delta\rho_1}{\rho} = \pi_l\tau_l + \pi_t\tau_t. \qquad (8.19)$$

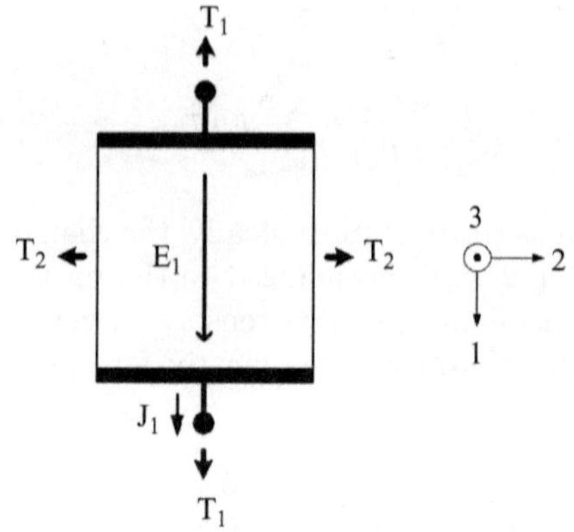

Fig. 8.2. Special case for piezoresistor sensing observable electric field, and current density colinear in same "1" direction for applied stresses longitudinal, "1", and transverse, "2", to the current density

It should be emphasized that this applies for the special case when the current is colinear with the observable electric field which primarily holds for piezoresistive transducer applications.

8.2.3 Coordinate Transformation to Arbitrary Directions

While the resistivity of an isotropic conductor is independent of orientation, the piezoresistivity of an anisotropic semiconductor such as strained Si is strongly dependent on the specific crystal direction. In cubic semiconductors, the material properties are defined in a coordinate system aligned with the principal $\langle 100 \rangle$ directions of the crystal unit cell. Longitudinal and transverse π coefficients in arbitrary directions can be obtained by transforming the piezoresistance and elastic coefficients to the new coordinate system (Mason and Thurston, 1957). Let $\hat{1}$, $\hat{2}$, and $\hat{3}$ be unit vectors along the 1, 2, 3 Cartesian axes aligned to the principal $\langle 100 \rangle$ directions of the crystal unit cell and $\hat{1}'$, $\hat{2}'$, and $\hat{3}'$ be unit vectors in the new coordinates of the rotated coordinate system as illustrated in Fig. 8.3. The new coordinates can be expressed in terms of the old coordinates using coordinate transforming rotations,

$$
\begin{aligned}
\hat{1}' &= l_1\hat{1} + m_1\hat{2} + n_1\hat{3}, \\
\hat{2}' &= l_2\hat{1} + m_2\hat{2} + n_2\hat{3}, \\
\hat{3}' &= l_3\hat{1} + m_3\hat{2} + n_3\hat{3},
\end{aligned}
\tag{8.20}
$$

where the direction cosines are given in (8.21) for rotation about the $2'$ axis,

$$
\begin{pmatrix} l_1 & m_1 & n_1 \\ l_2 & m_2 & n_2 \\ l_3 & m_3 & n_3 \end{pmatrix} = \begin{pmatrix} \cos\phi\cos\theta & \sin\phi\cos\theta & -\sin\theta \\ -\sin\phi & \cos\phi & 0 \\ \cos\phi\sin\theta & \sin\phi\sin\theta & \cos\theta \end{pmatrix}. \tag{8.21}
$$

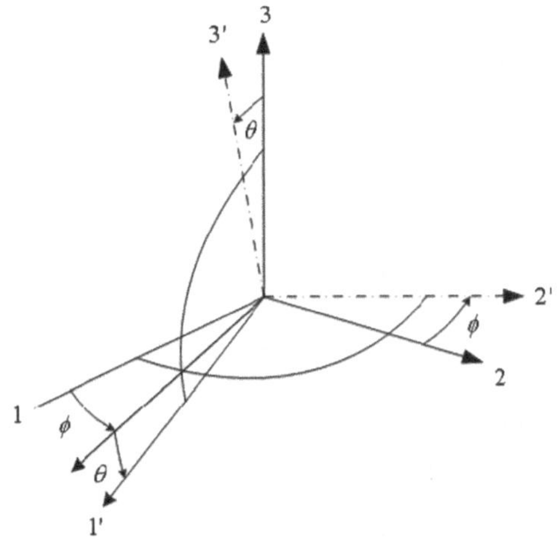

Fig. 8.3. θ-ϕ coordinate transformation with rotation about $2'$ axis

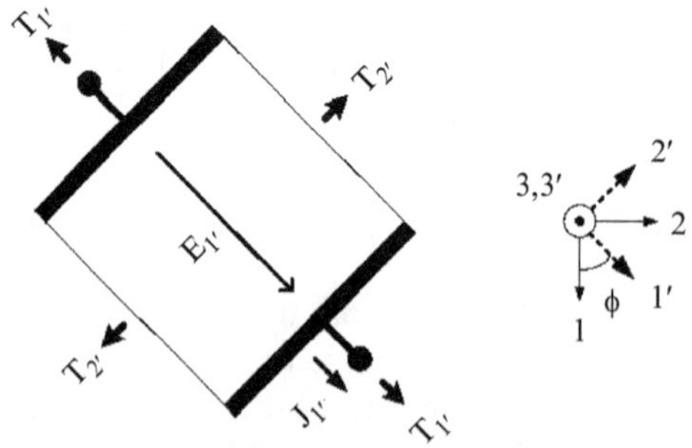

Fig. 8.4. Piezoresistive transducer rotated in the 1–2 plane while maintaining 3 and 3′ aligned with the [001] crystal axis ($\theta = 0$)

Since an important substrate for silicon logic and transducer devices is the (100) plane, we consider a special case where both the 3 and 3′ axes are aligned with the [001] crystal direction, i.e., $\theta = 0$, and the piezoresistive transducer is rotated in the 1–2 plane as shown in Fig. 8.4. By aligning the new coordinate 1′ to the direction of current flow in the rotated device shown in Fig. 8.3, the effective longitudinal and transverse piezoresistance coefficients can be obtained as follows (Mason and Thurston, 1957; Pfann and Thurston, 1961),

$$\begin{aligned} \pi_l' &= \pi_{1'1'} = \pi_{11} + 2(\pi_{44} + \pi_{12} - \pi_{11})(l_1^2 m_1^2 + l_1^2 n_1^2 + m_1^2 n_1^2), \\ \pi_t' &= \pi_{1'2'} = \pi_{12} - (\pi_{44} + \pi_{12} - \pi_{11})(l_1^2 l_2^2 + m_1^2 m_2^2 + n_1^2 n_2^2). \end{aligned} \tag{8.22}$$

Using the values of the π-coefficients obtained by Smith on bulk crystalline samples and summarized in Table 8.1 (Smith, 1954), the values of the rotated longitudinal and transverse π-coefficients are plotted in successive figures for n-type and p-type Si and Ge. In each polar plot, the rotation angle is scanned

Table 8.1. π-Coefficients for bulk Si and Ge (units 10^{-11} Pa^{-1}) (Smith 1954)

	ρ_0 (Ωcm)	π_{11}	π_{12}	π_{44}
n-Si	11.7	−102.2	53.7	−13.6
p-Si	7.8	6.6	−1.1	138.1
n-Ge	16.6	−5.2	−5.5	−138.7
p-Ge	15.0	−10.6	5.0	98.6

only from 0 to 180°. Using the convention of positive and increasing polar angle in the counter-clockwise direction beginning from 3 o'clock, curves in the top half of the polar plot correspond to positive values of the coefficient while curves in the bottom half correspond to negative values. The graphical representation of the longitudinal and transverse piezoresistive coefficients (Kanda, 1982) provides useful insight on how uniaxial stress affects carrier mobility via the piezoresistive effect and prospective directions to orient piezoresistive strain sensors to achieve a particular desired sensor response.

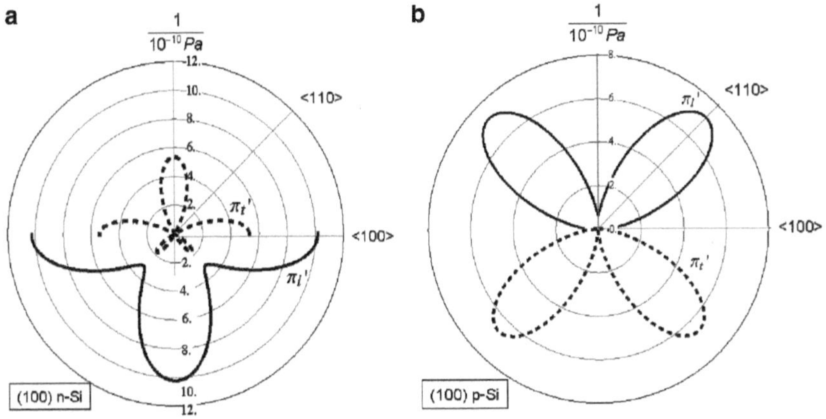

Fig. 8.5. Longitudinal and transverse π-coefficients for (**a**) n-type and (**b**) p-type Si in the (100) plane, i.e., $\theta = \angle 33' = 0$, and $\phi = \angle 11'$ from 0 to 180°

Figure 8.5 plots the longitudinal and transverse π-coefficients in the (100) plane for n-type and p-type Si. The negative polarity of the longitudinal n-Si π-coefficient illustrates qualitatively the rationale for employing tensile uniaxial stress to achieve negative resistance change (or equivalently, mobility enhancement) in n-type silicon devices. Similarly, the positive polarity of the longitudinal p-Si coefficient is consistent with the use of compressive uniaxial stress in (100) substrate p-type silicon devices, particularly in the $\langle 110 \rangle$ channel direction. The graphical results are qualitative for modeling the piezoresistance of the MOSFET inversion layer since bulk π-coefficients are used, but the insight is nonetheless useful. Recently, the piezoresistance coefficients in both n- and p-type bulk and MOSFET inversion layers on (100) and (110) Si have been summarized by (Chu et al, 2008). For piezoresistive sensor applications, the occurrence of nearly equal and opposite polarity longitudinal and transverse π-coefficients on (100) p-type Si in the $\langle 110 \rangle$ direction is useful for piezoresistive sensor bridge applications.

Similarly, the rotated longitudinal and transverse π-coefficients are shown in Fig. 8.6 for n-type and p-type Ge in the (100) plane. In contrast to Si, the longitudinal and transverse piezoresistance coefficients for n-type Ge and p-type Ge possess similar rotational symmetry. The polarity of the longitudinal π-coefficient is again negative for n-Ge and positive for p-Ge.

With the addition of a third rotation angle, ψ, a completely general coordinate transformation may be achieved according to the Euler rotation theorem (Weisstein, 1999–2009). The coordinate transformation described by the three Euler angles (θ, ϕ, ψ) is illustrated in Fig. 8.7. There are several conventions for the Euler rotations which require some care in their application. For a θ, ϕ, and ψ rotation order, the direction cosines using the "x-convention", i.e., θ rotation about the $1'$ axis, are given by (Weisstein, 1999–2009),

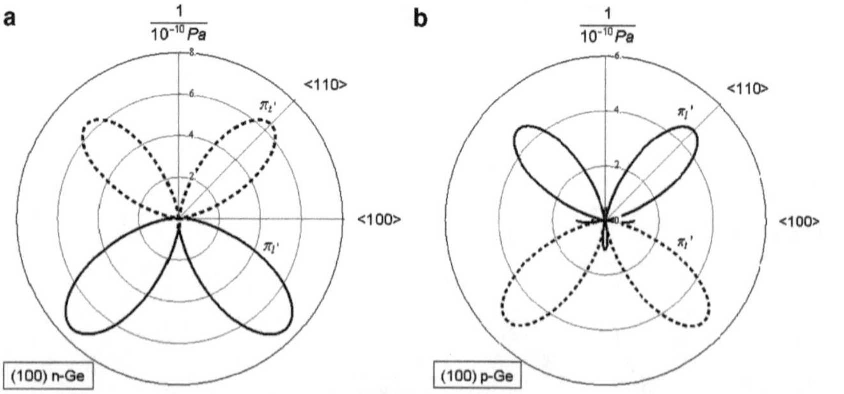

Fig. 8.6. Longitudinal and transverse π-coefficients for (**a**) n-type and (**b**) p-type Ge in the (100) plane, i.e., $\theta = \angle 33' = 0$, and $\phi = \angle 11'$ from 0 to 180°

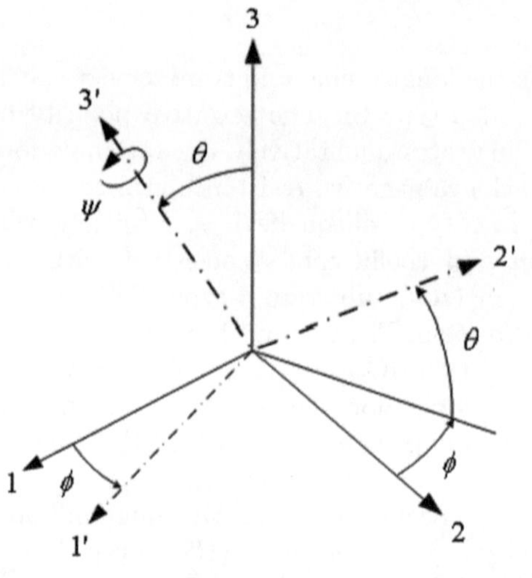

Fig. 8.7. General Euler angle coordinate transformation for rotation about the $1'$ axis (x-convention)

$$\begin{pmatrix} -\sin\phi\cos\theta\sin\psi+\cos\phi\cos\psi & \cos\phi\cos\theta\sin\psi+\sin\phi\cos\psi & \sin\theta\sin\psi \\ -\sin\phi\cos\theta\cos\psi-\cos\phi\sin\psi & \cos\phi\cos\theta\cos\psi-\sin\phi\sin\psi & \sin\theta\cos\psi \\ \sin\phi\sin\theta & -\cos\phi\sin\theta & \cos\theta \end{pmatrix}. \tag{8.23}$$

Kanda's graphical representation of the piezoresistance coefficients (Kanda, 1982) employs the "y-convention" Euler transformation, which is equivalent to the x-convention if $\phi_x = \phi_y + \pi/2$, $\psi_x = \psi_y - \pi/2$ is used (Weisstein, 1999–2009). The direction cosines in this case are given by

$$\begin{pmatrix} \cos\phi\cos\theta\cos\psi - \sin\phi\sin\psi & \sin\phi\cos\theta\cos\psi + \cos\phi\sin\psi & -\sin\theta\cos\psi \\ -\cos\phi\cos\theta\sin\psi - \sin\phi\cos\psi & -\sin\phi\cos\theta\sin\psi + \cos\phi\cos\psi & \sin\theta\sin\psi \\ \cos\phi\sin\theta & \sin\phi\sin\theta & \cos\theta \end{pmatrix}. \tag{8.24}$$

Using the Euler transformation, the longitudinal and transverse π-coefficients are computed for the (110) surface for Si and Ge and plotted in Figs. 8.8 and 8.9. A large negative longitudinal π-coefficient is evident for the $\langle 001 \rangle$ direction on (110) n-Si and a large positive longitudinal π-coefficient in the $\langle 111 \rangle$ direction on (110) p-Si. For (110) Ge, the maximum in the longitudinal π-coefficient occurs in the $\langle 111 \rangle$ direction for both doping types with negative polarity for n-type and positive polarity for p-type.

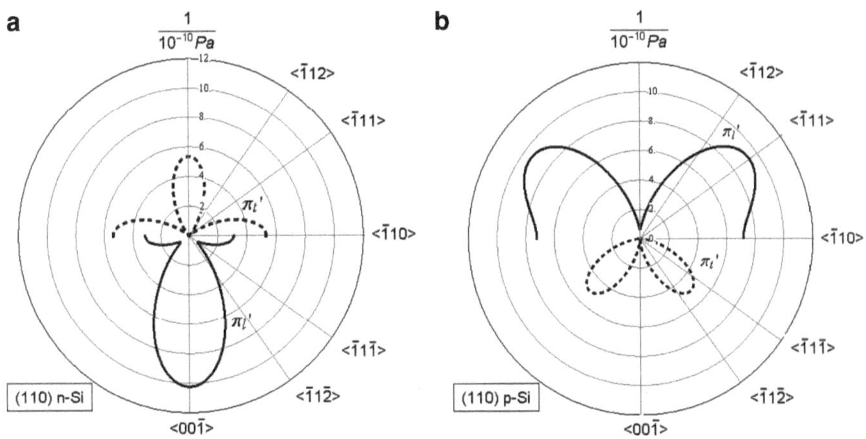

Fig. 8.8. Longitudinal and transverse π-coefficients for (**a**) n-type and (**b**) p-type Si in the (110) plane for ψ rotation from 0 to 180°

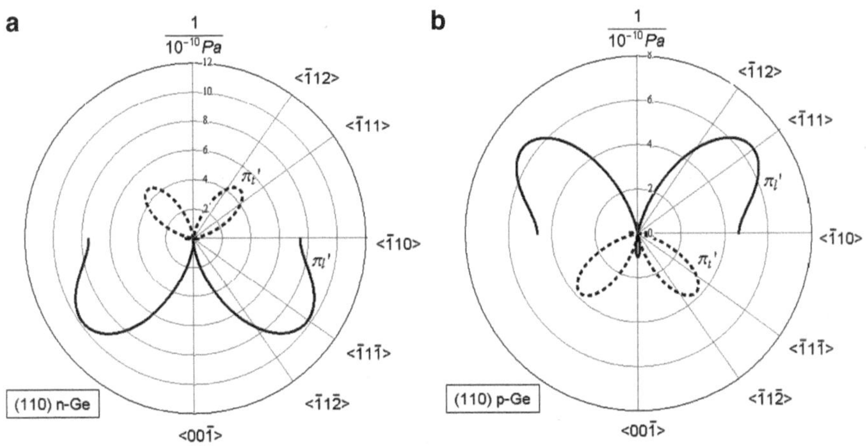

Fig. 8.9. Longitudinal and transverse π-coefficients for (**a**) n-type and (**b**) p-type Ge in the (110) plane for ψ rotation from 0 to 180°

Returning to the gauge factor, it was remarked that the gauge factor for semiconductors is up to two orders of magnitude larger than the gauge factor for metals due to the stress–strain effect on the electronic resistivity. By substituting the stress–strain relationship, (8.11) and (8.12), using the stiffness matrix,

$$C_{ij} = \begin{pmatrix} C_{11} & C_{12} & C_{12} & 0 & 0 & 0 \\ C_{12} & C_{11} & C_{12} & 0 & 0 & 0 \\ C_{12} & C_{12} & C_{11} & 0 & 0 & 0 \\ 0 & 0 & 0 & C_{44} & 0 & 0 \\ 0 & 0 & 0 & 0 & C_{44} & 0 \\ 0 & 0 & 0 & 0 & 0 & C_{44} \end{pmatrix}, \tag{8.25}$$

into the piezoresistance equation, (8.15), the normalized resistivity change can be related to strain via

$$\frac{\Delta\rho_i}{\rho} = \sum_{j,k=1}^{6} \pi_{ik} C_{kj} \tau_j = \sum_{j=1}^{6} m_{ij} \varepsilon_j \tag{8.26}$$

for direct estimation of the GF for some simple cases. Here, m_{ij} is the elastoresistance tensor,

$$m_{ij} = \sum_{k=1}^{6} \pi_{ik} C_{kj}, \tag{8.27}$$

which has the same form as the piezoresistance tensor. Using the stiffness coefficients in Table 8.2 (Brantley, 1973) and the piezoresistance coefficients previously listed in Table 8.1, the unique elastoresistance coefficients are computed and tabulated in Table 8.3.

Table 8.2. Stiffness[a] and compliance[b] coefficients for bulk Si and Ge in units of (a) 10^{11} Pa and (b) 10^{-11} Pa^{-1} (Brantley 1973)

	C_{11}	C_{12}	C_{44}	S_{11}	S_{12}	S_{44}
Si	1.657	0.639	0.7956	0.768	−0.214	1.26
Ge	1.292	0.479	0.670	0.964	−0.260	1.49

Table 8.3. Elastoresistance coefficients for bulk Si and Ge (non-dimensional)

	ρ_0 ($\Omega\cdot$cm)	m_{11}	m_{12}	m_{44}
n-Si	11.7	−100.7	58.0	−10.8
p-Si	7.8	9.5	1.7	109.9
n-Ge	16.6	−12.0	−12.2	−92.9
p-Ge	15.0	−8.9	3.8	66.1

Consider the special case where the longitudinal strain is in the principal [100] direction and originates from simple uniaxial stress and where the resistance change is measured in the same direction. Then, the gauge factor is given by

$$\mathrm{GF}|_{[100]\text{semiconductors}} = \frac{\Delta R_1/R}{\varepsilon_1} = (1+2\nu) + \frac{\Delta\rho_1/\rho}{\varepsilon_1}. \tag{8.28}$$

The normalized resistivity change follows from (8.26) (Pfann and Thurston, 1961),

$$\frac{\Delta\rho_1}{\rho} = m_{11}\varepsilon_1 + m_{12}\varepsilon_2 + m_{13}\varepsilon_3 = (m_{11} - 2\nu m_{12})\varepsilon_1. \tag{8.29}$$

Hence, the gauge factor for the silicon strain gauge becomes

$$\mathrm{GF}|_{[100]\text{semiconductors}} = \left(1 - 2\frac{S_{12}}{S_{11}}\right) + \left(m_{11} + 2\frac{S_{12}}{S_{11}} m_{12}\right), \tag{8.30}$$

where the definition of the Poisson coefficient in terms of the compliance coefficients is used,

$$\nu = -\frac{S_{12}}{S_{11}}. \tag{8.31}$$

For example, in the [100] direction, the value of ν is 0.279 for silicon.

The gauge factor of the silicon strain gauge in an arbitrary direction may be obtained by applying the coordinate transformation used previously for the π-coefficients. By aligning the new coordinate $1'$ to the direction of current flow in the rotated device as shown in Fig. 8.3, a similar transformation may be applied to obtain the elastoresistance coefficients, $m_{1'1'}$ and $m_{1'2'}$, in arbitrary directions (Pfann and Thurston, 1961).

$$\begin{aligned} m_{1'1'} &= m_{11} + 2(2m_{44} + m_{12} - m_{11})(l_1^2 m_1^2 + l_1^2 n_1^2 + m_1^2 n_1^2) \\ ,m_{1'2'} &= m_{12} - (2m_{44} + m_{12} - m_{11})(l_1^2 l_2^2 + m_1^2 m_2^2 + n_1^2 n_2^2) \end{aligned} \tag{8.32}$$

The rotated elastoresistance coefficients are plotted in Fig. 8.10 for n-Si and p-Si in the (100) plane.

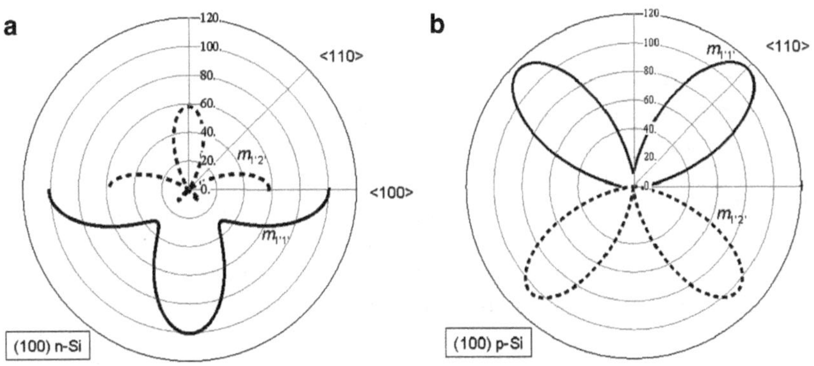

Fig. 8.10. Elastoresistance coefficient, $m_{1'1'}$, for (**a**) n-type and (**b**) p-type Si in the (100) plane

Similar transformations may be applied to the compliance coefficients. The compliance coefficients, $S_{1'1'}$ and $S_{1'2'}$, in arbitrary directions, are given by (Brantley, 1973),

$$
\begin{aligned}
S_{1'1'} &= S_{11} + 2(\frac{1}{2}S_{44} + S_{12} - S_{11})(l_1^2 m_1^2 + l_1^2 n_1^2 + m_1^2 n_1^2), \\
S_{1'2'} &= S_{12} - (\frac{1}{2}S_{44} + S_{12} - S_{11})(l_1^2 l_2^2 + m_1^2 m_2^2 + n_1^2 n_2^2),
\end{aligned}
\tag{8.33}
$$

and are plotted in Fig. 8.11 for (100) Si.

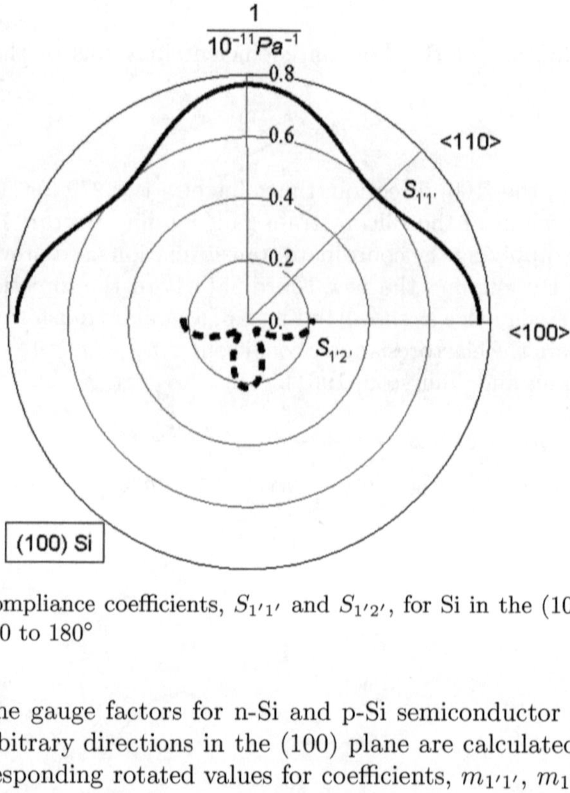

Fig. 8.11. Compliance coefficients, $S_{1'1'}$ and $S_{1'2'}$, for Si in the (100) plane for ψ rotation from 0 to 180°

Finally, the gauge factors for n-Si and p-Si semiconductor strain gauges aligned to arbitrary directions in the (100) plane are calculated using (8.30) and the corresponding rotated values for coefficients, $m_{1'1'}$, $m_{1'2'}$, $S_{1'1'}$, and $S_{1'2'}$, and are shown in Fig. 8.12. Note that the gauge factor is greater than 100 for $\langle 100 \rangle$ n-Si and $\langle 110 \rangle$ p-Si strain gauges, much larger than the gauge factor of metallic strain gauges, which was and continues to be a key motivation for the development of high-sensitivity crystalline semiconductor strain gauges and, as will be seen next, integrated piezoresistive stress transducers.

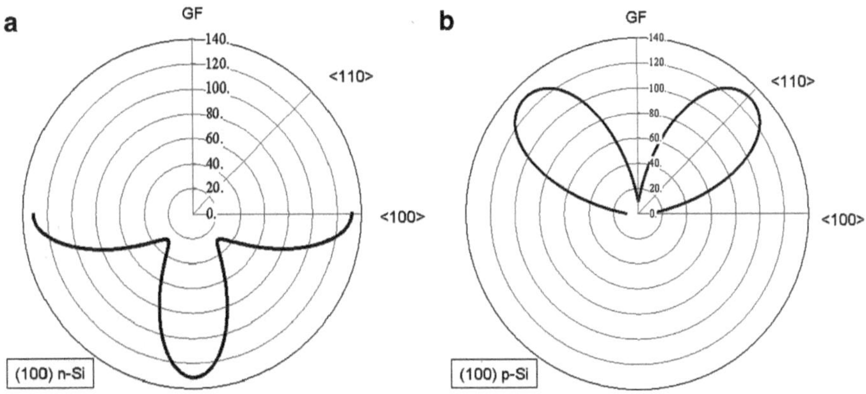

Fig. 8.12. Gauge factors for (**a**) n-Si and (**b**) p-Si silicon strain gauges for arbitrary directions in the (100) plane for ψ rotation from 0 to 180°

8.3 INTEGRATED PIEZORESISTIVE STRESS TRANSDUCERS

In contrast to discrete strain gauge sensors that are assumed to measure the local strain without significantly affecting the stiffness of the structure in question, integrated stress transducers are devices that integrate the piezoresistive strain gauge within a sensing structure. The combination of microelectromechanical systems (MEMS) and semiconductor strain gauges has enabled the development of integrated stress transducers. A conventional discrete strain transducer is contrasted with a MEMS piezoresistive pressure stress transducer (microphone) and a fixed stress-enhanced p-channel MOSFET in Fig. 8.13.

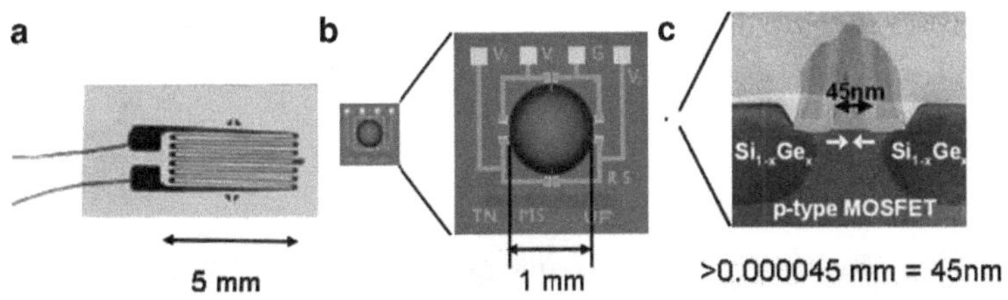

Fig. 8.13. Applications of strain and stress: (**a**) Discrete strain gauge (Omega.com, 2003–2009), (**b**) Si MEMS piezoresistive variable stress transducer with four integrated Si piezoresistors (Arnold et al, 2001), and (**c**) fixed stress-enhanced transistor (Thompson et al, 2004a)

The gauge factor described in the previous section defines the sensitivity of discrete strain gauges. The input is a local strain transferred to the strain gauge and the output is the resistance change incurred. The local strain is equal to the actual strain if a perfect bond exists between the strain gauge and the structural component being measured. The strain gauge is generally not the same as the structural component whose strain is being measured. In particular, metal films can be made sufficiently thin so that the strain gauge is much more compliant than the structural component so that the gauge itself does not affect the strain of the structural component.

Although it is possible to construct a discrete thin and compliant silicon strain gauge in the same manner as a metal film strain gauge, the vast silicon integrated circuit manufacturing knowledge base coupled with the fortuitous mechanical properties of silicon (Peterson, 1982) have enabled the fabrication of MEMS transducers that integrate silicon piezoresistors with a mechanical structure made of the same silicon material. We will consider two silicon transducers. The first example is a simple canonical cantilever-based force transducer with four integrated piezoresistors. The second example is a more elaborate circular diaphragm-based piezoresistive MEMS microphone.

8.3.1 Canonical Cantilever-Based Piezoresistive Force Transducer

A cantilever, defined as a beam with one end clamped and one end free, provides a powerful structure for transducing forces since the bending moment is proportional to the length of the lever arm between the point load and the clamp (Senturia, 2001). For the same applied force, a longer cantilever creates a larger bending moment, which produces a larger curvature. Curvature of the beam causes opposing strain on the top and bottom surfaces of the cantilever, which sandwich a neutral axis within the cantilever. The strain and hence the stress, $T(x)$, is maximum furthest away from the neutral axis, which is at the top and bottom surfaces of the cantilever. Consider a cantilever with length, L, width, W, and a height, H, as illustrated in Fig. 8.13. The position along the length of the unbent cantilever is defined as x where $x = 0$ at the clamped end and $x = L$ at the free end. With application of a point load, F, at the free end, the resulting moment causes the beam to bend downward as shown. Each point of a bent cantilever encounters a transverse displacement, $y(x)$, shown as a curve in Fig. 8.14.

Structure Interactions

In order to employ the cantilever structure in an integrated piezoresistive force transducer, the location and value of maximum stress is needed as a function of the applied force. The bending equation for the clamped-free cantilever, assuming small displacement, is given by (Senturia, 2001)

$$\frac{\mathrm{d}^2 y}{\mathrm{d}x^2} = \frac{F}{EI}(L - x), \tag{8.34}$$

where $I = WH^3/12$ is the moment of inertia of the beam cross section. Here, the small displacement assumption is valid when the bending angle, $\theta(x)$, is small enough such that $\cos\theta \sim 1$. By definition of an ideal clamped end, the

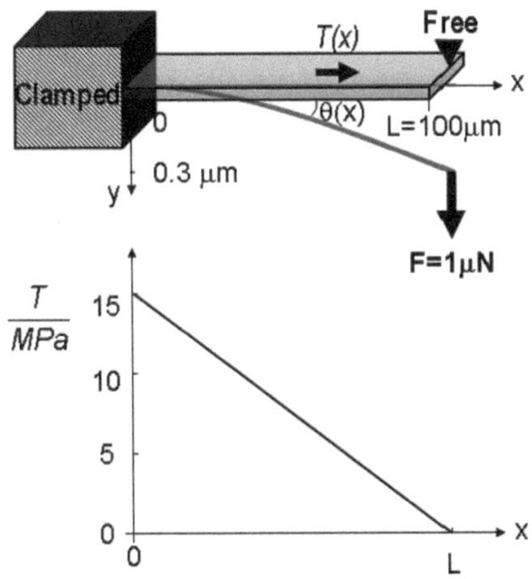

Fig. 8.14. Schematic of a clamped-free cantilever beam with a point force applied to the free end. The dimensions of the cantilever beam are $L = 100$ μm, $W = 10$ μm (into the page), and $H = 2$ μm (thickness). Also shown is the deflection, $y(x)$, and the longitudinal stress, $T(x)$, as a function of position, x, for an applied 1 μN force and a Young's modulus of 160 GPa. (not to scale)

boundary conditions, $y = 0$ and $\frac{dy}{dx}|_{x=0} = 0$, must hold. With these boundary conditions, the displacement as a function of position is obtained,

$$y(x) = \frac{FL}{2EI}x^2\left(1 - \frac{x}{3L}\right). \tag{8.35}$$

In the example shown in Fig. 8.14, the tip displacement for a 1 μN point load is approximately 0.3 μm.

For a piezoresistive force transducer or sensor, we wish to place the piezoresistor near the optimum location where the strain and corresponding stress are a maximum. The strain at the surface of the cantilever is equal to the product of the curvature at that point times the transverse distance from the neutral axis or half of the beam thickness, $H/2$, for a homogenous beam. Since the curvature under small displacement is equal to d^2y/dx^2 given by (8.34), the strain at the top surface, $\varepsilon_{top}(x)$, at any point x is given by

$$\varepsilon_{top}(x) = \frac{F}{EI}(L - x)\left(\frac{H}{2}\right) \tag{8.36}$$

and the corresponding uniaxial stress, $T(x)$, by the product of the strain and the Youngs modulus, E

$$T(x) = \frac{F}{I}(L - x)\left(\frac{H}{2}\right). \tag{8.37}$$

The uniaxial stress is approximately in the x direction and is tensile on the top surface and compressive on the bottom surface for the direction of applied point load shown in Fig. 8.14. As expected, the maximum stress occurs at the clamped end, $x = 0$,

$$T_{max} = F\left(\frac{LH}{2I}\right), \tag{8.38}$$

which is the location of the piezoresistor as seen in Fig. 8.15. The next consideration is the number and orientation of the piezoresistors.

Fig. 8.15. Top view of clamped-free cantilever with two longitudinal and two transverse piezoresistors located in the stress concentration region

Piezoresistor Configuration

The basis for the configuration of the piezoresistors is the ease of measuring the resulting resistance change from the stress induced by the applied force. As a force sensor, a key figure of merit is the sensitivity which in this case is the measured output voltage per input force. Therefore, the resistance change needs to be converted into an output voltage. As seen in (8.15), the normalized resistivity change may be resolved into two components, that induced by the longitudinal stress and a second component due to the transverse stress. Here, longitudinal refers to the case where the direction of the electric field and current is collinear with the applied stress, and transverse is where the direction of the electric field and current is perpendicular with the applied stress.

For crystalline semiconductors, since the resistivity change dominates the resistance change, i.e. the geometric contribution is negligible, (8.15) may be approximated as

$$\frac{\Delta R}{R} \approx \pi_l T_l + \pi_t T_t. \tag{8.39}$$

Since the unstressed resistance is R, the total resistance in general is equal to

$$R_{\text{total}} = R + \Delta R = R(1 + \pi_l T_l + \pi_t T_t). \tag{8.40}$$

If only one piezoresistor is employed, an output voltage may be measured by forcing a constant current, I, through the piezoresistor. However, the voltage has an offset that is usually much larger than the change in voltage due to the stress-induced change in resistance,

$$V_{\text{total}} = R_{\text{total}} I = RI + \Delta RI = V_{\text{offset}} + \Delta V. \tag{8.41}$$

Another question is the orientation of the piezoresistor. For some orientations, the stress parallel to the current flow may be much larger than the stress perpendicular to the current flow and vice versa, simplifying (8.39) and (8.40). In the simple case of the cantilever shown in Fig. 8.14, the tension is uniaxial along the length of the cantilever and the transverse tension is zero. Hence, (8.39) and (8.40) greatly simplify for the two orientations illustrated in Fig. 8.16. The total resistance formulations for the two orientations are given by,

$$\begin{aligned} R_{\text{total,longitudinal}} &= R(1+\pi_l T) = R + R\pi_l T, \\ R_{\text{total,transverse}} &= R(1+\pi_t T) = R + R\pi_t T. \end{aligned} \tag{8.42}$$

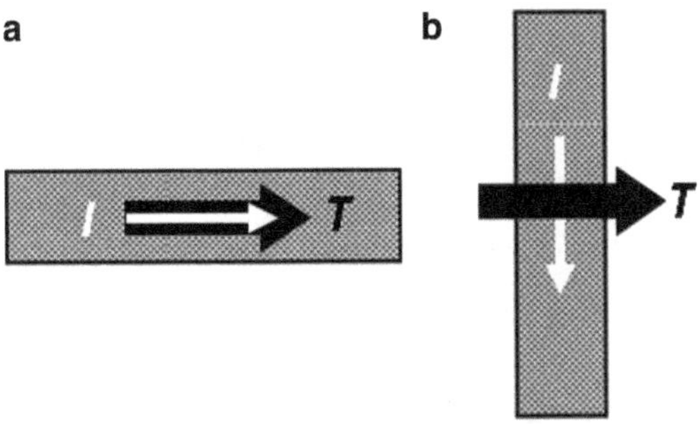

Fig. 8.16. Two piezoresistor orientations with respect to stress, T: (**a**) Current parallel (longitudinal) to stress, (**b**) current perpendicular (transverse) to stress

Of course, if both longitudinal and transverse stresses exist, the general formulation holds. However, part of the design of piezoresistive stress transducers involves designing the structure and placing the piezoresistors such that (8.42) is a valid approximation.

If a crystalline direction exists where π_l and π_t are equal and opposite in polarity, then the piezoresistors can be configured in a bridge circuit, which ideally has zero offset voltage and the output voltage is proportional to the input force. Examining the rotated π_l' and π_t' curves in Figs. 8.5 and 8.6, we find that (100) p-Si and (100) n- and p-Ge approximately meet this condition in the $\langle 110 \rangle$ directions. Furthermore, the aforementioned bridge circuit is realized in a Wheatstone bridge and is shown in Fig. 8.17. The circuit on the left is a general Wheatstone bridge where the four legs of the bridge are distinct in value.

Consider the ideal Wheatstone bridge. The ideal Wheatstone bridge, shown on the right in Fig. 8.17(b), occurs when the nominal unstressed

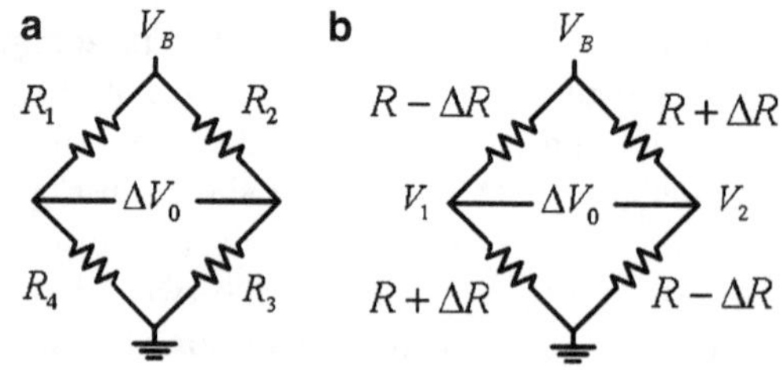

Fig. 8.17. (**a**) General Wheatstone bridge configuration of four piezoresistors. (**b**) Ideal Wheatstone bridge

resistance values of the four piezoresistors are equal in magnitude and opposite in polarity. The voltages at the two nodes are given by

$$\begin{aligned} V_1 &= \frac{R+\Delta R}{2R} V_B, \\ V_2 &= \frac{R-\Delta R}{2R} V_B, \end{aligned} \tag{8.43}$$

and the output voltage is their difference,

$$\Delta V_0 = \frac{\Delta R}{R} V_B. \tag{8.44}$$

Sensitivity

To find the overall sensitivity of the piezoresistive force transducer, the expression for the normalized resistance change in (8.39) is entered into (8.44).

$$\Delta V_0 = \pi_l T V_B. \tag{8.45}$$

Finally, since the piezoresistor is located near the clamped end, we assume that the stress is equal to the maximum stress given by (8.38), resulting in a relationship between the output voltage and the applied force,

$$\Delta V_0 = F\left(\frac{6\pi L V_B}{WH^2}\right). \tag{8.46}$$

The proportionality factor multiplying the input force is the overall sensitivity or the mechanical-to-electrical voltage sensitivity, S_{me}, of this canonical cantilever-based piezoresistive force transducer,

$$S_{me} = \frac{\Delta V_0}{F} = \frac{6\pi L V_B}{WH^2}. \tag{8.47}$$

The sensitivity is proportional to the bias voltage and to the length of the cantilever and inversely proportional to the cantilever width and height squared. For an applied bias voltage of 1 V and the dimensions given in Fig. 8.14 assuming a $\langle 110 \rangle$ cantilever on (100) p-Si, the estimated sensitivity is $S_{me} = 10.8\frac{\text{mV}}{\mu\text{N}}$.

Some Ramifications of Piezoresistive Stress Transducers

We shall now consider some ramifications of piezoresistive transducers. First, the sensitivity is proportional to the applied bias voltage, V_B. Unlike energy-conserving transduction techniques such as piezoelectric or capacitive, piezoresistive transducers dissipate power in order to sense. The dissipation of energy in the resistance has the consequence that the piezoresistor also introduces thermal noise. Second, piezoresistive MEMS transducers integrate the device or sensor and the structure, hence the mechanical-to-electrical transduction or sensitivity depends on how the input force "effort" "flows" through the mechanical structure and results in stress at the site of the piezoresistor. The analysis above is based on a small-displacement approximation and assumes an ideal clamped boundary. To maintain a linear response, the input force cannot exceed an upper limit set by the valid range of the small-displacement assumption. Naturally, a more detailed nonlinear finite element numerical analysis can be used to investigate 2nd order effects. In addition, bulk π-coefficients were used that were obtained for samples with relatively low doping concentrations. At high doping concentrations above 10^{18} cm^{-3}, the π-coefficients decrease as summarized by (Harley and Kenny, 2000) and shown in Fig. 8.18. Despite the limitations, the above analytical discussion provides insight into the key relevant physics of piezoresistive transducers.

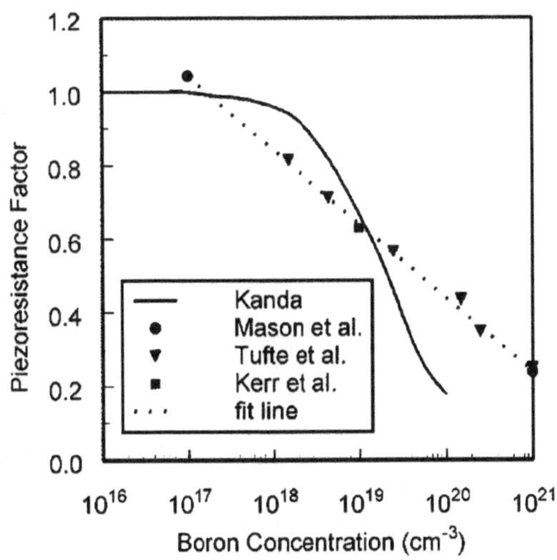

Fig. 8.18. Attenuation factor for bulk p-type Si piezoresistance coefficients as a function of boron concentration (Harley and Kenny, 2000)

The integrated piezoresistive cantilever device may be realized by fabricating p-type silicon piezoresistors in the ⟨110⟩ directions on a cantilever micromachined out of a (100) n-type Si wafer. Details of the MEMS fabrication process and methods of electrically isolating the piezoresistors from each other and interconnecting the piezoresistors in a Wheatstone bridge are covered elsewhere such as in (Senturia, 2001).

8.3.2 Circular Diaphragm MEMS Piezoresistive Microphone

A similar approach may be applied to other mechanical structures to derive the mechanical-to-electrical transducer sensitivity. The previous example considered a point load. Another general load of interest is a uniform distributed load such as a pressure for a wide range of applications from pressure transducers to acoustic microphones. Analogous to biological systems, an appropriate structure for a pressure load is a diaphragm. An axially symmetric circular diaphragm is desirable due to the absence of nonuniform stress regions found, for example, near the corners of square diaphragms. Only a qualitative discussion is presented here.

The general structural mechanics of a circular diaphragm under a uniform pressure load with built-in in-plane tension load is given by (Sheplak and Dugundji, 1998). The inclusion of a built-in load, N_0, models the presence of residual stress that typically occurs in deposited thin films. A top view of an unstressed diaphragm and the transverse displacement under uniform pressure load is shown in Fig. 8.19.

Key results of interest for the design of an integrated piezoresistive pressure transducer on a circular diaphragm are, similar to the canonical cantilever-based transducer, the linear diaphragm displacement as a function of input pressure under the small displacement approximation and the radial and tangential stresses as a function of radius. For placement of the piezoresistors, the location of maximum stress in the case of the circular diaphragm is near the perimeter as illustrated in Fig. 8.20.

Finally, to achieve a Wheatstone bridge with four active piezoresistors, two arc and two taper piezoresistors are placed as shown in Fig. 8.21 in the stress concentration region of the circular diaphragm, analogous to the transverse and longitudinal piezoresistors in the rectangular cantilever beam.

Two generations of a circular membrane piezoresistive MEMS microphone have been developed (Sheplak et al, 1998; Arnold et al, 2001) to date. The second-generation piezoresistive microphone is shown in Fig. 8.22(a) and compared to the earliest reported piezoresistive microphone, which employed a bulk Ge cantilever bimorph (Burns, 1957). Although much progress has been made, there is still ample room for improvement.

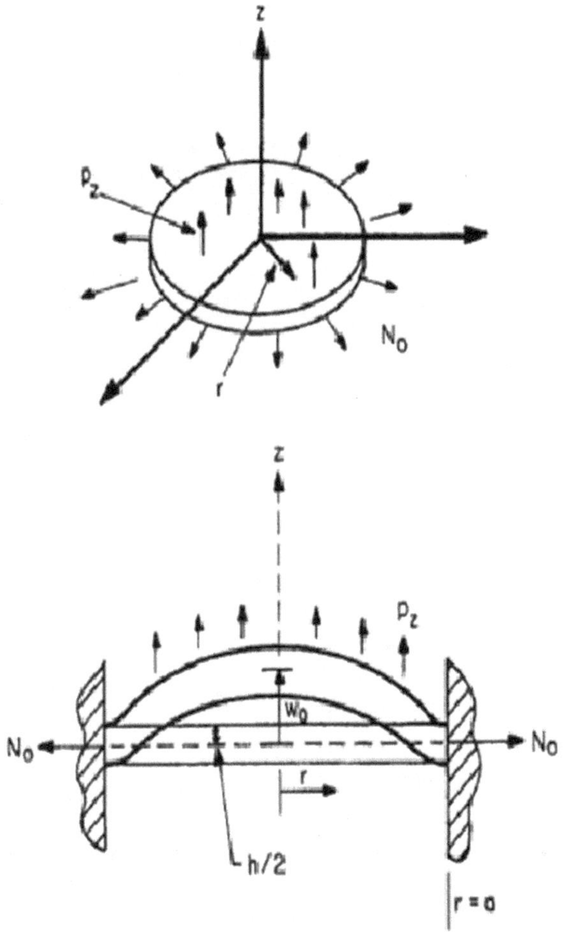

Fig. 8.19. Schematic of (**a**) circular diaphragm under in-plane tension load and (**b**) deflection under uniform pressure, $P = P_z$. The membrane is under in-plane stress, N_0 (Sheplak and Dugundji, 1998)

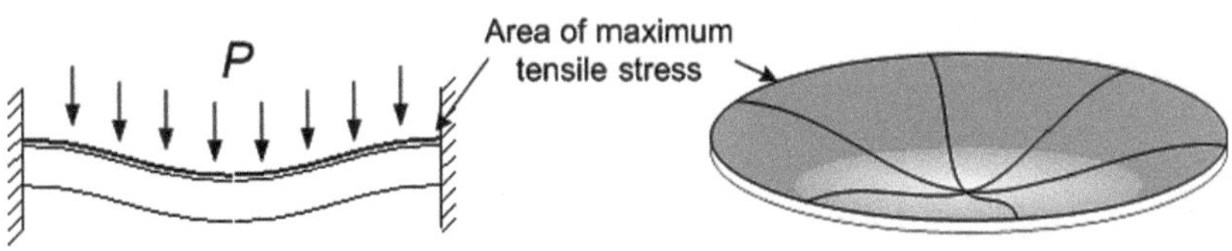

Fig. 8.20. Stress concentration region of circular membrane under uniform pressure load (Homeijer et al, 2006)

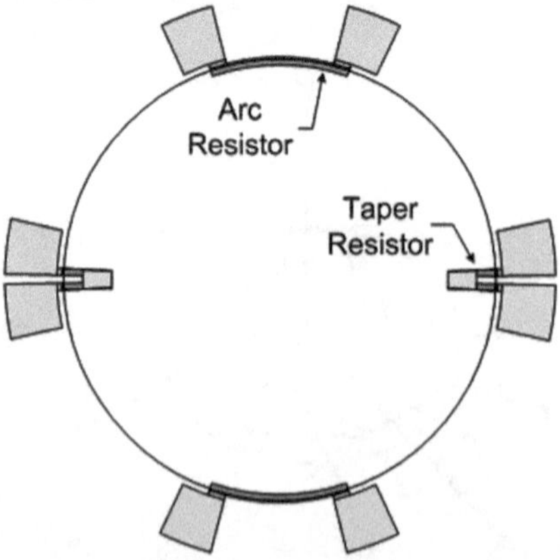

Fig. 8.21. Schematic of placement of arc and taper piezoresistors for a circular membrane piezoresistive transducer (Homeijer et al, 2006)

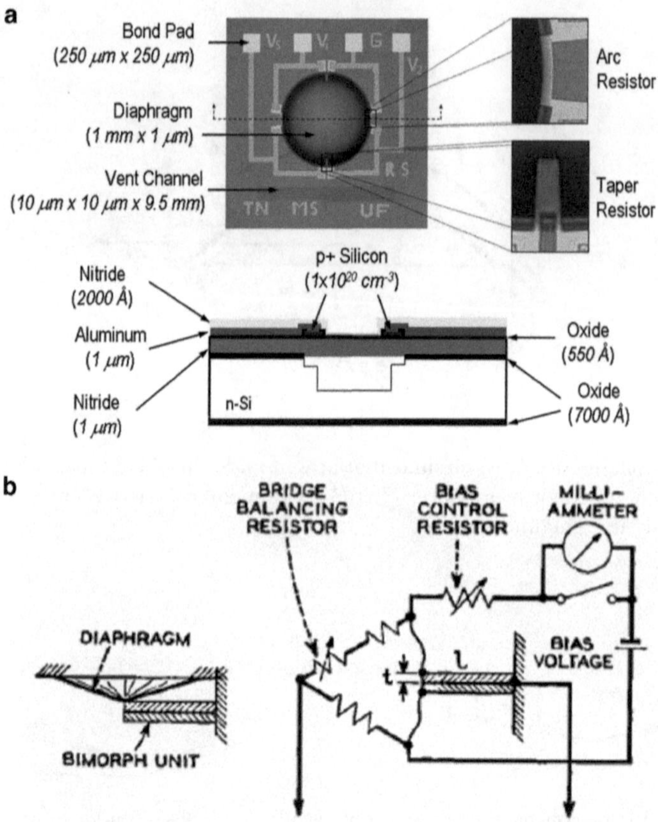

Fig. 8.22. (**a**) Photograph of second-generation Florida MEMS silicon piezoresistive microphone with circular diaphragm (Arnold et al, 2001). (**b**) First piezoresistive microphone with metallic diaphragm and bulk Ge bimorph piezoresistive cantilever transducer designed for type 500 telephone (Burns, 1957)

9

Strain Effects on Optoelectronic Devices

9.1 INTRODUCTION

Optoelectronics has become an essential part of modern lives in computer, consumer electronics, and communication. Optoelectronic devices either generate light or utilize light in their operation. Typical examples include light-emitting diodes in remote controls, battery chargers, even traffic lights, photodetectors in alarms, digital cameras, radars, and lasers in labs and bar-code scanners in stores. Optoelectronics is based on the quantum mechanical interactions between light and matter, which in most cases are semiconducting materials. In these interactions, photons as energy quanta of the light are either emitted or absorbed. These two basic quantum processes are similar to the phonon transition processes in semiconductors, involving two electronic states with conservation of both energy and momentum. The energy of a photon is determined by its wavelength λ, $E = hc/\lambda$, where h is the Plank constant and c is the speed of light, and its momentum is given by h/λ. Photon energies for the visible light are between 2 and 3 eV, which is the range of semiconductor bandgaps, with their wavelengths between 4000 and 7000 Å. Compared to the electron momentum hk in a solid, where k is the electron wave vector and in the order of π/a, where a is the lattice constant around the order of several Å, the typical momentum of a photon is very small. Therefore, the photon transitions between electronic states in solids are considered "vertical." That is to say, the electron momentum change for photon transitions in the Brillouin zone is often neglected, unlike the phonon processes. Also, phonons do not have angular momentum, but photons generally have definite polarization configuration and carry specific angular momentum, and thus angular momentum conservation has to be complied in photon transitions.

Photon emission is caused by radiative electron transitions in semiconductors, such as electron–hole recombination. Photon absorption is the reverse process. Since photon transition involves two electronic states, the transition probability is determined by the electronic state properties such as energy and wave function. The collective processes of photons, i.e., the light emission

Y. Sun et al., *Strain Effect in Semiconductors: Theory and Device Applications*, DOI 10.1007/978-1-4419-0552-9_9,

and absorption then depend on the semiconductor band structures. Strain can alter the band structure, then will also have effects on optoelectronic device performance. For solid-state lasers based on semiconductor heterostructures, strain is often inevitably induced. Study on strain effects on the gain, linewidth, quantum efficiency, and so on is an important subject. In optoelectronics, especially for quantum well lasers, strain engineering is already extensively employed to improve device performance.

In this chapter, we are going to investigate strain effects on optoelectronic devices. For this purpose, we first qualitatively illustrate how light interacts with semiconductor by the quantum processes of light emission and absorption in various optoelectronic devices and introduce the important material parameters that determine the device performance and could be varied by strain. Following that, we will quantitatively formulate the optical processes in semiconductors and use quantum well lasers as a typical optoelectronic device example to study how strain changes the device parameters.

9.2 STRAIN EFFECTS IN OPTOELECTRONIC DEVICES: AN OVERVIEW

9.2.1 Photon Emission and Absorption

Light is electromagnetic waves in a specific range of wavelength. When light interacts with matter, the electromagnetic field of the light couples two separate electronic states. The energy of the light can only be absorbed or emitted by quanta, which are called photons whose energy is related to the light wavelength λ or frequency ω by, $\hbar\omega$. An electromagnetic wave is a transverse wave in homogeneous and isotropic media, so that light is also a transverse wave. That is to say, the electromagnetic field vector is perpendicular to the propagation direction of the light. Photon absorption and emission in solids are illustrated in Fig. 9.1. An electron absorbs a photon and transits from the initial state labeled as E_a to the final state E_b in the absorption process. If this transition is across the bandgap of the semiconductor, then an electron–hole pair is generated by the photon absorption. But photon absorption can only take place between two electronic states if energy conservation is conserved and selection rule is complied. Photon emission is a radiative process such as electron–hole recombination. For photon emission, there are two cases, spontaneous and stimulated emission. In the spontaneous emission, electron–hole pairs that are thermally excited or injected recombine to emit photons with random phase and direction. However, a single photon traveling through a semiconductor with a proper energy is able to generate an identical second photon by stimulating the recombination of an electron–hole pair. The second photon has the same wavelength and the same phase as the first photon. When this multiplication process continues, it leads to strong light amplification. This is the working principle of laser (Light Amplification by Stimulated Emission of Radiation).

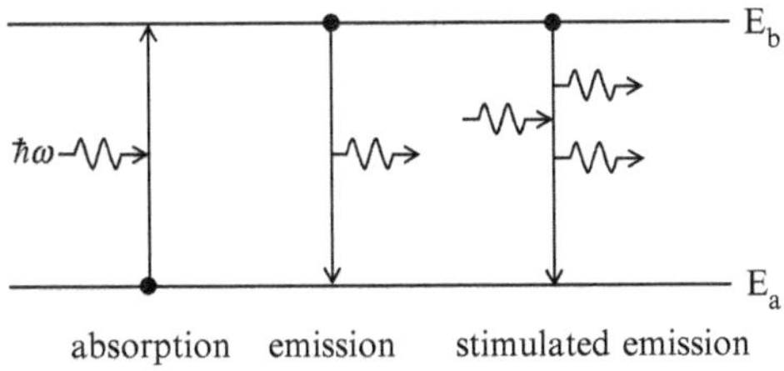

Fig. 9.1. Photon absorption, spontaneous photon emission and stimulated photon emission. In either photon absorption or photon emission process, two electronic states are involved

Photon transition selection rules are determined by the symmetry property of the electromagnetic wave. In coupling the two electronic states based on the dipole approximation, the electromagnetic wave acts like an odd operator such as **p**, the momentum operator. When photon transition occurs in atoms, the parities of the initial and final states must be different. In solids, this requires the symmetry of the initial and final states to be opposite (even and odd). The s and p states have even and odd symmetry, respectively, and thus photon transition can take place between conduction and valence bands. If the state symmetry is the same, photon transition is forbidden. In semiconductors, photon transition occurs either between two bands with definite symmetry such as s and p symmetry, or between states that contain, for example, p-components in the initial state and s-components in the final state. This is the reason why light absorption can also occur within the conduction band itself, since away from the Γ point, some p-characteristic is also mixed into the electron wave function.

A photon can have definite angular momentum depending on its polarization. Two basic configurations are the left circular and right circular, or σ^- and σ^+ polarizations, which carry $-\hbar$ and $+\hbar$ angular momentum, respectively, in the direction of the quantization axis that is parallel to $\mathbf{k}$, where $\mathbf{k}$ is the light wavevector. The polarization is described by a polarization vector $\hat{e}$. It depicts the phase variation patterns with elapsing time. More in detail, if we decompose the electric field vector into two orthogonal components in the perpendicular plane to the propagation direction, these two components have exactly the same amplitude and are 90° out of phase. Then the polarization vector is $(\hat{x} - \mathrm{i}\hat{y})/\sqrt{2}$ for σ^- photons and $(\hat{x} + \mathrm{i}\hat{y})/\sqrt{2}$ for σ^+ photons assuming $\mathbf{k}$ is along z. A linearly polarized photon can be considered as a coherent superposition of a left and right circularly polarized one with equal frequencies and wave vector k. The term coherent means that two light beams have a fixed-phase relation relative to each other. In the photon transition processes, angular momentum conservation has also to be obeyed. Thus, photon transition can only take place between two states with angular momentum differed by $\hbar$.

9.2.2 Working Principles for Photodiodes and Quantum Well Lasers

Photodiode and quantum well laser can serve as two examples for illustration of photon transitions in solids. Photodiodes are often used in photodetectors. The purpose of a photodiode is to converts light into electric current or voltage, in contrast to another more familiar optoelectronic device, the light-emitting-diode (LED), which convert electric power into light. Photodiode is the core part of many photodectors, charge-coupled devices (CCD), and CMOS image sensors, etc. In Fig. 9.2a we show the simplest structure of a passive CMOS image sensor pixel (Fossum, 1995), which is composed of a photodiode and a transfer transistor. The photodiode is normally operated in a reverse-biased mode. The active region for absorbing the light is the junction depletion region. When a photon is absorbed, it generates an electron–hole pair across the bandgap and then the electron and hole are separated by the strong electric field in the depletion region and create a reverse current or voltage, which is sensed by the auxiliary circuitry. When the transfer transistor is pulsed, the charges collected by the photodiode are sensed and amplified. One intrinsic layer is often added between the n- and p-doped layer to increase the dimension of the depletion region and thus increase the absorption of the light. For increasing responsivity, the photodiode can be operated with a much higher reverse bias and then allows the generated carriers to be multiplied by an avalanche breakdown. This type of photodiodes are called avalanche photodiodes, which have an internal gain. Photodiode is also the essential part of an phototransistor where the photodiode is integrated in the base. A light-induced current can then be amplified by the transistor's current gain.

Typical key parameters of a photodiode are 1) the responsivity, which is related to quantum efficiency, defined by the photocurrent divided by optical power; 2) the active area, where the photodiode absorbs the light; 3) the dark current, which is the current with the absence of light, with sources possibly being the background radiation and p-n diode reversely biased saturation current; 4) the bandwidth, which is determined by the rise and fall time of the photocurrent, influenced by the diode capacitance, and determines the photodiode speed. Traditional photodiodes are made from bulk semiconductors including Si, Ge, InGaAs, etc. Recently, quantum wells and superlattices are developed and employed to make new generation photodiodes for higher sensitivity, higher speed, and lower noise. These heterostructures include GaAs/$Al_{1-x}Ga_xAs$, $In_xGa_{1-x}As$/In_xAl_xAs, InSb/$InAs_{1-x}Sb_x$, InAs/$In_xGa_{1-x}As$, $In_xGa_{1-x}N$/GaN, GaN/$Al_xGa_{1-x}N$, $Si_{1-x}Ge_x$/Si, etc. Most importantly, in contrast to photodiodes made of bulk semiconductors whose bandgaps are a definite value and the photodiode responsivity only peaks at a specific wavelength, when using semiconductor heterostructures to make photodiodes, the wavelength of light that can be

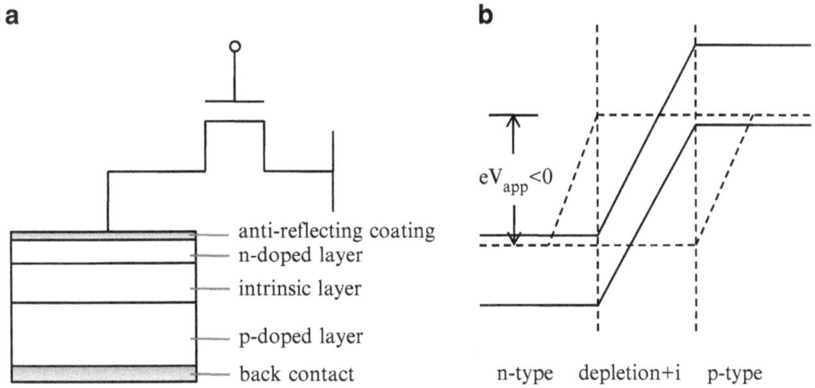

Fig. 9.2. (**a**) Diagram for a passive CMOS image sensor pixel (From Fossum (Fossum, 1995)), and (**b**) the band structure of the photodiode shown in (**a**) under a strong reverse bias

detected can be tuned by changing the material component x, from far infrared to ultra violet, and thus covers the application of light detection for a wide range of wavelength.

Laser products are now extensively manufactured and commonly used in everyday life, and they play an important role especially in information technology area. For example, in computer and consumer electronic applications, every CD-ROM or DVD player has a laser built in as a reading device, and in communication field, information is sent by modulated laser lights through optical fibers. As opposed to the photodiode we introduced earlier, a laser emits light. The theoretical basis for laser was developed by Einstein in 1916, and the concept of a semiconductor laser was introduced by Basov et al. (Nicolay G. Basov, Nobel Prize laureate in physics in 1964) in 1961 who suggested that stimulated light emission could occur through the recombination of carriers injected across a p–n junction. So basically, the active regions of a photodiode and a semiconductor laser are both p-n junctions, while in the photodiode electron–hole pairs are generated and in laser the electron–hole pairs are recombined and photon is emitted. For laser application purpose, Si and Ge are not suitable material candidates, since they both have indirect bandgap, and the light emission efficiency is poor. Rather, direct bandgap materials are requisite. The semiconductor laser structure from the device point of view is pretty similar to a photodiode. We show a very simplified laser structure diagram based on bulk semiconductors in Fig. 9.3a, where a p–n junction (or p–i–n structure) is sandwiched by the electrodes. In operation mode, the p–n junction is forwardly biased, and current is injected via the electrodes, one of which is connected to the heat sink shown in Fig. 9.3. Lasing occurs in the p–n junction where electrons and holes injected from the electrodes recombine and photons are emitted. Actual laser devices are much more complicated than shown in the figure and have auxiliary devices such as optical

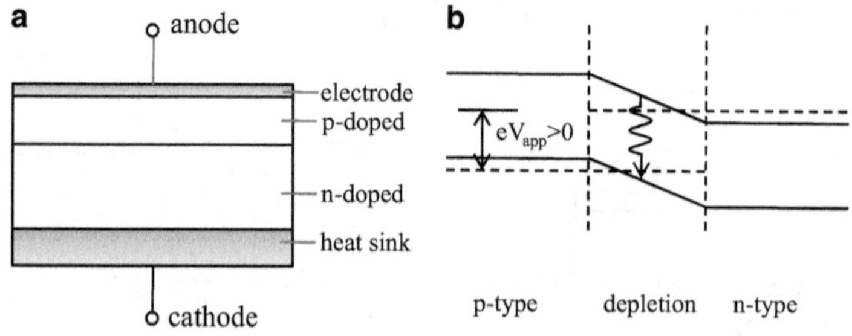

Fig. 9.3. (**a**)A simple structure of a semiconductor laser device, and (**b**) the band structure of the semiconductor laser device in working mode. Photons are emitted in the depletion region

resonators, etc. First generation of semiconductor lasers based on bulk materials suffered from low efficiency, which led to high heat dissipation and high threshold currents, and could only emit laser pulses. High efficiency and continuous wave lasers were not designed and produced until semiconductor heterostructures were successfully introduced and grown. That is to say, semiconductor lasers are only technologically and economically feasible by using quantum well structures.

A multiquantum well laser structure based on $GaAs/Al_xGa_{1-x}As$ heterostructure is shown in Fig. 9.4a, and the corresponding band line-up is shown in Fig. 9.4b. Instead of achieving lasing in the p–n junction depletion region in the bulk semiconductor lasers, quantum well lasers realize lasing in the quantum wells confined by larger bandgap barrier materials. Different heterostructures that can be used for laser applications also include $In_xGa_{1-x}As/In_xAl_xAs$, $InAs/In_xGa_{1-x}As$, $GaN/Al_xGa_{1-x}N$, etc. Comparing to the bulk semiconductor lasers, the advantages of quantum well lasers are obvious, because both electrons and holes are confined in the same region and the recombination rate is greatly enhanced compared to bulk p–n junction. The trademark parameter of a laser is certainly its emission wavelength. The other key parameters include power efficiency, threshold current, gain, linewidth, etc. By adjusting the barrier or well width, or the x value, the separation between the conduction and valence ground subbands can be almost continuously tuned, and thus the output light wavelength can be easily controlled. Because of the step-like DOS, which leads to narrow energy distribution of carriers, narrower luminescence spectra and higher optical gain can be achieved. Higher efficiency, higher gain, and small active volumes of quantum well lasers result in one order of magnitude reduction of threshold current compared to bulk devices. Also, quantum well lasers exhibit reduced

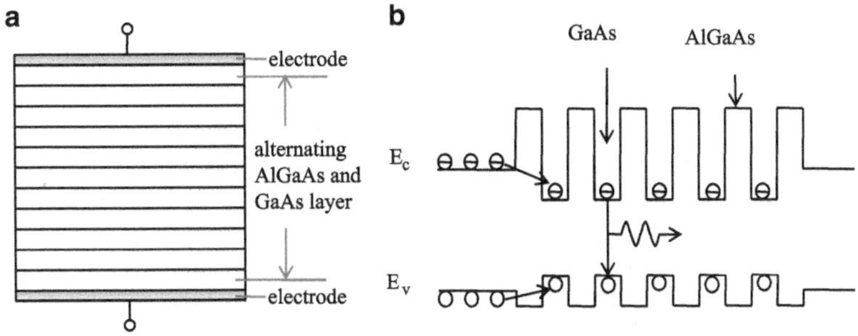

Fig. 9.4. (**a**) A laser device based on $GaAs/Al_xGa_{1-x}As$ heterostructure, and (**b**) the band structure of (**a**) in working mode. Photons are emitted in the multiple quantum wells

temperature sensitivity while retaining superior performance characteristic. Only based on the quantum well structures, does the room temperature semiconductor laser become possible.

Accompanying light emission in the laser operation, there is always the competing process, the light absorption. To have laser output, the optical gain must be positive, i.e., the emission strength is larger than the absorption strength. Photon transition rate, either for absorption or for emission, depends on three factors: 1) the photon–electron interaction strength; 2) the availability of electrons to be excited in the initial states; 3) availability of the final states. Assuming that the photon–electron interaction strengths are the same for both absorption and emission, then the rate difference is determined by the availability of the electrons in the initial states and final state vacancies. Under thermal equilibrium distribution, the electron occupation is always higher in the valence bands than in the conduction bands, as shown in Fig. 9.5a. This only favors photon absorption. To have a proper lasing

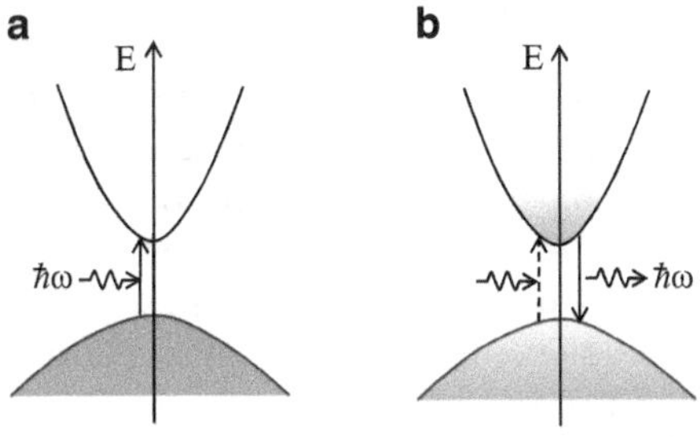

Fig. 9.5. (**a**) Electron distribution under thermal equilibrium state and, (**b**) electron distribution under population inverted state. The electron occupation is always higher in the valence bands than in the conduction bands in the thermal equilibrium state, but a population inversion can be realized by some particular approaches

condition, the electron population around the conduction band edge must be higher than that around the valence band edge, and thus form a state called "population inversion," as shown in Fig. 9.5b. For this end, external energy sources such as current injection and optical pumping are needed. Population inversion is a prerequisite condition for lasing not only for semiconductor lasers, but for all types of lasers. Except population inversion, a positive optical feedback and the confinement of photons in an optical resonator are also required because the stimulated emission depends on the number of photons present in the device to trigger the stimulated emission process. This is realized by different but similar ways. Traditionally, two reflecting facets (can be polished wafer surfaces, e.g., the p- and n-doped terminal ends) are put at opposite ends of the optical waveguide or cavity and thus the light is oscillating in the junction plane. This is called edge-emitting lasers. Vertical-cavity surface-emitting lasers are also designed whose emission direction is perpendicular to the in-plane direction. In quantum well lasers, proper selection of material systems can help confine the photons in the quantum well active region. The larger bandgap material usually has a lower refractive index. When light travels from material with a high refractive index to material with a low refractive index, it tends to be reflected back and thus the photons are effectively confined in the well region. Therefore, the barrier material also acts like a waveguide to the laser field, and it is transparent to the laser field. Cladding layers with even wider bandgap and lower refractive index can be used for further optical confinement. It is these special properties of the semiconductor heterostructures that make the room-temperature lasers into practical devices.

9.2.3 Strain Applications in Optoelectronic Devices

In the last subsection, we illustrated the optical processes in semiconductors including photon absorption and emission. The typical devices are photodiode and semiconductor lasers, respectively. Because of the merits we mentioned earlier, these semiconductor optoelectronic devices are now almost all made by heterostructures, which are generally accompanied with strain. In earlier time, people endeavored to diminish strain in lattice-matched systems to eliminate the possibility of device degradation caused by strain such as defects created by strain relaxation. With the maturation of film growth technology and scaling of device size, strain relaxation is now better controlled and becomes a less concern. On the other hand, strain effects were found to be of great potential to change the optoelectronic device characteristic and enhance their performance. Just like in electron devices, strain changes optoelectronic device performance through its effects on the band structure. However, due to different operation mechanisms, the emphasis of strain effects on band structures for optoelectronic devices is different from that for electron devices, where what we care about are 1) band warping, which induces conductivity effective mass change and subsequently affects the carrier velocity; 2) band splitting, which affects the interband scattering; 3) shifts of energy level, which alters the barrier height and alter the leakage current. However, the strain effects on band structure that will affect optoelectronic devices are manifested in the following three aspects: 1) shifts of bandgap; 2) changes of energy level DOS; 3) electronic wavefunction variation or mixing. Shifts of bandgap depend on the relative shifts of both the conduction and valence band edges. The latter two aspects are mainly related to the valence bands, since for direct-gap III-V semiconductors, strain has negligible effects on the DOS and wavefunction of the conduction band. In the following, we will elaborate on these points one by one by using their applications in both photodiodes and quantum well lasers.

First, bandgap shifts directly affect the wavelength of the light that can be absorbed or emitted. In a ternary or even quaternary strained quantum well system, which also falls into a large range of selection with different bandgaps and band offsets between well and barrier systems, three factors affect the bandgap of the active quantum well layers, which are the quantum well thickness, composition x, and strain. Under pseudomorphic approximation, strain is entirely accommodated by thin films (normally the quantum well) with thickness under critical thickness grown on lattice-mismatched barrier. The adjustability of the well thickness then is limited for reliability issue. Shifts of bandgap due to the latter two factors are equally important but are not independent. Thus, in strained quantum well systems, by adjusting the well thickness and strain (or x), the bandgap can be tuned, and for a specific wavelength requirement, a number of combinations of quantum well thickness and strain can be chosen in order to reach an optimal device performance. for a photodiode, the resonance wavelength can be adjustable, and the spectral range can be extended by shifting bandgap. Shown in Fig. 9.6 is an example of avalanche photodiode based on strained $In_xGa_{1-x}As/InP$ multiple quantum wells (Gershoni et al, 1988). In lattice-matched $In_{0.53}Ga_{0.47}As/InP$ quantum well, the cutoff wavelength the photodiode can detect is 1.65 μm. With increase of x from 0.53 to 0.74, the cutoff wavelength increases from 1.65 μm to about 1.95 μm. The photocurrent also increases significantly at the same wavelength for the strained photodiode than that of the lattice-matched one. Compared to photodiodes made of bulk semiconductors, strained quantum well photodiodes exhibit smaller dark current and higher speed (Dries et al, 1998; E. Özbay et al, 1999).

The shifts of bandgap in strained quantum well lasers also have critical use to enhance the laser operating wavelength and add extra flexibility in designing a semiconductor laser with particular operating parameters and optimized performance. In some cases, use of strained quantum well is the only means to achieve the required wavelength. $In_xGa_{1-x}P/AlGaInP$ quantum well lasers cover the wavelength from around 600 (Kikuchi et al, 1991; Hamada et al, 1992; Bour et al, 1994; Tanaka et al, 1994) to 710 nm

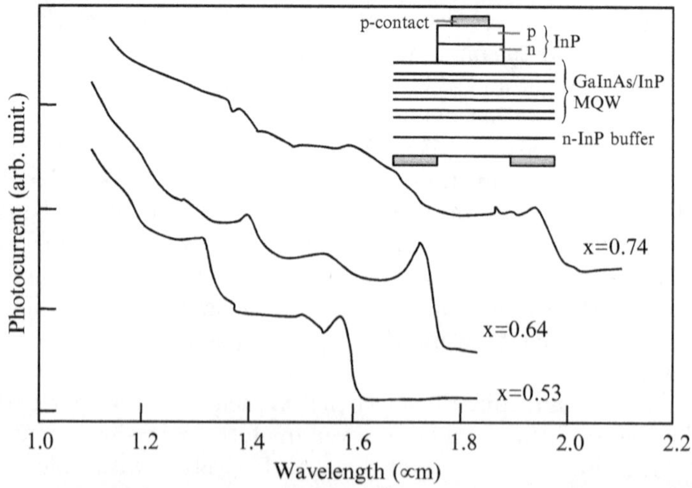

Fig. 9.6. Photocurrent vs. light wavelength in a $In_xGa_{1-x}As/InP$ avalanche photodiode with different x values. The inset shows the structure of the photodiode. When $x = 0.53$, InGaAs and InP are in lattice-matched condition. When $x > 0.53$, $In_xGa_{1-x}As$ layers are in compressive strain condition. From Gershoni (Gershoni et al, 1988)

(Ueno et al, 1993) where tensile and compressive strain is present, respectively, due to lattice mismatch. For a specific wavelength, the ability to choose from different combinations of quantum well thickness and strain provides significant advantage in a laser system design. For example, for a 633-nm strained quantum well laser based on $In_xGa_{1-x}P/(Al_yGa_{1-y})_{0.52}In_{0.48}P$ material system, minimum threshold current two times lower than the lattice-matched case was experimentally observed at a particular strain value for both tensile and compressive strains (Valster et al, 1992). On the other way, electron and hole ground energy level shift by strain pronouncedly affects carrier loss in laser operation, especially at room temperature and above, since the thermal excitation is the dominant carrier loss mechanism. The threshold current increases monotonically with the increase of Ga fraction in $In_xGa_{1-x}P/AlGaInP$ quantum well lasers at room temperature due to up-shifted ground subband and increased thermal leakage (Arkwright et al, 1994). Combined with the energy level shift by varying x, 1% compressive strain reduces the intrinsic current by about 50%.

Second, changes of the DOS affect the absorption efficiency in photodiode and the population inversion condition for quantum well lasers. Normally, the mismatch strain in quantum wells is biaxial since most quantum wells are grown on (100) surface. For bulk systems, strain is the only cause for the valence band splitting. With the degeneracy lifting of the HH and LH bands, the DOS at the valence edge is reduced. This effect can be large if the alignment of HH and LH is reversed so that the LH band is the ground valence band. For the case that HH is the ground valence band after splitting, there is the other cause for DOS reduction. When the HH and LH bands are degenerate, band warping is strong due to strong interband interaction. However, after they split, the HH band at the band edge tends to become parabolic and isotropic (for biaxial strain), and the warping is greatly reduced within the strain splitting energy range. This causes the DOS at the band edge to decrease further. In quantum well systems, DOS is determined by the in-plane subband structure. If we say that the bandgap shift in a quantum well is created by both quantum confinement and strain effect, the DOS change (in a quantum well that is already confined) can be largely ascribed to strain effect alone, since the spatial confinement always separates the HH and LH bands and makes the HH band the ground subband, while proper strain can warp and even reverse the HH and LH alignment. The valence band DOS dependence on strain of GaInAs/GaInAsP quantum well laser is shown in Fig. 9.7 (Silver and O'Reilly, 1994). For photodiodes, change of valence band DOS is not arduously pursued, since in general the conduction band DOS in III–V semiconductors is much smaller than that of the valence band, and the light absorption coefficient is proportional to the joint DOS, which is whereas mostly determined by the small DOS. However, for quantum well lasers, a LH ground valence subband is greatly desired to reduce the valence band DOS. Reduction of the valence band DOS makes it easier to achieve population inversion, and thus significantly reduces the threshold current and increases the gain. Gain increase due to strain in the same InGaAs/InGaAsP quantum well laser shown in Fig. 9.7 is shown in Fig. 9.8.

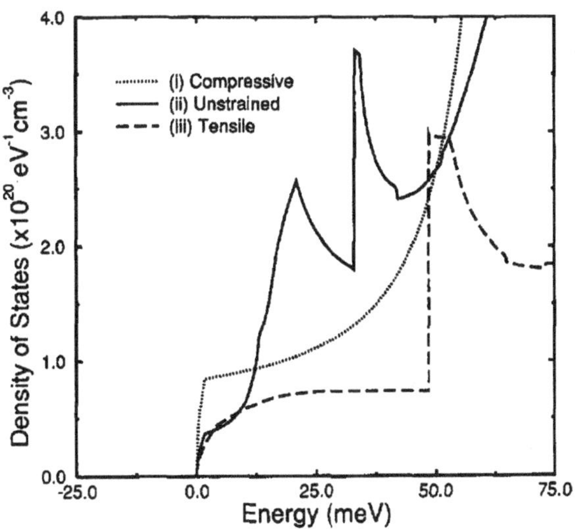

Fig. 9.7. Valence band DOS of InGaAs well in $In_{0.8}Ga_{0.2}As_{0.45}P_{0.55}$/GaInAs/$In_{0.8}Ga_{0.2}As_{0.45}P_{0.55}$ layers with well strain of (1) 1.2% compressive (25 Å), (2) unstrained, and (3) 1.2% tensile strain. DOS is reduced under either strain case. From Silver (Silver and O'Reilly, 1994)

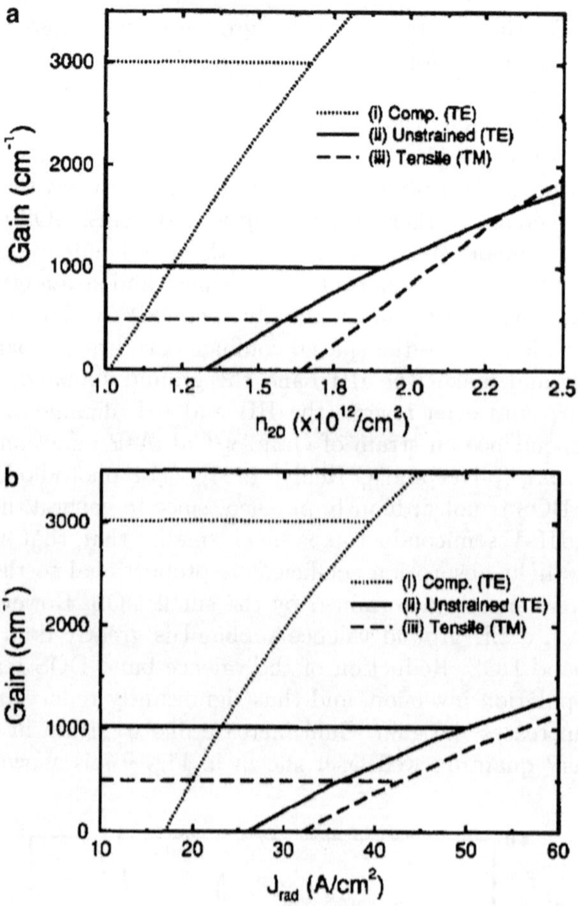

Fig. 9.8. Material gain of GaInAs quantum well laser vs. (**a**) 2D carrier density, and (**b**) radiative current density for the three structures shown in Fig. 9.7. The operating wavelength of the laser is at 1.5μm. The horizontal lines represent the threshold material gain required for lasing. From Silver (Silver and O'Reilly, 1994)

Both tensile and compressive-strained laser device greatly outperforms the unstrained one. Furthermore, with a reduced DOS and low current injection, the nonradiative Auger process (Beattie and Landsberg, 1958) and the inter-valence band absorption (IVBA) (Asada et al, 1981), which are responsible for temperature sensitivity of lasing characteristic for long-wavelength lasers, are significantly suppressed. Auger and IVBA processes are two dominant mechanisms for intrinsic charge loss in quantum well lasers. In Auger recombination, an electron–hole pair recombines to excite a third electron either in the conduction band or in the split-off band to higher energy states instead of emitting a photon. This process rate is proportional to $n^3 \exp(-E_a/k_BT)$ where n is the carrier density, E_a is the energy involved in an Auger process,

k_B is Boltsmann constant, and T is temperature. IVBA occurs when an electron in the split-off band absorbs a photon emitted by radiative recombination of an electron–hole pair and transits into an injected hole state in the HH band. Because in the III–V materials used for optoelectronic purpose the bandgap is generally much larger than the split-off energy, IVBA occurs usually well away from the zone center where the hole occupation probability depends strongly on temperature. Reduced DOS at the valence band edge and consequently the reduction of threshold injection current also reduce the available hole states, and thus the Auger and IVBA rates are greatly suppressed.

Lastly, electronic wavefunction variation with shift of the valence bands or electronic wavefunction mxing with strain alters the optical transition rate across the bandgap. Here, we only use light absorption for illustration. Under the dipole approximation, optical matrix element takes the form, $\langle\psi_f|\mathbf{p}\cdot\hat{e}|\ \psi_i\rangle = \mathbf{p}_{if}\cdot\hat{e}$, where $\mathbf{p}$ is the momentum operator and $\hat{e}$ is the photon polarization. For light propagation direction parallel to the z direction, $\mathbf{p}\cdot\hat{e} = \frac{1}{\sqrt{2}}(p_x \pm \mathrm{i}p_y)$ for σ^+ and σ^- photons, respectively. The possible light absorption processes from the HH and LH bands to the conduction bands are schematically shown Fig. 9.9. Let us concentrate on, for example, the σ^+ photon absorption process. Complied with the angular momentum conservation, two transition processes can occur. One is from $|\frac{3}{2},-\frac{3}{2}\rangle$ (HH $\downarrow$) to $|\frac{1}{2},-\frac{1}{2}\rangle$ (S $\downarrow$), and the other is from $|\frac{3}{2},-\frac{1}{2}\rangle$ (LH $\downarrow$) to $|\frac{1}{2},\frac{1}{2}\rangle$ (S $\uparrow$). In both processes, the electron angular momentum increases by $\hbar$. For light absorption across the bandgap, the final state is the conduction band which has the s symmetry. Because of symmetry, only matrix element in the form like $\langle S|p_x|X\rangle$ does not vanish. Recalling that $|\frac{3}{2},-\frac{3}{2}\rangle = \frac{1}{\sqrt{2}}|(X-\mathrm{i}Y)\downarrow\rangle$, and $|\frac{3}{2},-\frac{1}{2}\rangle = \frac{1}{\sqrt{6}}|(X-\mathrm{i}Y)\uparrow\rangle + \sqrt{\frac{2}{3}}|Z\downarrow\rangle$, and the absorption strength is proportional to $|\mathbf{p}_{if}\cdot\hat{e}|^2$, then the coupling strength of the HH band is three times as strong as that of the LH band. If parabolic band approximation can be applied, then when strain shifts the valence bands and changes the alignment of the HH and LH bands, the absorption rate from the valence band edge to the conduction band edge is strongly affected by this coupling

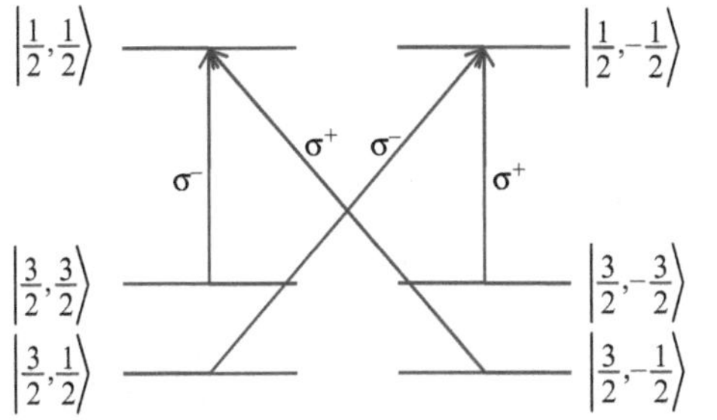

Fig. 9.9. Schematic of photon transition from the valence bands to the conduction band. In the process of photon transition, because the angular momentum has to be conserved, the transition probability is different for photons with different polarization

strength variation in addition to the DOS change. However, in quantum wells, the breaking of symmetry due to both confinement and strain strongly couples the carrier's in-plane (x-y) and z motion immediately away from the Γ point. As a result, the electronic wavefunction mixing is very strong. The valence bands show very pronouncedly nonparabolicity, and some bands have the electron-like dispersion, as shown in Fig. 9.10 where the subband structure of a 100 Å thick GaAs/$Ga_{0.6}Al_{0.4}As$ quantum well is presented (Sanders and Chang, 1985). In such cases, the HH and LH may be only defined at $k = 0$,

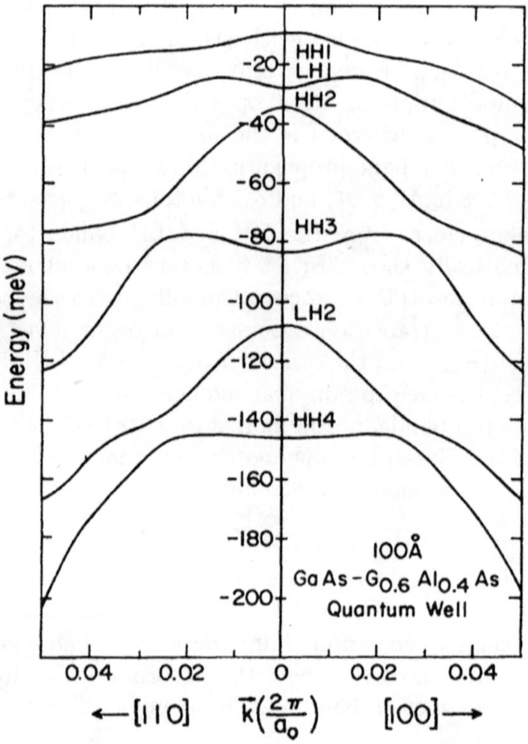

Fig. 9.10. Valence subbands of a 100-Å-thick GaAs/$Ga_{0.6}Al_{0.4}As$ quantum well. Strong nonparabolicity is present due to strong band mixing. From Sanders (Sanders and Chang, 1985)

and the distinction between the HH and LH characters away from $k = 0$ is obscure. Strain can even mix the HH states and the split-off hole states at $k = 0$, whereas the split-off energy is usually large for III–V materials, so that this mixing is not strong. Electronic wave function mixing leads to coupling of bands with different angular momentum, and the photon transition probability between a valence band to the conduction band needs the precise mixing behavior of the wavefunctions.

Light absorption or laser emission in quantum wells has strong polarization dependence in contrast to bulk materials, and this dependence is pronouncedly affected by strain. Exhibited in strained quantum well lasers, the emission light polarization is critically determined by strain condition. This effect is related to strain-shifted valence subband structures. For edge-emitting lasers, the light is emitted parallel to the quantum well surfaces and photons are confined to the well by reflection. Typically there are two polarization modes of the photons, the TE (transverse electric) mode and the TM (transverse magnetic) mode, as shown in Fig. 9.11, where the light incident angle to the surface is greatly amplified for clarity purpose. In real case, the angle is almost zero, and thus the magnetic field direction in Fig. 9.11a and the electric field direction in Fig. 9.11b are pointing to the z direction (quantum well growth direction). "Transverse" refers that the specified field is transverse to the plane constructed by the propagating direction and the surface normal. Photons resulted from the recombination of electrons with HHs favor the

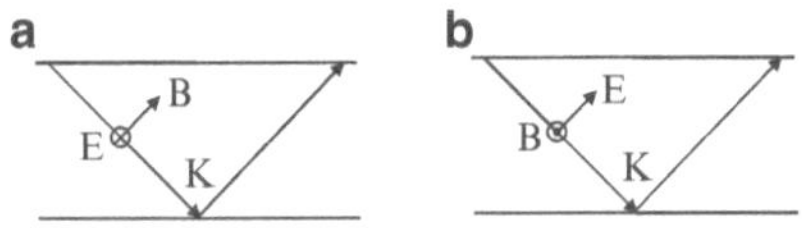

Fig. 9.11. Diagrams of (**a**) TE and (**b**) TM mode light polarization

TE polarization, and those from electron-LH recombination favor the TM polarization. The fundamental reason for this is the wave function of the HHs and LHs. Because the only nonvanishing optical matrix element is in the form $\langle S|p_x|X\rangle$, the electron-HH transition is strong for TE photons, while the electron-HH transition is actually forbidden for TM photons, since there is no in-plane electric field components, but the electron-LH transition is strong for TM photons. TE/TM gain ratio can be changed by strain by lifting of HH and LH band degeneracy and shifting their ordering (Patel et al, 1973; Ahn and Chuang, 1988; Boermans et al, 1990). Predominance of TE or TM emission depends on whether the lowest occupied band is the HH band or the LH band, respectively, and desired HH and LH band ordering can be obtained by choosing a suitable mismatch between the lattices of the active layer and the substrate. For example, for GaInP lattice-matched to a GaAs substrate, the band structure has degenerate HH and LH valence bands at $k = 0$. TE polarized light is emitted when an excited electron in the conduction band recombines with a hole in the heavy hole valence band, whereas TM polarized light is emitted when an electron combines with a light hole. Because of the predominant HH DOS, a laser with a lattice-matched InGaP active region will generally lase in the TE mode. An $In_xGa_{1-x}P$ layer with $x < 0.53$ will be under tensile strain and will emit TM polarized light. However, a compressively strained layer ($x > 0.53$) will emit TE polarized light due to the dominance of the HH band. Prior laser diodes emitting at the shorter 0.62–0.65 μm wavelengths have tensile-strained quantum wells and thus operate in the TM mode, while those emitting at the longer 0.65–0.69 μm wavelengths are compressively strained and operate in the TE mode.

9.3 OPTICAL PROCESSES IN SEMICONDUCTORS

In the last section, qualitative picture is given for basic optical processes in semiconductors and their heterostructures, and strain effects on photodiodes and quantum well lasers are briefly introduced. In this section, we will introduce the quantum interaction between light and matter, and the formalism of quantitatively calculating the light absorption and gain for a particular quantum structure. Since the band structure calculation methods for both bulk semiconductors and quantum wells have already been introduced in earlier chapters, so here we assume that the band structures are already known and elaborate on the optical processes, such as light absorption and gain. Physically, light absorption and light gain are two different terms to describe the same process for different application purposes. For lasers, light gain is defined as the negative of the absorption. We have to also distinguish "gain" used in avalanche photodiodes, where it is often to refer to the electron–hole pair (or current) multiplication factor. In this chapter, we only concentrate on the light gain. When the light gain is positive, i.e., the absorption is negative, the light strength is enhanced when traveling along the optical resonator, and the light amplification is achieved. Only when the gain is positive, will a stable laser power output be possible.

Light absorption is characterized by the absorption coefficient. Going through a slab of material with thickness of $\mathrm{d}x$, the reduction of the light strength, $\mathrm{d}I$, is proportional to the slab thickness and the light strength I,

$$\mathrm{d}I = -\alpha I \mathrm{d}x, \tag{9.1}$$

where α is the light absorption coefficient. Generally, it is a function of the photon energy (or light wavelength). The light strength then decreases exponentially with traveled distance,

$$I = I_0 \exp(-\alpha x), \tag{9.2}$$

where I_0 is the initial light strength at $x = 0$. In literature, the unit for α is usually cm^{-1}. The reverse of it measures the average penetration depth of the light.

Light absorption can be induced by many different mechanisms such as free carrier or impurity absorption. But what is really important for optoelectronics is the light absorption across the bandgap by electron transitions from the valence bands to the conduction band. This is also called the intrinsic absorption. Measurements of the absorption spectra have been the major means to characterize the electronic properties of semiconductors. In this section, we discuss how to calculate the absorption coefficient.

9.3.1 Light Absorption Coefficient

In optoelectronic applications, direct gap semiconductors are of overwhelming importance. Since the conduction band valence band edges are at the same k point (the Γ point), the discussion is greatly simplified. The indirect gap photon transition also involves phonon transition, and the process is much more complicated. Theoretically, the photon transition in direct gap semiconductors can also be phonon involved. However, this type of transition has a much lower probability and does not have significant effect of the light absorption coefficient.

The absorption coefficient for monochromatic light with angular frequency ω due to cross-bandgap transition can be obtained as follows.

If we assume the light energy flux is S, then the photon flux is $S/\hbar\omega$. The photon absorption rate per unit volume, T, is related to the photon flux by $T\mathrm{d}x = \alpha \mathrm{d}x S/\hbar\omega$, thus we have

$$\alpha = \frac{\hbar\omega T}{S}. \tag{9.3}$$

There are two contributions to T in optical transition: 1) T_1, due to the electron transition from the valence bands to the conduction band. In this process, photons are absorbed; 2) T_2, due to the stimulated photon emission process where an excited electron by a photon emits a photon and transits from the conduction band to the valence bands. The photons emitted by stimulated emission are not distinguishable from the injected photons, so this process is a reverse process of absorption and has a negative contribution to photon absorption. Accompanying the stimulated emission, there also exists spontaneous emission. However, photons emitted by spontaneous emission have random phase and random direction, and thus they are not additive to the injected light strength and are considered a part of optical power loss. Thus, only stimulated photon emission process is considered as one intrinsic part of light absorption, and the net photon absorption is the combined contribution of the photon absorption and stimulated emission processes, $T = T_1 - T_2$.

T_1 is the total possible transition from the valence bands to the conduction band divided by the volume of the semiconductor,

$$T_1 = \frac{1}{V}\sum_{a,b} W_{ab} f_a (1 - f_b), \tag{9.4}$$

where a, b label the energy states in the valence bands and conduction band, and the sum runs over all possible states in the conduction and valence bands. W_{ab} is the transition rate from state a to b, and f_a and f_b are the fermi functions for energies at levels a and b,

$$f(E, E_F) = \frac{1}{1 + \exp(\frac{E - E_F}{k_B T})}, \tag{9.5}$$

where E_F is the Fermi energy and k_B is the Boltzmann constant. Here, $f_a(1 - f_b)$ is exactly what we have stated previously: the transition rate is proportional to the availability of electrons in the initial states and availability of the empty final states. Similarly, T_2 is given by

$$T_2 = \frac{1}{V}\sum_{b,a} W_{ba} f_b (1 - f_a). \tag{9.6}$$

Next we study optical transition rate W_{ab}.

Photon transition in semiconductor is also treated using perturbation theory. The procedure can be found in standard quantum mechanic textbooks. Here we give a simple introduction. In an electromagnetic field, the total Hamiltonian for an electron in a crystal is

$$H = \frac{1}{2m_0}(\mathbf{p} - e\mathbf{A})^2 + V(\mathbf{r}), \tag{9.7}$$

where m_0 is the free electron mass, e is the electron charge (negative for electrons), $\mathbf{A}$ is the vector potential of the electromagnetic field, and $V(\mathbf{r})$ is the periodic crystal potential. Note that in quantum mechanics, $\mathbf{p} = -i\hbar\nabla$, is a differential operator. Expanding this Hamiltonian, we get

$$\begin{aligned} H &= \frac{\mathbf{p}^2}{2m_0} + V(\mathbf{r}) - \frac{e}{2m_0}(\mathbf{p}\cdot\mathbf{A} + \mathbf{A}\cdot\mathbf{p}) + \frac{e^2\mathbf{A}^2}{2m_0}, \\ &= H_0 + H', \end{aligned} \tag{9.8}$$

where $H_0 = \frac{\mathbf{p}^2}{2m_0} + V(\mathbf{r})$ is the unperturbed Hamiltonian and H' is the perturbation due to the electromagnetic field of light. We choose to work in the Coulomb gauge (also called the radiation gauge, corresponding to transverse electromagnetic wave), then $\nabla\cdot\mathbf{A} = 0$, and $\mathbf{p}\cdot\mathbf{A} = \mathbf{A}\cdot\mathbf{p}$, then the perturbation becomes

$$\begin{aligned} H' &= -\frac{e}{2m_0}(\mathbf{p}\cdot\mathbf{A} + \mathbf{A}\cdot\mathbf{p}) + \frac{e^2\mathbf{A}^2}{2m_0} \\ &\simeq -\frac{e}{m_0}\mathbf{A}\cdot\mathbf{p}. \end{aligned} \tag{9.9}$$

where in the last step we neglected the second order term of $\mathbf{A}$. This can be justified by the following argument. The vector potential is related to the optical power by the equation

$$A^2 = \frac{2S}{n_r c\epsilon_0\omega^2}, \tag{9.10}$$

where S is the optical power, n_r is the refractive index of the material, c is the speed of the light, and ϵ_0 is the dielectric constant in vacuum. For an optical power of 1 W/cm^2, the ratio of $eA/p \sim 10^{-5}$ for an electron with momentum $p = \hbar k$ with k at the order of Å^{-1}. Even for an optical power of 1 MW/cm^2, $eA/p \sim 10^{-2}$. This fact not only justifies the neglecting of the second order term of A, but also justifies the implementation of the perturbation theory to treat the interaction between light and matter.

Assuming that the vector potential for the incident electromagnetic wave takes the form

$$\mathbf{A} = A_0\hat{e}\cos(\mathbf{k_L}\cdot\mathbf{r} - \omega t), \tag{9.11}$$

where $\hat{e}$ is the photon polarization vector and $\mathbf{k_L}$ is the light wave vector, then

$$\begin{aligned} H'(\mathbf{r},t) &= -\frac{e}{m_0}\mathbf{A}\cdot\mathbf{p} \\ &= \frac{eA_0}{2m_0}\hat{e}\cdot\mathbf{p}\{\exp[i(\mathbf{k_L}\cdot\mathbf{r}-\omega t)]+\exp[-i(\mathbf{k_L}\cdot\mathbf{r}-\omega t)]\} \\ &= H'(\mathbf{r})\exp(-i\omega t)+H'^{\dagger}\exp(i\omega t), \end{aligned} \tag{9.12}$$

where † signifies the Hermitian conjugation operation. From the time-dependent perturbation theory, the first term in (9.12) corresponds to photon absorption, and the second term corresponds to stimulated photon emission. Then for the absorption transition rate, the Fermi's golden rule gives (only consider the first term):

$$\begin{aligned} W_{ab} &= \frac{2\pi}{\hbar}|\langle b|H'(\mathbf{r})|a\rangle|^2\delta(E_b-E_a-\hbar\omega) \\ &= \frac{2\pi}{\hbar}|H'_{ba}|^2\delta(E_b-E_a-\hbar\omega). \end{aligned} \tag{9.13}$$

The δ-function complies with the conservation of energy in the transition. Identically, when we only consider photon emission state b to a, the second term in (9.12) gives

$$W_{ba} = \frac{2\pi}{\hbar}|H'_{ab}|^2\delta(E_a-E_b+\hbar\omega). \tag{9.14}$$

Because of the Hermitian property of H', $|H'_{ba}|^2 = |H'_{ab}|^2$. A δ-function is an even function, $\delta(x)=\delta(-x)$, then we have $\delta(E_a-E_b+\hbar\omega)=\delta(E_b-E_a-\hbar\omega)$. Thus, the photon absorption rate is

$$T = T_1 - T_2 = \frac{1}{V}\frac{2\pi}{\hbar}\sum_{a,b}|H'_{ba}|^2\delta(E_b-E_a-\hbar\omega)(f_a-f_b). \tag{9.15}$$

Note this equation does not already take into account the spin degeneracy. If in the band calculation formalism spin states are degenerate, then the right-hand side of this equation should be multiplied by 2.

For an electromagnetic field with the vector potential in (9.11), the energy flux is given by the time-averaged Poynting vector,

$$S = \frac{n_r\omega^2A_0^2}{2\mu c} = \frac{n_r c\epsilon_0\omega^2A_0^2}{2}, \tag{9.16}$$

where n_r is the refractive index of the material, μ is the permeability, which is close to the permeability of the vacuum μ_0 in semiconductors, and ϵ_0 is the vacuum dielectric constant. Therefore, from (9.3), we obtain the absorption coefficient

$$\alpha(\omega) = \frac{1}{V}\frac{2\mu c\hbar}{n_r\omega A_0^2}\frac{2\pi}{\hbar}\sum_{a,b}|H'_{ba}|^2\delta(E_b-E_a-\hbar\omega)(f_a-f_b). \tag{9.17}$$

In the expression for H', there is also a phase factor $\exp(i\mathbf{k_L}\cdot\mathbf{r})$, so in photon transition, the electron wave vectors, between the initial and final states shall be differed by $\mathbf{k_L}$. However, as we discussed earlier, k_L is much smaller than typical electron wave vectors, and is generally neglected. Thus, $\exp(i\mathbf{k_L}\cdot\mathbf{r}) \simeq 1$. Then the optical matrix element takes the form

$$H'_{ba} = -\frac{eA_0}{2m_0}\langle b|\hat{e}\cdot\mathbf{p}|a\rangle = -\frac{eA_0}{2m_0}\hat{e}\cdot\mathbf{p}_{ba}, \tag{9.18}$$

and the absorption coefficient becomes

$$\alpha(\omega) = \frac{1}{V}\frac{\pi e^2}{n_r c\epsilon_0 m_0^2\omega}\sum_{a,b}|\hat{e}\cdot\mathbf{p}_{ba}|^2\delta(E_b - E_a - \hbar\omega)(f_a - f_b). \tag{9.19}$$

The right-hand side of this equation manifests the factors we mentioned before that affect the absorption process. First is the optical transition matrix element $\mathbf{p}_{ba}$. First, because of the differential characteristic of the operator $\mathbf{p}$, the symmetry of the initial and final states has to be opposite to have unforbidden transition. In III–V material, the cross-gap transition certainly is allowed due to the s and p nature of the conduction band valence bands. Also, the photon polarization $\hat{e}$ selects what components of the vector operator $\mathbf{p}$ take part in the transition, so it determines the transition strength between any two s and p bands. In bulk, because of the cubic symmetry of the III–V materials, there is no absorption dependence on photon polarization. But in low-dimensional structures, the cubic symmetry is broken, and the absorption has a strong dependence on photon polarization. We will come back to this point later. Second, the light absorption strongly depends on the DOS of the band structure through the δ-function. Since there are two energy levels involved, the DOS for light absorption is also determined by the combined behavior of these two energy levels, which is called the joint DOS. Third, the fermi energies term $(f_a - f_b)$ describes the absorption dependence on availability of the electrons at the initial states and vacancies for the final states. Generally when considering the light absorption for medium to wide gap intrinsic materials, $f_a = 1$ and $f_b = 0$ are good approximations. However, for doped or narrow-gap materials, f_a and f_b have to be carefully determined.

Although in the derivation of (9.19) we assumed valence to conduction cross-gap photon transition, however, this assumption is not requisite. Equation (9.19) can be used to calculate light absorption between any two bands, if the summation of states is changed properly. Because of the wave function mixing, light absorption within the valence bands does occur.

The absorption mechanism we discussed above is based on so-called *electric dipole approximation.* That means, the interaction between light and matter is from the diploe field the light creates inside the material which subsequently induces photon transition. This can be better understood if we transfer the perturbation Hamiltonian (9.12) from the momentum space into real space. In an electromagnetic field without current and charge sources, the electric field is related to the vector potential by $\mathbf{E} = -\frac{\partial \mathbf{A}}{\partial t}$. The momentum $\mathbf{p}$ and the electron displacement vector $\mathbf{r}$, which is caused by the force of the electromagnetic field, are related by $\mathbf{p} = m_0 \frac{\partial \mathbf{r}}{\partial t}$. In a monochromatic electromagnetic wave with angular frequency ω, $\mathbf{E}$, $\mathbf{A}$, and the charge displacement $\mathbf{r}$ are all varying with time with phase factor $\exp(-\mathrm{i}\omega t)$, so from (9.12) we have

$$\begin{aligned} H' &= -\frac{e}{m_0}\mathbf{A} \cdot \mathbf{p} = -e\mathbf{A} \cdot \frac{\partial \mathbf{r}}{\partial t} \\ &= -e\mathbf{A} \cdot (-\mathrm{i}\omega)\mathbf{r} = -e\frac{\partial \mathbf{A}}{\partial t} \cdot \mathbf{r} \\ &= e\mathbf{E} \cdot \mathbf{r} = -\mathbf{E} \cdot \mathbf{d}, \end{aligned} \tag{9.20}$$

where $\mathbf{d} = -e\mathbf{r}$ is the light-induced electric dipole moment. Furthermore, the electric field of the incident wave can be written as the same form of $\mathbf{A}$ as

$$\begin{aligned} \mathbf{E} &= E_0 \cos(\mathbf{k_L} \cdot \mathbf{r} - \mathrm{i}\omega t) \\ &= \frac{\hat{e}E_0}{2}\{\exp[\mathrm{i}(\mathbf{k_L} \cdot \mathbf{r} - \omega t)] + \exp[-\mathrm{i}(\mathbf{k_L} \cdot \mathbf{r} - \omega t)]\}. \end{aligned} \tag{9.21}$$

For the usual light we encounter, $\mathbf{k_L} \cdot \mathbf{r} \simeq a/\lambda \ll 1$ in the dimension of a unit cell, where a is the lattice constant, so the change of the electric field in the atomic scale is very small, and we can expand

$$\exp(\mathrm{i}\mathbf{k_L} \cdot \mathbf{r}) = 1 + \mathrm{i}\mathbf{k_L} \cdot \mathbf{r} + \cdots \tag{9.22}$$

Electric dipole approximation is to keep only the lowest order term of this expansion, $\exp(\mathrm{i}\mathbf{k_L} \cdot \mathbf{r}) = 1$, and thus the optical transition is induced by the electric dipole only. The next term in the expansion, $\mathrm{i}\mathbf{k_L} \cdot \mathbf{r}$, induces the quadrupole interaction, which is usually not important in most optical transitions we encounter. Only when the light wavelength becomes so small as comparable to the lattice constant will the quadrupole transition become important.

9.3.2 Joint Density of States

In discussions of light absorption in semiconductors, especially under low temperature, the involved electrons in the optical transition are generally confined in a small k-range, normally around the Γ point. Under such a condition, the electron wave function is a slow varying function of k compared to energy, and thus the optical matrix element $\mathbf{p}_{ba}$ is also slowly varying or even a constant. Therefore, it is the joint DOS that largely determines the absorption coefficient trend. For $f_a = 1$ and $f_b = 0$, the joint DOS is the integral of the δ-function in (9.19),

$$g_{cv} = \sum_{a,b} \delta(E_b - E_a - \hbar\omega). \tag{9.23}$$

Here, we only consider a simple parabolic case, where the conduction and valence bands both have a well-defined effective mass, m_n, and m_p, respectively. Then (9.23) can be rewritten as

$$g_{cv} = \sum_{\mathbf{k}} \delta(E_b - E_a - \hbar\omega) = \frac{V}{8\pi^3} \int \delta(E(\mathbf{k}) - E_g - \hbar\omega) \mathrm{d}\mathbf{k} \tag{9.24}$$

where we used $E(\mathbf{k}) - E_g$ to represent $E_c(\mathbf{k}) - E_v(\mathbf{k}) = E_b - E_a$. For parabolic bands, we have

$$E(\mathbf{k}) = \frac{\hbar^2 k^2}{2m_n} + \frac{\hbar^2 k^2}{2m_p} = \frac{\hbar^2 k^2}{2m_r}. \tag{9.25}$$

with the reduced effective mass defined as

$$\frac{1}{m_r} = \frac{1}{m_n} + \frac{1}{m_p}. \tag{9.26}$$

Substituting (9.25) and (9.26) to (9.24), we find that the joint DOS that determines the photon transition is the DOS of a band with dispersion $E(\mathbf{k})$ at the energy at $\hbar\omega - E_g$, $g_{cv}(\hbar\omega - E_g)$, where

$$g_{cv}(E) = \frac{V}{4\pi^2} \left(\frac{2m_r}{\hbar^2} \right)^{3/2} E^{1/2}. \tag{9.27}$$

Thus, the absorption coefficient has a square root relation with photon energy for bulk semiconductors,

$$\alpha(\omega) = \frac{e^2}{2\pi n_r c \epsilon_0 m_0^2 \omega} \left(\frac{2m_r}{\hbar^2} \right)^{3/2} |\hat{e} \cdot \mathbf{p}_{cv}|^2 (\hbar\omega - E_g)^{1/2}, \tag{9.28}$$

where we already took into account the spin degeneracy and assumed fully filled valence bands and empty conduction band. Shown in Fig. 9.12a is the absorption coefficient for intrinsic GaAs at room temperature (Moss and Hawkins, 1961). Above bandgap, the absorption coefficient follows the square root relation very well. In lower energy region, the absorption coefficient does not fall as steep as predicted by theory. This absorption region is called the absorption band-tail, whose origin is complicated and beyond the scope of current discussion.

Parabolic approximation works well for medium to wide gap bulk semiconductors at low temperature. At high temperature or quantum well structures, nonparabolicity strongly affects the $\alpha - \hbar\omega$ relation. Also, for quantum well structures, because of the strong band mixing, the electron wave function changes abruptly with k. Optical matrix element dependence on k must also be considered.

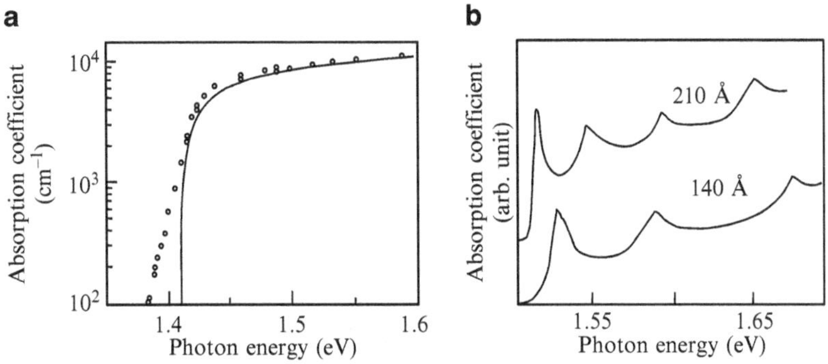

Fig. 9.12. Absorption coefficient vs. photon energy of (**a**) bulk GaAs, and (**b**) GaAs quantum wells. (**a**) is reproduced after Moss (Moss and Hawkins, 1961), and (**b**) is reproduced after Dingle (Dingle et al, 1974)

9.3.3 Optical Transitions in Quantum Wells

When the discussion of the light absorption changes from bulk to quantum wells, some new features arise. Because of the quantum confinement in quantum wells, an electron is localized in the z direction and has a z-dependent probability distribution. The light absorption in quantum wells does not depend only on the optical matrix element, but also on the overlap of the electron wave function in the z direction (envelope function). There are chances that when the energy difference and in-plane wave vectors between two states satisfy the requirement for photon transition, but their wave functions in the z direction do not overlap, and thus the photon transition in fact cannot occur. As we know, a 2D subband is the superposition of a set of basis states, which we label as u_i. The envelope function $f^i(z)$ of each component u_i is a function of z. Therefore, the optical matrix element in bulk case $\mathbf{p}_{ba} = \langle b|\mathbf{p}|a\rangle$ becomes in 2D case

$$\mathbf{p}_{nm}^{2D} = \sum_{i,j} \langle f_{en}^i | f_{hm}^j \rangle \langle u_{en}^i | \mathbf{p} | u_{hm}^j \rangle, \tag{9.29}$$

where en and hm label the nth electron subband and mth hole subband, respectively, and

$$\langle f_{en}^i | f_{hm}^j \rangle = \int f_{en}^{i*}(z) f_{hm}^j(z) \mathrm{d}z. \tag{9.30}$$

In above derivation, we assume that the electron wave function in the conduction band is also mixed with p states. Generally for medium to wide gap semiconductors, this mixing is small. If it is neglected, then the summation in the above equation can run over j only.

The DOS in 2D quantum wells is a step function of energy if we assume a parabolic subband structure. So naturally, the joint DOS for quantum wells is also a step function of energy under this assumption and is given by

$$g_{eh}^{2D}(\omega) = \frac{m_r}{2\pi\hbar^2} \sum_{m,n} \langle f_{en} | f_{hm} \rangle \Theta(\hbar\omega - E_{nm}), \tag{9.31}$$

where $\Theta(E)$ is a step function of E, $E_{nm} = E_{en} - E_{hm}$, and

$$\langle f_{en} | f_{hm} \rangle = \sum_{i,j} \langle f_{en}^i | f_{hm}^j \rangle. \tag{9.32}$$

For symmetrical quantum wells, one has approximately

$$\langle f_{en} | f_{hm} \rangle \approx \delta_{nm}, \tag{9.33}$$

where for valence bands, the HH and LH subbands are numbered individually. The absorption coefficient across the bandgap in quantum wells is then

$$\begin{aligned} \alpha_{2D}(\omega) = & \frac{1}{W} \frac{\pi e^2}{n_r c \epsilon_0 m_0^2 \omega} \times \\ & \sum_{n,m} \sum_{i,j} |\langle f_{en}^i | f_{hm}^j \rangle|^2 |\langle u_{en}^i | \hat{e} \cdot \mathbf{p} | u_{hm}^j \rangle|^2 \delta(\hbar\omega - E_{nm}), \end{aligned} \tag{9.34}$$

where W is the quantum well thickness. The light absorption for the GaAs/AlGaAs quantum well absorption shown in Fig. 9.12b clearly shows the step behavior (Dingle et al, 1974), and the absorption for GaInAs/InP quantum well structure shown in Fig. 9.6 also exhibits the step-like spectra.

It is useful to study the optical transition using a two-band model, which is enlightening since first in many cases the optical transition between the ground conduction subband and the ground valence subband is most important, and second, the optical transition between multiple valence to conduction subbands can always be studied as superposition of two-band transition. For the two-band model based on a parabolic approximation, substituting the 2D joint DOS to (9.34), the absorption coefficient is obtained as

$$\alpha_{hm \to en}(\omega) = \frac{1}{W} \frac{\pi e^2}{n_r c \epsilon_0 m_0^2 \omega} \frac{m_r}{2\pi\hbar^2} |\langle f_{en} | f_{hm} \rangle|^2 |\hat{e} \cdot \mathbf{p}_{nm}|^2 \Theta(\hbar\omega - E_{nm}). \tag{9.35}$$

9.3.4 Optical Matrix Elements

In this subsection, let us investigate the optical matrix elements and discuss their dependence on photon polarization. For near band edge transitions, we will assume that is $u_{\mathbf{k}}$ is given by their zone center expressions. These basis functions are given in Chap. 4 for quantization along the z direction. Here for convenience, we list them in the following:

- Conduction band:

$$u_{c0} = |S \uparrow\rangle, \quad \text{and,} \quad |S \downarrow\rangle \tag{9.36}$$

- Valence bands:
  1. HH states:

$$|HH\uparrow\rangle = -\frac{1}{\sqrt{2}}|(X+iY)\uparrow\rangle,$$
$$|HH\downarrow\rangle = \frac{1}{\sqrt{2}}|(X-iY)\downarrow\rangle. \tag{9.37}$$

  2. LH states:

$$|LH\uparrow\rangle = -\frac{1}{\sqrt{6}}(|(X+iY)\downarrow\rangle - 2|Z\uparrow\rangle),$$
$$|LH\downarrow\rangle = \frac{1}{\sqrt{6}}(|(X-iY)\uparrow\rangle + 2|Z\downarrow\rangle). \tag{9.38}$$

From symmetry, the only nonvanishing matrix element takes the form

$$p_{cv} = \langle S|p_x|X\rangle = \langle S|p_x|X\rangle = \langle S|p_x|X\rangle\,, \tag{9.39}$$

which is related to Kane's parameter by

$$P = \frac{-i\hbar}{m_0}p_{cv}, \tag{9.40}$$

and a more common used band parameter E_P in unit of eV,

$$E_P = \frac{2m_0}{\hbar^2}P^2. \tag{9.41}$$

Next, we list the band-to-band optical matrix elements. Since the deferential operator $\mathbf{p}$ does not operate on spin states, so optical transition only occurs between the states (or more accurately, state components) with the same spin. Thus we have:

$$\langle HH\uparrow|p_x|S\uparrow\rangle = \langle HH\uparrow|p_y|S\uparrow\rangle = \frac{1}{\sqrt{2}}p_{cv},$$
$$\langle HH\uparrow|p_z|S\uparrow\rangle = 0 \tag{9.42}$$

and,

$$\langle LH\uparrow|p_x|S\downarrow\rangle = \langle LH\uparrow|p_y|S\downarrow\rangle = \frac{1}{\sqrt{6}}p_{cv},$$
$$\langle LH\uparrow|p_z|S\uparrow\rangle = \frac{2}{\sqrt{6}}p_{cv}. \tag{9.43}$$

Changing the spin states (the up- and down-arrows) for the initial and final states simultaneously gives the same results. With these matrix elements, it is easy to find out the transition strength between any valence band to the conduction band for a specific photon polarization. First, for polarization $\hat{e} = \hat{z}$, there is no coupling between HH band and conduction band. For either polarization $\hat{e} = \hat{x}$ or $\hat{e} = \hat{y}$, the coupling strength between HH and conduction band is three times as strong as that between the LH band and conduction band, since the coupling is proportional to $|p_{ba}|^2$. However, no matter which polarization the photon has, the transition strength from the valence band to conduction band is the same. That means, there is no polarization dependence for photon transition in bulk semiconductors. This result is expected, because any polarization can be decomposed to some addition of $\hat{x}$, $\hat{y}$, and $\hat{z}$, and for cubic semiconductors, these three directions are equivalent.

Interesting polarization dependence of photon transition occurs in quantum well structures. This may be ascribed to symmetry breaking effect of the quantum well structure, and physically manifested by the split HH and LH bands. For quantum wells grown along the $\langle 001 \rangle$ direction, which is simultaneously selected as the Hamiltonian quantization axis, the basis functions to expand each subband are the same as listed above. Growth on the other direction is a little more complicated, but situations are similar. So we always define the growth direction as the z direction and discuss only the $\langle 001 \rangle$ quantum wells. Since the electron motion is quantized in the z direction unlike the motion in the x and y direction, the optical transition then also depends on the light polarization since as we mentioned earlier, the polarization selects which vector components of the momentum operator $\mathbf{p}$ for subband coupling and then only couples some particular subbands. The polarization configurations for TE is $\hat{e} = \hat{x}$ or $\hat{e} = \hat{y}$, and for TM polarization, $\hat{e} = \hat{z}$. The optical coupling strength for each polarization can be obtained following the discussion above for bulk. We immediately see that the TE photons couple with the HH band more strongly than with the LH band, and TM photons couple only with the LH band. Thus, the HH and LH splitting in the quantum well structures has a strong effect on light transition. However, quantum well band mixing is very strong away from the Γ point, and the subband character is smeared out at finite k. If the optical transition is not confined only at very small k, the optical matrix elements have finite strength for both TE and TM polarizations. Biaxial strain mixes the HH and split-off band even at the Γ point. However, the split-off energy is usually large for III–V materials; this coupling is then weak.

The most important strain effect on optical matrix elements in quantum wells is the relative shifts of the hole subbands. We shall concentrate on the subbands around the zone center, since for quantum well lasers, the contribution to the optical gain from the states near the zone center is very important, and these zone center optical momentum matrix elements play dominant roles in determining the optical gain in addition to the contribution of the subband DOS. For compressive biaxial strained quantum wells, the ground subband is always formed by the HH subband, and in contrast, the ground subband is the LH band for tensile biaxial strain. Chang and Chuang (Chang and Chuang, 1995) studied the polarization-dependent optical matrix elements in strained $In_{1-x}Ga_xAs$/InGaAsP quantum wells. Figure 9.13 shows the TE and TM polarization coupling strength between the ground hole subband and the ground electron subband for zero, tensile, and compressive strain. Different well width has been selected to make the edge to edge energy difference to be the same, and optical transition only occurs between two subbands, the ground hole subband, and the ground electron subband. Not surprisingly, the

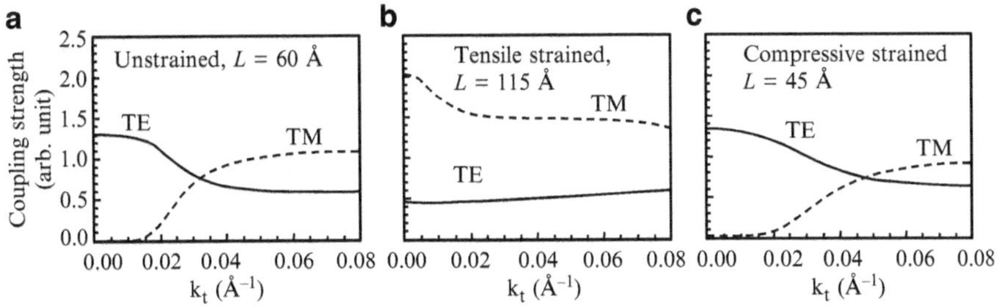

Fig. 9.13. TE and TM coupling strength vs. wave vector for unstrained, tensile strained and compressive strained $In_{1-x}Ga_xAs/InGaAsP$ quantum wells. From Chang (Chang and Chuang, 1995)

TE and TM coupling strengths for unstrained and compressive-strained cases are very similar, because in both cases the ground hole subband is the HH band. In these two cases, TM transition is forbidden at the zone center, while for tensile-strained case, the TM coupling strength is very strong. In all three cases, the coupling strength varies with k_t, the in-plane electron wave vector, due to band mixing at finite k_t.

9.4 NONEQUILIBRIUM CARRIER DISTRIBUTION AND GAIN

If a ray of light is incident on a semiconductor system under thermal equilibrium, although there is stimulated and spontaneous emission, the absorption always dominates. This is obvious from common sense and is easily evident from (9.19), where $f_a - f_b$ is always larger than zero. For optical output like in lasers, nonequilibrium carrier distribution has to be created. For laser, the population inversion is a typical case of nonequilibrium carrier distribution. Only under population inversion will a positive optical gain be possible.

Population inversion, implied by its name, simply means that the carriers in the upper level (e.g., the conduction band) are more than those in the lower level (e.g., the valence band). It can be achieved by current injection or optical pumping. In quantum well systems, current injection mechanism is simple and is realized by properly choosing the composite materials to form suitable band lineup. Shown in Figs. 9.4 and 9.14 are typical quantum well structures, which are widely used for achieving population inversion. Under a steady injection state, the electrons and holes are in quasi-equilibrium states,

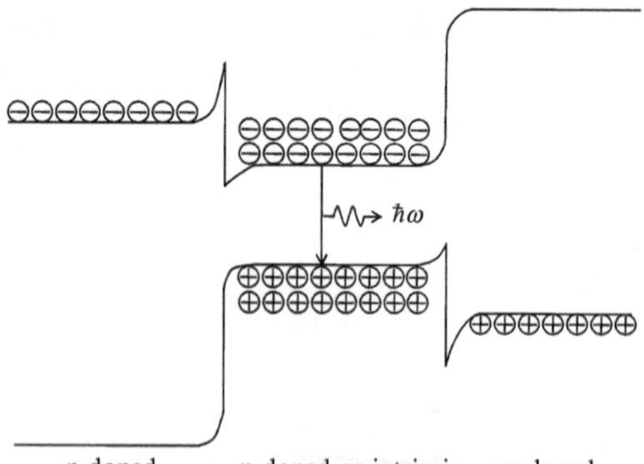

Fig. 9.14. A double heterostructure suitable for laser applications. The electrons are supplied from the n-doped electron barrier material, and the holes are supplied from the p-doped hole barrier material. Electron–hole recombination takes place in the quantum well sandwiched in the middle

and their distribution with energy can be described separately with two different Fermi energies, F_c F_v, respectively. The determination of quasi-Fermi energies is pretty simple. Suppose we have an injected electron density N, which occupies the conduction band (or subband labeled use "n"), then we have

$$N = 2\sum_{n}\sum_{\mathbf{k}_n} f_n(E_{\mathbf{k}}, F_c), \tag{9.44}$$

where the spin degeneracy is already considered. This is an implicit equation for F_c, and it can be numerically found by root-finding routines. The quasi-Fermi energy for holes, F_v can also be found by the same way, but with the hole concentration given by

$$P = 2\sum_{m}\sum_{\mathbf{k}_n}[1 - f_m(E_{\mathbf{k}}, F_c)], \tag{9.45}$$

where m runs over all the valence sunbands.

Next, let us inspect the absorption coefficient under population inversion. Repeat the equation for absorption again here as

$$\alpha(\omega) = \frac{1}{V}\frac{\pi e^2}{n_r c\epsilon_0 m_0^2\omega}\sum_{a,b}|\hat{e}\cdot\mathbf{p}_{ba}|^2\delta(E_b - E_a - \hbar\omega)(f_a - f_b), \tag{9.46}$$

but now we assume that there is only one subband for both electrons and holes, and the optical matrix element does not depend on k. Then the absorption coefficient becomes

$$\alpha(\omega) = \frac{1}{V}\frac{\pi e^2}{n_r c\epsilon_0 m_0^2\omega}|\hat{e}\cdot\mathbf{p}_{cv}|^2 \times \tag{9.47}$$
$$\int \frac{V}{4\pi^3}\delta[E_c(\mathbf{k}) - E_v(\mathbf{k}) - \hbar\omega][f_v(E_{v\mathbf{k}}, F_v) - f_c(E_{c\mathbf{k}}, F_c)]\mathrm{d}\mathbf{k}.$$

Under a parabolic approximation, we can write

$$\delta[E_c(\mathbf{k}) - E_v(\mathbf{k}) - \hbar\omega] = \delta(E_g + \frac{\hbar^2 k^2}{2m_r} - \hbar\omega), \tag{9.48}$$

and find that integration in (9.48) becomes

$$g_{cv}(\hbar\omega - E_g)[f_v(E_{vk_0}, F_v) - f_c(E_{ck_0}, F_c)], \tag{9.49}$$

where g_{cv} is the joint DOS given by (9.27), and

$$k_0 = \sqrt{\frac{2m_r}{\hbar^2}(\hbar\omega - E_g)}. \tag{9.50}$$

Thus, the two-band model light absorption coefficient becomes

$$\alpha(\omega) = \frac{e^2}{2\pi n_r c\epsilon_0 m_0^2 \omega}\left(\frac{2m_r}{\hbar^2}\right)^{3/2} |\hat{e}\cdot\mathbf{p}_{cv}|^2 \times \\ (\hbar\omega - E_g)^{1/2}(f_v(E_{vk_0}, F_v) - f_c(E_{ck_0}, F_c)). \tag{9.51}$$

By comparing this equation with (9.28), we find that they are only different by a multiplier $(f_v(E_{vk_0}, F_v) - f_c(E_{ck_0}, F_c))$. For all thermal equilibrium absorption, the absorption coefficient is positive. Recalling that the gain is defined as the negative absorption, so for positive gain, we have to have $f_v(E_{vk_0}, F_v) - f_c(E_{ck_0}, F_c) < 0$. That is,

$$\frac{\exp((E_{ck_0} - F_c)/k_BT) - \exp((E_{vk_0} - F_v)/k_BT)}{(1 + \exp[(E_{ck_0} - F_c)/k_BT])(1 + \exp[(E_{vk_0} - F_v)/k_BT])} < 0, \tag{9.52}$$

which leads to

$$\exp((E_{ck_0} - F_c)/k_BT) < \exp((E_{vk_0} - F_{vk_0})/k_BT) \tag{9.53}$$

or

$$F_c - F_v > E_{ck_0} - E_{vk_0} = \hbar\omega > E_g. \tag{9.54}$$

Then from this equation, the population inversion condition in such a two-band model is that the energy difference between the quasi-Fermi energies shall be larger than the photon energy. This is the Bernard-Duraffourg inversion condition (Bernard and Duraffourg, 1961). In quantum well lasers, this two-band model still applies if the 3D joint DOS is replaced by the 2D joint DOS, since in most cases, it is the ground electron and hole subbands that dominate the optical transition. Shown in Fig. 9.15 is the gain spectra for bulk GaAs as a function of energy with different carrier density (Corzine et al, 1993). Increased carrier density also increases the quasi-Fermi level separation, thus the gain spectra have a wider energy range above zero. Also, with the increased carrier density, gain also increases at the same energy. As we can see from (9.52), the absorption coefficient is related to carrier density

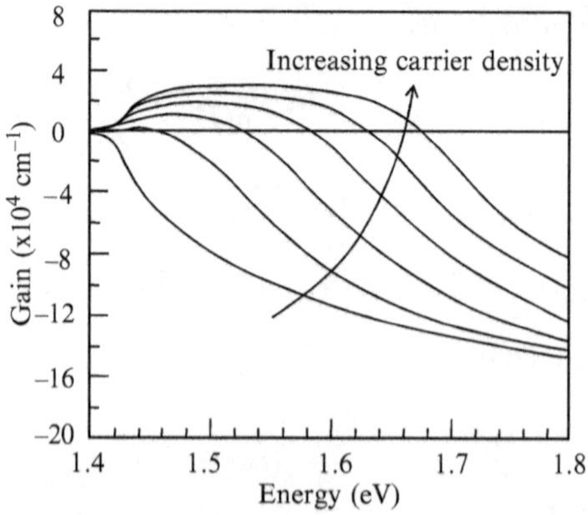

Fig. 9.15. Gain vs. photon energy for bulk GaAs with different carrier densities. From Corzine (Corzine et al, 1993)

only through the quasi-Fermi energy. So Fig. 9.15 indicates that increased quasi-Fermi level distance enhances the gain. Differential gain is the physics quantity to measure this enhancement. It is defined as the derivative of the optical gain to the increment of the injected carrier density. It measures the effectiveness of the laser to transform the input electric power into the luminous flow and translates the optical gain through carrier injection. Generally, a positive gain does not necessarily guarantee lasing because there is intrinsic power loss due to nonradiative recombination such as Auger process and external loss due to the reflection facets. Instead, the optical gain of the laser medium must be sufficient to balance the sum of all the losses experienced by the light in one round trip of the laser's optical resonator for lasing to start. The injection current to meet this condition is called the threshold current. Since the optical loss is nearly constant for any particular laser the optical gain must be larger than a finite positive value.

It is enlightening to look at 2D cases, because first 2D quantum wells are more realistic device structures for today's semiconductor lasers, and second the Fermi energy can be explicitly obtained in terms of the carrier density under the two-band model. Let us first solve (9.44) and (9.45) to obtain the quasi-Fermi levels for both electrons and holes assuming a carrier density $N = P = n$. Then using the 2D DOS for subbands we have for electrons

$$n = \frac{m_c k_B T}{\pi \hbar^2} \ln \left[1 + \exp\left(\frac{F_c - E_c}{k_B T} \right) \right], \tag{9.55}$$

and for holes,

$$n = \frac{m_v k_B T}{\pi \hbar^2} \ln \left[1 + \exp\left(\frac{E_v - F_v}{k_B T} \right) \right]. \tag{9.56}$$

Then F_c and F_v is given by

$$F_c = k_BT \ln \left[\exp\left(\frac{n\pi\hbar^2}{m_c k_B T} \right) - 1 \right] + E_c, \tag{9.57}$$

and

$$F_v = E_v - k_BT \ln \left[\exp\left(\frac{n\pi\hbar^2}{m_v k_B T} \right) - 1 \right]. \tag{9.58}$$

Then we have

$$F_c - F_v = E_g + k_BT \left[\ln \left(\exp\left(\frac{n\pi\hbar^2}{m_c k_B T} \right) - 1 \right) + \ln \left(\exp\left(\frac{n\pi\hbar^2}{m_v k_B T} \right) - 1 \right) \right). \tag{9.59}$$

From (9.59), we can see that for a given photon energy ($>E_g$), if the DOS of either band or both bands can be reduced, the inversion conduction can be realized by lower carrier density, or lower injection current, and higher gain can be achieved. Also smaller DOS can result in a larger quasi-Fermi level separation and thus enhance the gain shift for a fixed injection current increase, and a higher differential gain can be achieved, too. For shifting the DOS, strain can play a critical role.

9.5 STRAINED QUANTUM WELL LASERS

In this section, we will use the $In_xGa_{1-x}As/InP$ quantum well laser as a prototype to introduce the properties of strained quantum well lasers and to see how strain alters their operation parameters. Also, $In_xGa_{1-x}As$/InGaAsP/InP structures with InGaAsP lattice matched to InP substrate are extensively used. There is no difference in fundamental physics between these two types of heterostructures.

9.5.1 Subband Structure and Modal Gain

Consider $In_xGa_{1-x}As$ thin film grown on InP substrate. Both the ternary structure $In_xGa_{1-x}As$ and crystal InP are cubic. The lattice constant is 5.8697 Å for InP, 5.6533 Å for GaAs, and 6.0583 Å for InAs. For $In_xGa_{1-x}As$, the lattice constant can be obtained by linear extrapolation, $a(x) = 5.6533 + 0.405x$. From this relation, the lattice constants of $In_xGa_{1-x}As$ and InP match when $x = 0.53$. Assume the pseudomorphic strain condition, When $x < 0.53$, $In_xGa_{1-x}As$ film is under biaxial tensile strain, and when $x > 0.53$, $In_xGa_{1-x}As$ film is under biaxial compressive strain. As we discussed before, the film thickness shall be controlled under the critical thickness to prevent defect formation. The critical thickness for $In_xGa_{1-x}As$ grown on InP substrate is shown in Fig. 9.16 as a function of lattice mismatch using Matthews and Blakeslee's model. The well thicknesses for modern quantum well lasers

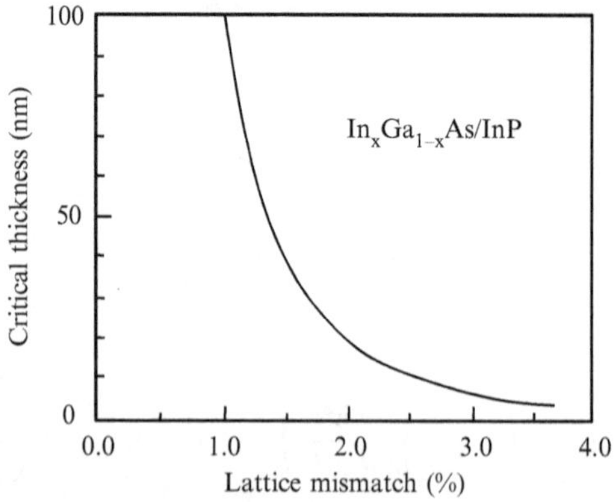

Fig. 9.16. Critical thickness vs. lattice mismatch for InGaAs layers grown on InP substrate

are generally very small and below the critical thickness. From the analysis for strain effects on band structures we did in earlier chapters, the hydrostatic strain component in this biaxial strain case alters the bandgap. In III–V direct gap materials, shear strain had negligible effect on conduction band. Thus, we may just look at the strain effect on valence bands. It lifts the HH and LH degeneracy and shifts the LH up to the top band in biaxial tensile case while shifts the HH up in biaxial compressive case. The subband structure diagram for a $In_xGa_{1-x}As$ quantum well under different strain conditions is shown in Fig. 9.17. At the left panel, the subband edges are shown. They reflect the band structure at the z direction. At the right panel, the in-plane valence subband structures are shown. They demonstrate the effects of both quantum confinement and strain on band curvatures and DOS. Without strain, the HH subband is always at the top. Under biaxial tensile stress, the LH shifts up and becomes the top subband when strain reaches a certain value, which also depends on the well width. Calculations show that for a well thickness of 12 nm, the LH and HH subbands cross over each other at $x = 0.42$. The quantum confinement and strain effects in quantum wells are generally not separable. For the same In composition $x = 0.7$, although they have the same strain, the DOS is smaller in $d_w = 3\,\text{nm}$ structure than the 5 nm structure due to the stronger confinement and consequent larger subband splitting. Overall, quantum confinement and strain combined together have very significant effects on the effective masses and DOS for holes. In III–V materials, at least in bulk, the hole effective masses are generally several times larger than those of electrons. So when carriers are injected into the laser, the quasi-Fermi level for electrons is much easier to get into the conduction band, while the quasi-Fermi level for holes is often above the valence band edge. The effective masses for the top subbands are also plotted as a function of the in-plane wave vector in Fig. 9.18, where we can see that both tensile and compressive strains reduce the top valence subband effective mass in a large k_t range; especially for the compressive strained case, the effective mass is much smaller than the

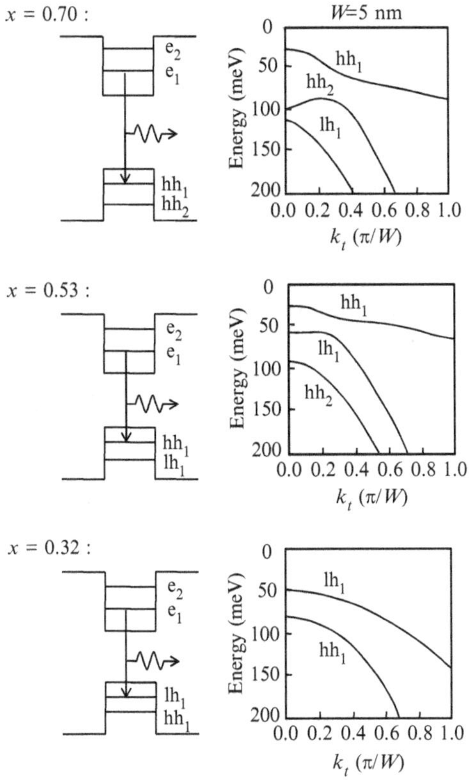

Fig. 9.17. Band structure of $In_xGa_{1-x}As/InP$ quantum well. (**a**) Subband structure at $k = 0$; (**b**) The valence subband dispersion in the transverse plane

unstrained case. Smaller mass leads to smaller DOS and subsequently makes it easier for the quasi-Fermi level for holes to enter the valence band. This means, to realize the population inversion, a lower carrier density is required compared to the lattice-matched device.

In tensile-strained quantum well, the LH subband is at the top, while in compressive-strained case, the HH subband is at the top. The dependence of the ordering of the subband structure on strain makes the quantum well lasers capable to emit light with desirable polarization by strain engineering. As shown in Fig. 9.13, the TE mode dominates the light emission in unstrained and compressive strain cases, while in tensile strain case, the TM mode dominates. The trend shown in Fig. 9.13 does not depend on the particular well material, but is universal (Chang and Chuang, 1995) for quantum well lasers designed using III–V direct gap materials. Taking into account both the DOS change and the optical matrix element dependence on strain, the modal gain for both tensile-strained and compressive-strained $In_xGa_{1-x}As/InP$ quantum well lasers is shown in Fig. 9.19. Compared to the unstrained device, the modal gain is larger in both tensile and compressive-strained devices.

Single quantum well lasers sometimes suffer from high carrier loss due to thin well thickness and subsequent less time staying in the well active region. For better employing the merits of quantum well structure and enhance the quantum efficiency at the same time, multiple quantum wells (MQW) have been developed where several identical single quantum wells are grown by multiple layer structures. In MQW, the injected carriers can be collected by these quantum wells by utilizing the quantum tunneling effect and then recombining to emit photons. The quantum efficiency in MWW systems can be as high as 100%. With the enhanced quantum efficiency, the threshold current for lasing can be lowered. But meanwhile, since every quantum well can be a emitting channel, the output optical power is greatly enhanced, and the threshold current will begin to increase after a certain number of integrated quantum wells. Shown in Fig. 9.20 is the threshold current dependence on a

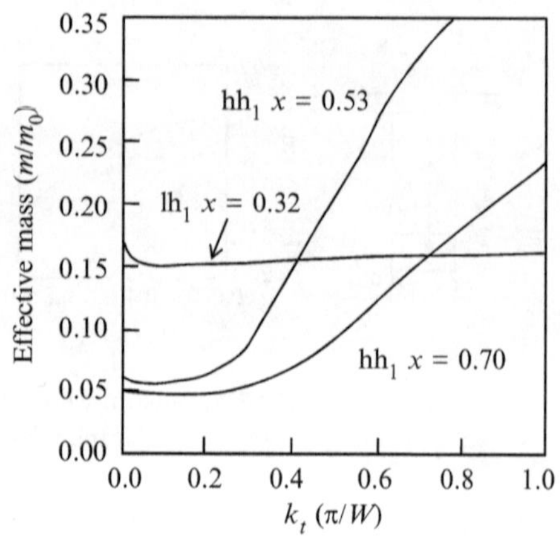

Fig. 9.18. Effective masses for top valence subbands of $In_xGa_{1-x}As/InP$ quantum wells as a function of in-plane wave vector

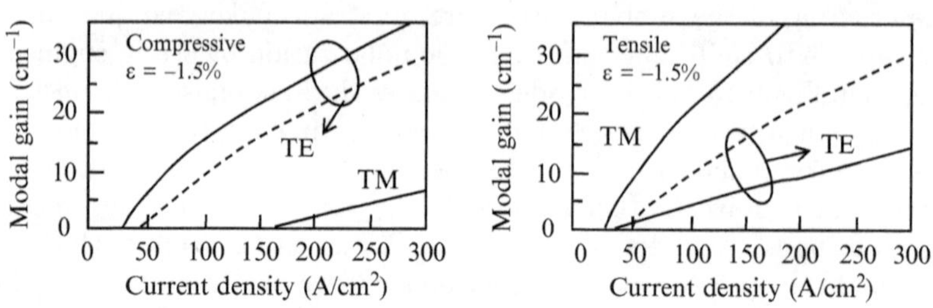

Fig. 9.19. Modal gain for $In_xGa_{1-x}As/InP$ quantum well lasers under both compressive and tensile strains

number of quantum wells in a $In_xGa_{1-x}As$/InGaAsP MQW laser assuming an optical loss of 20 cm^{-1}. Strain effect on the threshold current is shown by comparison between different curves. For TE mode emission, the optimal number of quantum wells is 3. Especially for the $x = 0.7$ compressive-strained MQW device, the threshold current is reduced by 100% when changing from single quantum well to three quantum wells. At the same time, the threshold current reduction by compressive strain for TE mode lasing is also beyond 100%.

In the past 40 years, from bulk semiconductor lasers to double heterostructure lasers, from lattice-matched quantum well lasers to strained multiple-quantum well lasers, strain has enabled a great development of semiconductor laser technology, which reduced the threshold current by tens of times, and has achieved a linewidth that is hundreds of times narrower. The substantial importance of semiconductor laser applications in communications makes

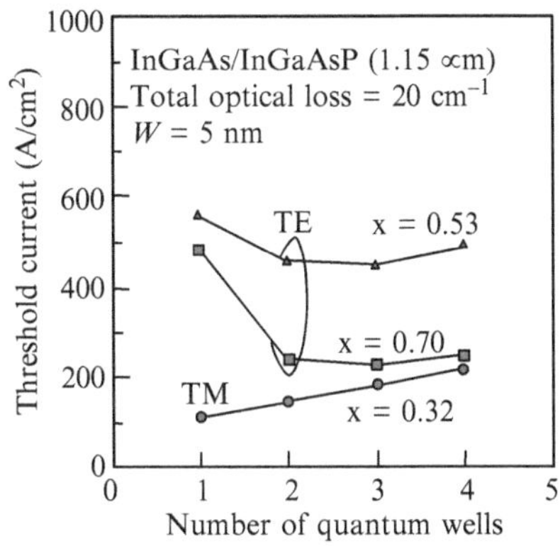

Fig. 9.20. Threshold current vs. number of quantum wells for unstrained, compressive and tensile strained InGaAs/InGaAsP multiple-quantum well lasers

it safe to say that the semiconductor optoelectronics, which pivots on the semiconductor laser technology, will continue to play a more important role in the twenty-first century, the information century.

Appendix: Effective Mass Theorem

For solving a Schrödinger equation with a slowly varying perturbation $\delta V(\mathbf{r})$ in a crystal,

$$[H_0 + \delta V(\mathbf{r})]\psi(\mathbf{r}) = E\psi(\mathbf{r}), \tag{A.1}$$

with $H_0 = \frac{p^2}{2m_0} + V(\mathbf{r})$ the unperturbed Hamiltonian, we may expand the new eigenstates as the linear superposition of the perfect bulk eigenstates $\psi_{j\mathbf{k}}(\mathbf{r})$, thus we look for a solution in the form of

$$\psi(\mathbf{r}) = \sum_j \int f_j(\mathbf{k})\psi_{j\mathbf{k}}(\mathbf{r})\mathrm{d}\mathbf{k}, \tag{A.2}$$

where

$$f_j(\mathbf{k}) = \int_\Omega \psi^*_{j\mathbf{k}}(\mathbf{r})\psi(\mathbf{r})\mathrm{d}\mathbf{r} \tag{A.3}$$

is the projection of $\psi(\mathbf{r})$ on the bulk eigenstates $\psi_{j\mathbf{k}}(\mathbf{r})$. We integrate over k because that with the breaking of the translational symmetry, now the eigenstates are the mixture from the original eigenstates over the whole Brillouin zone. The summation is over the band we include in our calculation. Sometimes one band is not very bad, such as for the donor states for GaAs, which are generally on the magnitude of meV below the conduction band edge. On the other occasions such as for the acceptor states or the valence subband structures in a quantum well, we need to include the HH, LH, and split-off hole states based on the Luttinger Hamiltonian, and sometimes we shall also put the conduction band into consideration in addition.

Before proceeding, we first consider the integral

$$I = \int_\Omega F(\mathbf{r})u(\mathbf{r})\mathrm{d}\mathbf{r}, \tag{A.4}$$

where Ω is the volume of the crystal, $F(\mathbf{r})$ is a slowly varying function at the scale of the primitive cell, and $u(\mathbf{r})$ is a periodic function with the periodicity of the lattice. Varying "slowly" here means that in the scale of a primitive cell, the quantities it refers to can be assumed unchanged. There are typically two length scales in our considered systems. One is a relatively macroscopic scale such as the quantum well width, on which the perturbation and the envelope function vary. The other is the scale of the crystal primitive cell, or about the lattice constant, on which the periodic part of the Bloch function varies. Using these spatial properties of the functions, we can factorize an integral over the whole crystal into a product of two integrals as in the following,

$$\begin{aligned} I &= \int_{\Omega} F(\mathbf{r})u(\mathbf{r})d\mathbf{r} = \sum_{\mathbf{R_n}} \int_{\Omega_0} F(\mathbf{r}-\mathbf{R_n})u(\mathbf{r}-\mathbf{R_n})d\mathbf{r} \\ &= \sum_{\mathbf{R_n}} F(\mathbf{r}-\mathbf{R_n}) \int_{\Omega_0} u(\mathbf{r}-\mathbf{R_n})d\mathbf{r} \\ &= \frac{1}{\Omega} \int_{\Omega} F(\mathbf{r})d\mathbf{r} \int_{\Omega_0} u(\mathbf{r})d\mathbf{r}, \end{aligned} \tag{A.5}$$

where Ω_0 is the volume of the primitive cell. In the above derivation, the periodicity of $u(\mathbf{r})$ is used.

According to the discussions in the last chapter, the bulk eigenstates $\psi_{j\mathbf{k}}(\mathbf{r})$ are Bloch functions with the periodic part being superpositions of the band edge function u_{i0}, i.e.,

$$\psi_{j\mathbf{k}}(\mathbf{r}) = \mathrm{e}^{i\mathbf{k}\cdot\mathbf{r}} \sum_i c_{ji}(\mathbf{k}) u_{i0}(\mathbf{r}). \tag{A.6}$$

Inserting the above equation into (A.2), we obtain

$$\begin{aligned} \psi(\mathbf{r}) &= \sum_{i,j} \int \mathrm{e}^{i\mathbf{k}\cdot\mathbf{r}} c_{ji}(\mathbf{k}) f_j(\mathbf{k}) u_{i0}(\mathbf{r}) \mathrm{d}\mathbf{k} \\ &= \sum_i \left(\int \mathrm{e}^{i\mathbf{k}\cdot\mathbf{r}} F_{i\mathbf{k}} \mathrm{d}\mathbf{k} \right) u_{i0}(\mathbf{r}) \\ &= \sum_i F_i(\mathbf{r}) u_{i0}(\mathbf{r}), \end{aligned} \tag{A.7}$$

where we define

$$F_{i\mathbf{k}} = \sum_j c_{ji}(\mathbf{k}) f_j(\mathbf{k}) \tag{A.8}$$

and

$$F_i(\mathbf{r}) = \int \mathrm{e}^{i\mathbf{k}\cdot\mathbf{r}} F_{i\mathbf{k}} \mathrm{d}\mathbf{k}. \tag{A.9}$$

So eventually we expressed the eigenstates of the perturbed system as a sum of modulated periodic band edge functions. The $F_i(\mathbf{r})$s are the envelope functions.

Inserting (A.7) into (A.1), then multiplying $e^{-i\mathbf{k}'\cdot\mathbf{r}}u_{i0}^*(\mathbf{r})$, and then integrating over the volume of the crystal, we obtain

$$\sum_j \frac{1}{\Omega}\int_\Omega \mathrm{d}\mathbf{r} \int \mathrm{d}\mathbf{k}\ \ F_{j\mathbf{k}} e^{i(\mathbf{k}-\mathbf{k}')\cdot\mathbf{r}} u_{i0}^*(\mathbf{r}) \times \left[\frac{\hbar^2 k^2}{2m_0} + E_j - E + \frac{\hbar}{m_0}\mathbf{k}\cdot\mathbf{p} + \delta V(\mathbf{r})\right] u_{j0}(\mathbf{r}) = 0. \tag{A.10}$$

Here we can use (A.5) to greatly reduce the complexity of the above equation. Following the same procedure, we have

$$\frac{1}{\Omega}\int_\Omega \mathrm{d}\mathbf{r}\ \ e^{i(\mathbf{k}-\mathbf{k}')\cdot\mathbf{r}} u_{i0}^*(\mathbf{r}) \frac{\hbar}{m_0}\mathbf{k}\cdot\mathbf{p} u_{j0}(\mathbf{r}) = \delta_{\mathbf{k},\mathbf{k}'}\frac{\hbar}{m_0}\mathbf{k}\cdot\mathbf{p}_{ij}, \tag{A.11}$$

where the identity

$$\frac{1}{\Omega}\int_\Omega \mathrm{d}\mathbf{r}\ \ e^{i(\mathbf{k}-\mathbf{k}')\cdot\mathbf{r}} = \delta_{\mathbf{k},\mathbf{k}'} \tag{A.12}$$

is used, and $\mathbf{p}_{ij}$ is the momentum matrix element between states $u_{i0}(\mathbf{r})$ and $u_{j0}(\mathbf{r})$. Similarly,

$$\frac{1}{\Omega}\int_\Omega \mathrm{d}\mathbf{r}\ \ e^{i(\mathbf{k}-\mathbf{k}')\cdot\mathbf{r}} u_{i0}^*(\mathbf{r})\delta V(\mathbf{r}) u_{j0}(\mathbf{r}) = \delta V_{\mathbf{k}-\mathbf{k}'}\delta_{i,j}, \tag{A.13}$$

where

$$\delta V_{\mathbf{k}-\mathbf{k}'} = \frac{1}{\Omega}\int_\Omega \mathrm{d}\mathbf{r}\ \ e^{i(\mathbf{k}-\mathbf{k}')\cdot\mathbf{r}}\delta V(\mathbf{r}) \tag{A.14}$$

is actually the Fourier transformation of $\Delta V(\mathbf{r})$, and we used the property that periodic Bloch states are orthonormal in a primitive cell, i.e.,

$$\frac{1}{\Omega}\int_\Omega u_{j0}^* u_{i0}\mathrm{d}\mathbf{r} = \frac{1}{\Omega_0}\int_{\Omega_0} u_{j0}^* u_{i0}\mathrm{d}\mathbf{r} = \delta_{ij}. \tag{A.15}$$

Using these results, (A.10) finally becomes

$$\left(\frac{\hbar^2 k^2}{2m_0} + E_i\right) F_{i\mathbf{k}} + \frac{\hbar}{m_0}\sum_j \mathbf{k}\cdot\mathbf{p}_{ij} F_{j\mathbf{k}} + \int \mathrm{d}\mathbf{k}'\ \ \delta V_{\mathbf{k}-\mathbf{k}'}(\mathbf{r}) F_{i\mathbf{k}'} = E F_{i\mathbf{k}}. \tag{A.16}$$

This equation clearly shows that δV mixes the states with different k, but does not couple one state with the other. Next, we deal with the coupling from the other bands using perturbation theory. One approximation is to neglect the effects of the perturbative potential to the other bands. Then if we consider a singly degenerate band i, and assume that one band j only couples with band i, then for all the other bands, (A.16) becomes

$$\left(\frac{\hbar^2 k^2}{2m_0} + E_j\right) F_{j\mathbf{k}} + \frac{\hbar}{m_0}\mathbf{k}\cdot\mathbf{p}_{ij} F_{i\mathbf{k}} = E F_{j\mathbf{k}}. \tag{A.17}$$

By approximating $E \simeq E_i + \frac{\hbar^2 k^2}{2m_0}$, we have

$$F_{j\mathbf{k}} = \frac{\hbar}{m_0}\frac{\mathbf{k}\cdot\mathbf{p}_{ij}}{E_i - E_j} F_{i\mathbf{k}}. \tag{A.18}$$

Inserting the above equation into (A.16), we obtain the equation for band i,

$$\left(\frac{\hbar^2 k^2}{2m_0} + E_i + \frac{\hbar^2}{m_0^2}\sum_{j\neq i}\frac{|\mathbf{k}\cdot\mathbf{p}_{ij}|^2}{E_i - E_j}\right) F_{i\mathbf{k}} + \int \mathrm{d}\mathbf{k}' \;\; \delta V_{\mathbf{k}-\mathbf{k}'}(\mathbf{r}) F_{i\mathbf{k}'} = E F_{i\mathbf{k}}. \tag{A.19}$$

Comparing with (4.116), we can write the above equation as

$$\left(\frac{\hbar^2 k^2}{2m^*} + E_i\right) F_{i\mathbf{k}} + \int \mathrm{d}\mathbf{k}' \;\; \delta V_{\mathbf{k}-\mathbf{k}'}(\mathbf{r}) F_{i\mathbf{k}'} = E F_{i\mathbf{k}}, \tag{A.20}$$

where m^* is exactly the effective mass for band i in the prefect crystal. Taking a Fourier transform of the above equation from the reciprocal space to the real space, the above equation becomes

$$\left[-\frac{\hbar^2 \nabla^2}{2m^*} + \delta V(\mathbf{r})\right] F_i(\mathbf{r}) = E F_i(\mathbf{r}), \tag{A.21}$$

which is an equation for the envelope function $F_i(\mathbf{r})$. Following the same procedure, a coupled second-order differential equation set can be obtained for the envelope functions for a degenerate set, written in a matrix form as $(H + \delta V)F = EF$, where F is the envelope function in a form of $1 \times n$ vector, if the band considered is n-fold degenerate, and H is in the same form as for a perfect crystal with k_i replaced by $-i\frac{\partial}{\partial x_i}$, where $x_i = x, y, z$.

References

Adachi A (1994) GaAs and related materials: bulk semiconducting and superlattice properties. World Scientific, New York

Ahn D, Chuang SL (1988) Optical gain in a strained-layer quantum-well laser. IEEE J Quantum Electron 24:2400–2406

Alieu J, Bouillon P, Gwoziecki R, Moi D, Bremond G, Skotnicki T (1998) Optimisation of $Si_{0.7}Ge_{0.3}$ channel heterostructures for 0.15/0.18 μm CMOS process. In: Proceedings of ESSDERC, Bordeaux, France, pp 144–147

Ando T (1982a) Self-consistent results for a gaas/al_xga_{1-x}as heterojunction. II. Low temperature mobility. J Phys Soc Japan 51:3900–3907

Ando T (1982b) Self-consistent results for a GaAs/Al_xGa_{1-x}As heterojunction. I. subband structure and light-scattering spectra. J Phys Soc Japan 51:3893–3899

Andrieu F, Ernst T, Romanjek K, Weber O, Renard C, Hartmann JM, Toffoli A, Papon AM, Truche R, Holliger P, Brévard L, Ghibaudo G, Deleonibus S (2003) SiGe channel p-MOSFETS scaling-down. In: Proceedings of ESSDERC, Grenoble, France, pp 267–270

Ang KW, Chui KJ, Tung CH, Balasubramanian N, Li MF, Samudra GS, Yeo YC (2007) Enhanced strain effects in 25-nm gate-length thin-body NMOSFETS with silicon-carbon source/drain and tensile-stress liner. IEEE Electron Dev Lett 28:301–304

Arkwright J, Wu B, Skinner I, Chu P (1994) Resonantly enhanced nonlinearity using stimulated down-pumping in neodymium doped twin-core fibre. Electron Lett 30:235–236

Arnold DP, Gururaj S, Bhardwaj S, Nishida T, Sheplak M (2001) A piezoresistive microphone for aeroacoustic measurements. In: ASME International Mechanical Engineering Congress and Exposition, New York, pp 23,841-1–23,841-8

Asada M, Adams AR, Stubkjaer KE, Suematsu Y, Itaha Y, Arai S (1981) The temperature dependence of the threshold current of GaInAsP/InP DH lasers. IEEE J Quantum Electron 17:611–619

Ashcroft NW, Mermin ND (1976) Solid state physics. Saunders College Publishing/Harcourt Brace College Publishers, Forth Worth, TX

Aspnes DE, Cardona M (1978) Piezoresistance and the conduction-band minima of GaAs. Phys Rev B 17:741–751

Assad F, Ren Z, Bendix P, Lundstrom MS (1999) Performance limits of Si MOSFETS. In: Tech. Dig. Int. Electron Device Meet., San Francisco, CA, pp 547–550

Assaderaghi F, Shahidi G (2000) SOI at IBM: current status of technology, modeling, design, and the outlook for the 0.1 μm generation. In: IEEE International SOI Conference, Wakefield, MA, pp 6–9

Bai P, Auth C, Balakrishnan S, Bost M, Brain R, Chikarmane V, Heussner R, Hussein M, Hwang J, Ingerly D, James R, Jeong J, Kenyon C (2004) A 65nm logic technology featuring 35 nm gate lengths, enhanced channel strain, 8 Cu interconnect layers, low-k ild and 0.57 μm^2 SRAM cell. In: Tech. Dig. Int. Electron Device Meet., San Francisco, CA, pp 657–660

Ball CA, van der Merwe JH (1983) Dislocations in solids. North-Holland Press, Amsterdam

Bardeen J (1938) An improved calculation of the energies of metallic Li and Na. J Chem Phys 6:367–371

Bardeen J, Shockley W (1950) Deformation potentials and mobilities in non-polar crystals. Phys Rev 80:72–80

Beattie AR, Landsberg PT (1958) Auger effect in semiconductors. Proc R Soc A 249:16–29

Bernard MGA, Duraffourg G (1961) Phys Status Solodi 1:699

Bernardini F, Fiorentini V (1997) Spontaneous polarization and piezoelectric constants of III–V nitrides. Phys Rev B 56:R10,024–R10,027

Berz F (1985) The bethe condition for thermionic emission near an absorbing boundary. Solid State Electron 28:1007–1013

Boermans MJB, Hagen SH, Valster A, Finke MN, der Heyden JMMV (1990) Investigation of TE and TM polarised laser emission in GaInp/AlGaInp lasers by growth-controlled strain. IEEE Electron Lett 26:1438–1439

Bonno O, Barraud S, Mariolle D, Andrieu F (2008) Effect of strain on the electron effective mobility in biaxially strained silicon inversion layers: an experimental and theoretical analysis via atomic force microscopy measurements and kubo-greenwood mobility calculations. J Appl Phys 103:063,715-1–063,715-9

Boresi AP, Schmidt RJ (2003) Advanced mechanics of materials. Wiley, New York

Bour DP, Treat DW, Beernink KJ, Krusor BS, Geels RS, Welch DF (1994) 610-nm band AlGaInp single quantum well laser diode. IEEE Photon Tech Lett 6:128–131

Braga N, Buczkowski A, Kirk HR, Rozgonyi GA (1994) Formation of cylindrical n/p junction diodes by arsenic enhanced diffusion along interfacial misfit dislocations in p-type epitaxial Si/Si(Ge). Appl Phys Lett 64:1410–1412

Brantley WA (1973) Calculated elastic-constants for stress problems associated with semiconductor devices. J Appl Phys 44:534–535

Bufler FM, Fichtner W (2003) Scaling of strained-Si n-MOSFETS into the ballistic regime and associated anisotropic effects. IEEE Trans Electron Devices 50:278–284

Burns FP (1957) Piezoresistive semiconductor microphone. J Acoust Soc Amer 29:248–253

Callaway J (1976) Quantum theory of the solid state, 2nd edn. Academic, Boston, MA

Chadi DJ, Cohen ML (1975) Tight-binding calculations of the valence bands of diamond and zincblende crystals. Phys Stat Sol (B) 68:405–491

Chan V, Rengarajan R, Rovedo N, Wei J, Hook T, Nguyen P, Chen J, Nowak E, Xiang-Dong C, Lea D, Chakravarti A, Ku V, Yang S, Steegen A, Baiocco C, Shafer P, Hung N, Shih-Fen H, Wann C (2003) High speed 45nm gate length CMOSFETS integrated into a 90nm bulk technology incorporating strain engineering. In: Tech. Dig. Int. Electron Device Meet., Washington, DC, pp 3.8.1–3.8.4

Chang CS, Chuang SL (1995) Universal curves for optical-matrix elements of strained quantum wells. Appl Phys Lett 66:795–797

Chang RPH, Coleman JJ (1978) A new method of fabricating gallium arsenide MOS devices. Appl Phys Lett 32:332–333

Chelikowsky JR, Cohen ML (1976) Calculations for the electronic structures of eleven diamond and zinc-blende semiconductors. Phys Rev B 14:556–582

Chidambaram PR, Smith BA, Hall LH, Bu H, Chakravarthi S, Kim Y, Samoilov AV, Kim AT (2004) 35 percent drive current improvement from recessed-SiGe drain extensions on 37 nm gate length PMOS. Hawaii, USA, pp 48–49

Chien-Hao C, Lee TL, Hou TH, Chen CL, Chen CC, Hsu JW, Cheng KL, Chiu YH, Tao HJ, Jin Y, Diaz CH, Chen SC, Liang MS (2004) Stress memorization technique (SMT) by selectively strained-nitride capping for sub-65nm high-performance strained-Si device application. Honolulu, HI, pp 56–57

Cho YK, Kwon SK, Jung HB, Kim J (2005) High Performance Power MOSFETs with Strained-Si Channel. Proceedings-ISPSD, Santa Barbara, CA

Chu M, Nishida T, Lv XL, Mohta N, Thompson SE (2008) Comparison between high-field piezoresistance coefficients of Si metal-oxide-semiconductor field-effect transistors and bulk Si under uniaxial and biaxial stress. J Appl Phys 103:113,704-1–113,704-7

Chuang SL (1995) Physics of optoelectronic devices. Wiley-Interscience, Wiley, New York

Corzine SW, Yan RH, Coldren LA (1993) Optical gain in III–V bulk and quantum well semiconductors. In: Zory PS (ed) Quantum Well Lasers. Academic, San Diego, CA

Cui Y, Zhong Z, Wang D, Wang W, Lieber CM (2003) High performance silicon nanowire field effect transistors. Nano Lett 3:149–152

Currie MT, Samavedam SB, Langdo TA, Leitz CW, Fitzgerald EA (1998) Appl Phys Lett 72:1718

Currie MT, Leitz CW, Langdo TA, Taraschi G, Antoniadis DA, Fitzgerald EA (2001) Carrier mobilities and process stability of strained Si n- and p-MOSFETS on SiGe virtual substrates. J Vac Sci Technol B 19:2268–2279

Datta S, Dewey G, Doczy M, Doyle BS, Jin B, Kavalieros J, Kotlyar R, Metz M, Zelick N, Chau R (2003) High mobility Si/SiGe strained channel mos transistors with HfO_2/TiN gate stack. In: Tech. Dig. Int. Electron Device Meet., Washington, DC, pp 653–656

Davies JH (1998) The physics of low-dimensional semiconductors – an introduction. Cambridge University Press, New York

Dingle R, Wiegmann W, Henry CH (1974) Quantum states of confined carriers in very thin $Al_xGa_{1-x}As$-GaAs-$Al_xGa_{1-x}As$ heterostructures. Phys Rev Lett 33:827–830

Dingle R, Stormer H, Gossard AC, Wlegmann W (1978) Selective absorption of solar energy in ultrafine chromium particles. Appl Phys Lett 31:665–666

Dismukes JP, Ekstrom L, Steigmeier EF, Kudman I, Beers SD (1964) Thermal and electrical properties of heavily doped Ge-Si alloys up to 1300 k. J Appl Phys 35

Dries JC, Gokhale MR, Thompson KJ, Forrest SR, Hull R (1998) Strain compensated $In_{1-x}Ga_xAs$(x ¡ 0.47) quantum well photodiodes for extended wavelength operation. Appl Phys Lett 73:2263–2265

Özbay E, Kimukin İ, Biyikli N, Aytür O, Gökkavas M, Ulu G, Ünlü MS, Mirin RP, Berthness KA, Christensen DH (1999) High-speed ¿90% quantum-efficiency *p–i–n* photodiodes with a resonance wavelength adjustable in the 795–835 nm range. Appl Phys Lett 74:1072–1074

Evans MH, Zhang XG, Joannopoulos JD, Pantelides ST (2005) First-principles mobility calculations and atomic-scale interface roughness in nanoscale structures. Phys Rev Lett 95:106,802-1–160,802-4

Fang FF, Howard WE (1966) Negative field-effect mobility on (100) Si surfaces. Phys Rev Lett 16:797–799

Ferrier M, Clerc R, Ghibaudo G, Boeuf F, Skotnicki T (2006) Analytical model for quantization on strained and unstrained bulk NMOSFET and its impact on quasi-ballistic current. Solid State Electron 50:69–77

Fiorenza JG, Braithwaite G, Leitz CW, Currie MT, Yap J, Singaporewala F, Yang VK, Langdo TA, Carlin J, Somerville M, Lochtefeld A, Badawi H, Bulsara MT (2004) Film thickness constraints for manufacturable strained silicon CMOS. Semicond Sci Technol 19:L4–L8

Fischetti MV, Laux SE (1996) Band structure, deformation potentials, and carrier mobility in strained Si, Ge, and SiGe alloys. J Appl Phys 80:2234–2252

Fischetti MV, Gámiz F, Hänsch W (2002) On the enhanced electron mobility in strained-silicon inversion layers. J Appl Phys 92:7320–7324

Fischetti MV, Ren Z, Solomon PM, Yang M, Rim K (2003) Six-band k·p calculation of the hole mobility in silicon inversion layers: dependence on surface orientation, strain, and silicon thickness. J Appl Phys 94:1079–1095

Fortuna V, Bournel A, Dollfus P, Retailleau S (2006) Ultra-short n-mosfets with strained Si: device performance and the effect of ballistic transport using monte carlo simulation. Semiconductor Sci Technol 21:422–428

Fossum ER (1995) CMOS image sensors: electronic camera on a chip. In: Tech. Dig. Int. Electron Device Meet., Washington, DC, pp 17–25

Fossum JG, Weimin Z (2003) Performance projections of scaled CMOS devices and circuits with strained Si-on-SiGe channels. IEEE Trans Electron Devices 50:1042–1049

Frank DJ, Laux SE, Fischetti MV (1992) Mente carlo simulation of a 30 nm dual-gate mosfet: how short can Si go? In: Tech. Dig. Int. Electron Device Meet., Washington, DC, pp 553–556

Froyen S, Harrison WA (1979) Elementary prediction of linear combination of atomic orbitals matrix elements. Phys Rev B 20:2420–2422

Ridley BK (1999) Quantum Processes in Semiconductors, frouth edn. Clarendon Press Oxford, New York

Gámiz F, Roldán JB, López-Villanueva JA, Cartujo-Cassinello P, Carceller JE (1999) Surface roughness at the Si–SiO_2 interfaces in fully depleted silicon-on-insulator inversion layers. J Appl Phys 86:6854–6863

Gannavaram S, Pesovic N, Ozturk C (2000) Low temperature (800°c) recessed junction selective silicon–germanium source/drain technology for sub-70 nm CMOS. In: Tech. Dig. Int. Electron Device Meet., San Francisco, CA, pp 437–440

Gaworzewski P, Tittelbach-Helmrich K, Penner U (1998) Electrical properties of lightly doped *p*-type silicon–germanium single crystals. J Appl Phys 83

Gershoni D, Temkin H, Panish MB (1988) Strained-layer $ga_{1-x}in_x$As/InP avalanche photodetectors. Appl Phys Lett 53:1294–1296

Ghani T, Armstrong M, Auth C, Bost M, Charvat P, Glass G, Hoffmann T, Johnson K, Kenyon C, Klaus J, McIntyre B, Mistry K, Murthy A, Sandford J, Silberstein M, Sivakumar S, Smith P, Zawadzki K, Thompson S, Bohr M (2003) A 90nm high volume manufacturing logic technology featuring novel 45nm gate length strained silicon CMOS transistors. In: Tech. Dig. Int. Electron Device Meet., Washington, DC, pp 11.6.1–11.6.3

Gnani E, Reggiani S, Gnudi A, Parruccini P, Colle R, Rudan M, Baccarani G (2007) Band-structure effects in ultrascaled silicon nanowires. IEEE Trans Electron Devices 54:2243–2254

Goo JS, Xiang Q, Takamura Y, Wang H, Pan J, Arasnia F, Paton EN, Besser P, Sidorov MV, Adem E, Lochtefeld A, Braithwaite G, Currie MT, Bulsara RHMT, Lin MR (2003) Scalability of strained-Si NMOSFETS down to 25 nm gate length. IEEE Electroc Dev Lett 24:351–353

Goodnick SM, Gann RG, Sites JR, Ferry DK, Fathy D, Krivanek OL (1983) Surface roughness scattering at the Si-sio_2 interface. J Vac Sci Technol B 1:803–808

Goodnick SM, Ferry DK, Wilmsen CW, Liliental Z, Fathy D, Krivanek OL (1985) Surface roughness at the Si(100)-SiO_2 interface. Phys Rev B 32:8171–8186

Hackbarth T, Herzog HJ, Rinaldi F, Soares T, Holländer B, Manti S, Luysberg M, Fichtner PF (2003) High frequency n-type modfets on ultrathin virtual SiGe substrates. Solid State Electronics 47:1179–1182

Hadjisavvas G, Tsetseris L, Pantelides ST (2007) The origin of electron mobility enhancement in strained MOSFETS. IEEE Electron Dev Lett 28:1018–1020

Hamada H, Tominaga K, Shono M, Honda S, Yodoshi K, Yamaguchi T (1992) Room-temperature cw operation of 610 nm band AlGaInp strained multiquantum well laser diodes with multiquantum barrier. Electron Lett 28:1834–1836

Harley JA, Kenny TW (2000) 1/F noise considerations for the design and process optimization of piezoresistive cantilevers. J Microelectromech Syst 9:226–235

Harrison WA (1976) Pseudopotential theory of covalent bonding. Phys Rev B 14:702–711

Harrison WA (1989) Electronic structure and the properties of solids. Dover, New York

Harrison WA (1999) Elementary electronic structure. World Scientific, Singapore

Hasegawa H (1963) Theory of cyclotron resonance in strained silicon crystals. Phys Rev 129:1029–1040

He R, Yang P (2006) Giant piezoresistance effect in silicon nanowires. Nat Nanotechnol 1:42–46

Hensel JC, Hasegawa H, Nakayama M (1965) Cyclotron resonance in uniaxially stressed silicon. II. nature of the covalent bond. Phys Rev 138:A225–A238

Herring C, Vogt E (1956) Transport and deformation-potential theory for many-valley semiconductors with anisotropic scattering. Phys Rev 101:944–961

Höck G, Kohn E, Rosenblad C, von Känel H, Herzog HJ, König U (2000) High hole mobility in $Si_{0.17}Ge_{0.83}$ channel metal-oxide-semiconductor field-effect transistors grown by plasma-enhanced chemical vapor deposition. Appl Phys Lett 76

Homeijer B, Griffin B, Nishida T, Cattafesta L, Sheplak M (2006) Design optimization of a microelectromechanical piezoresisitive microphone for use in aeroacoustic measurements. J Acoust Soc Am 120:3330

Hong M, Ren F, Kuo JM, Hobson WS, Kwo J, Mannaerts JP, Lothian JR, Chen YK (1997) Depletion mode GaAs metal-oxide-semiconductor field effect transistors with $Ga_2O_3(Gd_2O_3)$ as the gate oxide. J Vac Sci Technol B 16:1398–1400

Houghton DC, Gibbings CJ, Tuppen CG, Lyons MH, Halliwell MAG (1990) Equilibrium critical thickness for $Si_{1-x}Ge_x$ strained layers on (100) Si. Appl Phys Lett 56:460–462

Huang L, Chu JO, Goma SA, D'Emic CP, Koester SK, Canaperi SF, Mooney PM, Cordes SA, Speidell JL, Anderson RM, Wong HSP (2002) Electron and hole mobility enhancement in strained SOI by wafer bonding. IEEE Trans Electron Devices 49

Huet K, Chassat C, Nguyen DP, Retailleau S, Bournel A, Dollfus P (2007) Full band Monte Carlo study of ballistic effects in nanometer-scaled strained p channel double gate MOSFETS. Physica Status Solidi (c) 5:43–46

Inoue K, Sakaki H, Yoshino J, Hotta T (1985) Self-consistent calculation of electronic states in AlGaAs/GaAs/AlGaAs selectively doped double-heterojunction systems under electric fields. J Appl Phys 58:4277–4281

Irisawa T, Numata T, Toyoda E, Hirashita N, Tezuka T, Sugiyama N, Takagi S (2007) Physical understanding of strain effects on gate oxide reliability of MOSFETS. In: VLSI Symp. Tech. Dig., Hawaii, USA, pp 36–37

Ito S, Namba H, Yamaguchi K, Hirata T, Ando K, Koyama S, Kuroki S, Ikezawa N, Suzuki T, Saitoh T, Horiuchi T (2000) Mechanical stress effect of etch-stop nitride and its impact on deep submicron transistor design. In: Tech. Dig. Int. Electron Device Meet., Washington, DC, pp 247–250

Jogai B (1998) Effect of in-plane biaxial strains on the band structure of wurtzite GaN. Phys Rev B 57:2382–2386

Kahng AB, Sharma P, Topaloglu R, Rasit O (2007) In: Proc. IEEE Intl. Conf. on CAD

Kanda Y (1982) A graphical representation of the piezoresistance coefficients in silicon. IEEE Trans Electron Dev 29:64–70

Kane EO (1957) Band structure of indium antimonide. J Phys Chem Solids 1:249–261

Kash JA, Zachau M, Tischler MA, Ekenberg U (1994) Optical measurements of warped valence bands in quantum wells. Surf Sci 305:251–255

Kettel C (1996) Introduction to solid state physics, 7th edn. Wiley, New York

Kikuchi A, Kishino K, Kaneko Y (1991) 600 nm-range GaInP/AlInP multi-quantum-well (MQW) lasers grown on misorientation substrates by gas source molecular beam epitaxy (gs-mbe). Jpn J Appl Phys 30:3865–3872

Kim BW (2001) Piezoelectric-field effect on electronic and optical properties of [111] $In_xGa_{1-x}As$/GaAs superlattices. J Appl Phys 89:1197–1204

Kohn W, Luttinger JM (1954) Quantum theory of cyclotron resonance in semiconductors. Phys Rev 96:529–530

Krivanek OL, Sheng TT, Tsui DC (1978) A high-resolution electron microscopy study of the Si-SiO_2interface. Appl Phys Lett 32:437–439

Lammers D (2001) IBM banks on SOI, SiGe. EE Times: Columns

Lammers D (1998) IBM takes SOI technology to market. EE Times: Columns

Lee BH, Mocuta A, Bedell S, Chen H, Sadana D, Rim K, O'Neil P, Mo R (2002) Performance enhancement on sub-70 nm strained Si SOI MOSFETS on ultra-thin thermally mixed strained Si/SiGe on insulator (TM-SGOI)

substrate with raised S/D. In: Tech. Dig. Int. Electron Device Meet., San Francisco, CA, pp 946–948

Lee WH, Waite A, Nii H, Nayfeh HM, McGahay V, Nakayama H, Fried D, Chen H, Black L, Bolam R, Cheng J (2005) High performance 65 nm SOI technology with enhanced transistor strain and advanced-low-k BEOL. In: Tech. Dig. Int. Electron Device Meet., Washington, DC, pp 56–59

Leitz CW, Currie MT, Kim AY, Lai J, Robbins E, Fitzgerald EA (2001) J Appl Phys 90

Leu PW, Svizhenko A, Cho K (2008) *Ab initio* calculations of the mechanical and electronic properties of strained Si nanowires. Phys Rev B 77:235,305-1–235,305-14

Lim J, Thompson SE, Fossum JG (2004) Comparison of threshold-voltage shifts for uniaxial and biaxial tensile-stressed n-MOSFETS. IEEE Electron Device Lett 25:731–733

Lime F, Andrieu F, Derix J, Ghibaudo G, Boeuf F, Skotnicki T (2005) Low temperature characteriztion of effective mobility in uniaxially and biaxially strained n-MOSFETS. In: Proceedings of ESSDERC, Grenoble, France, pp 525–528

Lindgren AC, Hellberg PE, von Haartman M, Wu D, Menon C, Zhang S, Östling M (2002) Enhanced intrinsic gain (g_m/g_d) of pMOSFETS with a $Si_{0.7}Ge_{0.3}$ channel. In: Proceedings of ESSDERC, Firenze, Italy, pp 175–178

Liu YC, Pan JW, Chang TY, Liu PW, Lan BC, Tung CH, Tsai CH, Chen TF, Lee CJ, Wang WM, Chen YA, Shih HL, Tung LY, Cheng LW, Shen TM, Chiang SC, Lu MF, Chang WT, Luo YH, Nayak D, Gitlin D, Meng HL, Tsai CT (2005) Single stress liner for both NMOS and PMOS current enhancement by a novel ultimate spacer process. In: Tech. Dig. Int. Electron Device Meet., Washington, DC, pp 836–839

Lochtefeld A, Antoniadis DA (2001) On experimental determination of carrier velocity in deeply scaled NMOS: How close to the thermal limit? IEEE Electron Device Lett 22:95–97

Loo R, Collaert N, Verheyen P, Caymax M, Delhougne R, Meyer KD (2004) Fabrication of 50 nm high performance strained-SiGe pMOSFETS with selective epitaxial growth. Appl Surf Sci 224

Löwdin P (1951) A note on the quantum-mechanical perturbation theory. J Chem Phys 19:1396–1401

Luisier M, Schenk A, Fichtner W (2006) Atomistic simulation of nanowires in the $sp^3d^5s^*$ tight-binding formalism: From boundary conditions to strain calculations. Phys Rev B 74:205,323-1–205,334-12

Lundstrom MS (1997) Elementary scattering theory of the Si mosfet. IEEE Electron Device Lett 18:361–363

Lundstrom MS (2001) On the mobility versus drain current relation for a nanoscale mosfet. IEEE Electron Device Lett 22:293–295

Lundstrom MS, Ren Z (2002) Essential physics of carrier transport in nanoscale MOSFETS. IEEE Trans Electron Devices 49:133–141

Luttinger JM (1956) Quantum theory of cyclotron resonance in semiconductors: general theory. Phys Rev 102:1030–1041

Luttinger JM, Kohn W (1955) Motion of electrons and holes in perturbed periodic fields. Phys Rev 97:869–883

Manasevit HM, Gergis IS, Jones AB (1982) Electron mobility enhancement in epitaxial multilayer Si-$Si_{1-x}Ge_x$ alloy films on (100) Si. Appl Phys Lett 41:464–466

Mason WP, Thurston RN (1957) Use of piezoresistive materials in the measurement of displacement, force, and torque. J Acoust Soc Am 29:1096–1101

Matthews JW, Blakeslee AE (1974) Defects in epitaxial multilayers. I. Misfit dislocations. J Cryst Growth 27:118–125

Mheen B, Song YJ, Kang JY, Hong S (2005) Strained-SiGe complementary MOSFETS adopting different thickness of silicon cap layers for low power and high performance applications. ETRI J 27

Mimura T, Yokoyama N, Fukuta M (1979) Electrical characteristics of the plasma-grown native-oxide/GaAs interface. Appl Phys Lett 34:642–644

Mimura T, Hiyamizu S, Fujii T, Nanbu K (1980) A new field-effect transistor with selectively doped GaAs/n-$Al_xGa_{1-x}As$ heterojunctions. Jpn J Appl Phys 19:L225–L227

Mishima Y, Ochimizu H, Mimura A (2005) Strained-silicon formation on relaxed Silicon-germanium/Silicon-on-insulator substrate using laser annealing. Appl Phys Lett 86

Mistry K, Allen C, Auth C, Beattie B, Bergstrom D, Bost M, Brazier M, Buehler M, Cappellani A, Chau R, Choi CH, Ding G, Fischer K, Ghani T, Grover R, Han W, Hanken D, Hattendorf M, He J, Hicks J, Huessner R, Ingerly D, Jain P, James R, Jong L, Joshi S, Kenyon C, Kuhn K, Williams S, Zawadzki K (2007) A 45nm logic technology with high-k+metal gate transistors, strained silicon, 9 cu interconnect layers, 193nm dry patterning, and 100 percent pb-free packaging. In: Tech. Dig. Int. Electron Device Meet., Washington, DC, pp 247–250

Mizuno T, Sugiyama N, Tezuka T, Takagi S (2003) [110] strained-SOI n-MOSFETS with higher electron mobility. IEEE Electron Dev Lett 24

Moss TS, Hawkins TDF (1961) Infrared absorption in gallium arsenide. Infrared Phys 1:111–115

Murthy A, Chau R, Ghani T, Mistry K (2003) Semiconductor transistor having a stressed channel. Intel Corporation, USA

Nayak DK, Woo JCS, Park JS, Wang KL, MacWilliams KP (1993) High-mobility *p*-channel metal-oxide-semiconductor field-effect transistor on strained Si. Appl Phys Lett 62:2853–2855

Nicholas G, Grasby TJ, Parker EHC, Whall TE, Skotnicki T (2005) IEEE Electron Dev Lett 26: 684–686

Nishizawa J, Shiota I (1980) Inst Phys Conf Ser Chapt 4:289

Nye JF (1984) Physical properties of crystals: their representation by tensors and matrices. Oxford University Press, Oxford

Oberhuber R, Zandler G, Vogl P (1998) Subband structure and mobility of two-dimensional holes in strained Si/SiGe mosfets. Phys Rev B 58:9941–9948

Olsen SH, Escobedo-Cousin E, Varzgar JB, Agaiby R, Seger J, Dobrosz P, Chattopadhyay S, Bull SJ, O'Neill AG, Hellström PE, Edholm J, Östling M, Lyutovich KL, Oehme M, Kasper E (2006) Control of self-heating in thin virtual substrate strained Si MOSFETS. IEEE Trans Electron Devices 53:2296–2305

Omegacom OEI (2003–2009) Omega strain gauges. http://www.omega.com/tocasp/sectionsc.asp?book=pressure&section=e

Onsongo D, Kelly DQ, Dey S, Wise RL, Cleavelin CR, Banerjee SK (2004) Improved hot-electron reliability in strained-Si NMOS. IEEE Trans Electron Devices 51:2193–2199

Packan P, Cea S, Deshpande H, Ghani T, Giles M, Golonzka O, Hattendorf M, Kotlyar R, Kuhn K, Murthy A, Ranade P, Shifren L, Weber C, Zawadzki K (2008) High performance hi-k + metal gate strain enhanced transistors on (110) silicon. In: Tech. Dig. Int. Electron Device Meet., San Francisco, CA, pp 1–4

Passlack M, Hong M, Mannaerts JP, Chu SNG, Opila RL, Moriya N (1995) In-situ Ga_2O_3 process for GaAs inversion/accumulation device and surface passivation applications. In: Tech. Dig. Int. Electron Device Meet., Washington, DC, p 383

Passlack M, Hong M, Mannaerts JP, Opila RL, Chu SNG, Moriya N, Ren F, Kwo JR (1997) Low d_{it}, thermodynamically stable Ga_2O_3–GaAs interfaces: Fabrication, characterization, and modeling. IEEE Trans Electron Devices 44:214–225

Patel N, Ripper J, Brosson P (1973) Behavior of threshold current and polarization of stimulated emission of gaas injection lasers under uniaxial stress. IEEE J Quantum Electron 9:338–341

Persson C, Lindefelt U (1997) Relativistic band structure calculation of cubic and hexagonal SiC polytypes. J Appl Phys 82:5496–5508

Peterson KE (1982) Silicon as a mechanical material. Proc IEEE 70:420–457

Pfann WG, Thurston RN (1961) Semiconducting stress transducers utilizing transverse and shear piezoresistance effects. J Appl Phys 32:2008–2019

Phillips JC, Kleinman C (1959) New method for calculating wave functions in crystals and molecules. Phys Rev 116:287–294

Pidin S, Mori T, Inoue K, Fukuta S, Itoh N, Mutoh E, Ohkoshi K, Nakamura R, Kobayashi K, Kawamura K, Saiki T, Fukuyama S, Satoh S, Kase M, Hashimoto AK (2004) A novel strain enhanced CMOS architecture using selectively deposited high tensile and high compressive silicon nitride films. In: Tech. Dig. Int. Electron Device Meet., San Francisco, CA, pp 213–216

Price J (1981) Two-dimensional electron transport in semiconductor layers. I. phonon scattering. Ann Phys 133:217–239

Qiqing O, Min Y, Holt J, Panda S, Huajie C, Utomo H, Fischetti M, Rovedo N, Jinghong L, Klymko N, Wildman H, Kanarsky T, Costrini G, Fried DM, Bryant A, Ott JA, Meikei I, Chun-Yung S (2005) Investigation of CMOS devices with embedded SiGe source/drain on hybrid orientation substrates. Honolulu, HI, pp 28–29

Rajagopalan K, Drooopad R, Abrokwah J, Passlack M (2006) Compound semiconductor mosfet structure with high-κ dielectric. In: CS MANTECH Conference, Vancouver, BC, Canada, p 119

Rajagopalan K, Droopad R, Abrokwah J, Zurcher P, Fejes P, Passlack M (2007) 1-μm enhancement mode GaAs n-channel MOSFETS with transconductance exceeding 250 mS/mm. IEEE Electron Dev Lett 28:100–102

Reitz JR, Milford FJ, Christy RW (1979) Foundations of electromagnetic theory. Addison-Wesley, Reading, MA

Ren Z, Lundstrom MS (2000) Simulation of nanoscale MOSFETS: a scattering theory interpretation. Supperlattices Microstruct 27:177–189

Rim K, Gusev EP, D'Emic C, Kanarsky T, Chen H, Chu J, Ott J, Chan K, Boyd D, Mazzeo V, Lee BH, Mocuta A, Welser J, Cohen SL, Leong M, Wong HS (2002) Mobility enhancement in strained Si NMOSFETS with HfO_2 gate dielectrics. Honolulu, HI, pp 12–13

Rim K, Chan K, Shi L, Boyd D, Ott J, Klymko N, Cardone F, Tai L, Koester S, Cobb M (2003) Fabrication and mobility characteristics of ultrathin strained Si directly on insulator (SSDOI) MOSFETS. In: Tech. Dig. Int. Electron Device Meet., Washington, DC, pp 49–52

People R, Bean JC, Lang DV, Sergent AM, Stormer HL, Wecht KW, Lynch RT, Baldwin K (1984) Appl Phys Lett 45:1231

Sacconi F, Carlo AD, Lugli P, Morkoç H (2001) Spontaneous and piezoelectric polarization effects on the output characteristics of AlGaN/GaN heterojunction modulation doped FETs. IEEE Trans Electron Dev 48:450–457

Sah CT (1991) Fundamentals of Solid-State Electronics. World Scientific, Singapore

Samavedam SB, Taylor WJ, Grant JM, Smith JA, Tobin PJ, Dip A, Phillips AM (1999) Relaxation of strained Si layers grown on SiGe buffers. J Vac Sci Technol B 17:1424–1429

Sanders GD, Chang YC (1985) Optical properties in modulation-doped GaAs-$Ga_{1-x}Al_xAs$ quantum wells. Phys Rev B 31:6892–6895

Schulz M, Klansmann E (1979) Transient capacitance measurements of interface states on the intentionally contaminated Si-SiO_2 interface. J Appl Phys 18:169–175

Seitz F (1940) The modern theory of solids. McGraw Hill, New York

Semikolenova NA, Nesmelowa IM, Khabarov EN (1978) Investigation of the impurity interaction mechanism in indium arsenide. Sov Phys Semicond 12:1139–1142

Senturia SD (2001) Microsystem design. Kluwer Academic, Boston

Sheplak M, Breuer KS, Schmidt MA (1998) A wafer-bonded, silicon-nitride membrane microphone with dielectrically-isolated, single-crystal silicon piezoresistors. In: Technical Digest, Solid-State Sensor and Actuator Workshop, Hilton Head, SC, pp 23–26

Sheplak M, Dugundji J (1998) Large deflections of clamped circular plates under initial tension and transitions to membrane behavior. J Appl Mech-Trans Asme 65:107–115

Shima M (2002) <100> strained-SiGe-channel p-mosfet with enhanced hole mobility and lower parasitic resistance. FUJITSU Sci Tech J 39

Shimizu A, Hachimine K, Ohki N, Ohta H, Koguchi M, Nonaka Y, Sato H, Ootsuka F (2001) Local mechanical-stress control (LMC): a new technique for CMOS-performance enhancement. In: Tech. Dig. Int. Electron Device Meet., San Francisco, CA, pp 433–436

Shiri D, Kong Y, Buin A, Anantram MP (2008) Strain induced change of bandgap and effective mass in silicon nanowires. Appl Phys Lett 93:073,114-1–073,114-3

Shiue CC, Sah CT (1979) New mobility-measurement technique on inverted semiconductor surfaces near the conduction threshold. Phys Rev 9:2149–2162

Shockley W, Bardeen J (1950) Energy bands and mobilities in monatomic semiconductors. Phys Rev 77:407–408

Silver M, O'Reilly EP (1994) Gain and radiative current density in InGaAs/lnGaAsP lasers with electrostatically confined electron states. IEEE J Quantum Electron 30:547–553

Slater JC, Koster GF (1954) Simplified LCAO method for the periodic potential problem. Phys Rev 94:1498–1524

Smith CS (1954) Piezoresistance effect in germanium and silicon. Phys Rev 94:42–49

Smith DL, Mailhiot C (1988) Piezoelectric effects in strained-layer superlattices. J Appl Phys 63:2717–2719

Stern F (1972) Self-consistent results for n-type Si inversion layers. Phys Rev B 5:4891–4899

Sugano T, Chen JT, Hamana T (1980) Morphology of Si-SiO_2 interface. Surf Sci 98:154–166

Sugii N, Washio K (2003) Low-temperature electrical characterization of strained-Si MOSFETS. Jpn J Appl Phys 42

Sun G, Sun Y, Nishida T, Thompson SE (2007) Hole mobility in silicon inversion layers: stress and surface orientation. J Appl Phys 102:084,501-1–084,501-7

Suzuki N, Hariu T, Shibata Y (1978) Effect of native oxide on the interface property of GaAs MIS structures. Appl Phys Lett 33:761–762

Takagi S (2003) Re-examination of subband structure engineering in ultrashort channel MOSFETS under ballistic carrier transport. Honolulu, HI, pp 115–116

Takagi S, Mizuno T, Tezuka T, Sugiyama N, Numata T, Usuda K, Moriyama Y, Nakaharai S, Koga J, Tanabe A, Hirashita N, Maeda T (2004) Channel structure design, fabrication and carrier transport properties of strained-Si/SiGe-on-insulator (strained-SOI) MOSFETS. In: Tech. Dig. Int. Electron Device Meet

Takagi S, Hoyt JL, Welser JJ, Gibbons JF (1996) Comparative study of phonon-limited mobility of two-dimensional electrons in strained and unstrained Si metal-oxide-semiconductor field-effect transistors. J Appl Phys 80:1567–1577

Takagi S, Mizuno T, Sugiyama N, Tezuka T, Kurobe A (2001) Strained -Si-on-insulator (strained-SOI) MOSFETS – concept, structures and device characteristics. IEICE Trans Electron E84-C

Takagi S, Mizuno T, Tezuka T, Sugiyama N, Numata T, Usuda K, Moriyama Y, Nakaharai S, Koga J, Tanabe A, Maeda T (2004) Fabrication and device characteristics of strained-Si-on-insulator (strained-SOI) CMOS. Appl Surf Sci 224

Tanaka SI, Lundstrom MS (1994) A compact model HBT device model based on a one-flux treatment of carrier transport. Solid State Electron 37:401–410

Tanaka T, Yanagisawa H, Minagawa S (1994) Comparison between tensile-strained AlGalnp SQW and MQW LDS emitting at 615 nm. Electron Lett 30:566–568

Taraschi G, Pitera AJ, Fitzgerald EA (2004) Strained Si, SiGe, and Ge on-insulator: review of wafer bonding fabrication techniques. Solid State Electron 48:1297–1305

Taur Y, Ning TH (1998) Fundamentals of modern VLSI devices. Cambridge University Press, New York

Taur Y, Buchanan DA, Chen W, Ismail DJF, Lo SH, SaiHalasz GA, Viswanathan RG, Wann HJC, Wind SJ, Wong HS (1997) CMOS scaling into the nanometer regime. Proc IEEE 85:486–504

Thompson SE, Anand N, Armstrong M, Auth C, Arcot B, Alavi M, Bai P, Bielefeld J, Bigwood R, Brandenburg J (2002) A 90 nm logic technology featuring 50 nm strained silicon channel transistors, 7 layers of cu interconnects, low κ ild, and 1 μm^2 SRAM cell. In: Tech. Dig. Int. Electron Device Meet., San Francisco, CA, pp 61–64

Thompson SE, Amstrong M, Auth C, Cea S, Chau R, Glass G, Hoffman T, Klaus J, Ma Z (2004a) A logic nanotechnology featuring strained-silicon. IEEE Electroc Dev Lett 25:191–193

Thompson SE, Sun G, Wu K, Lim J, Nishida T (2004b) Key differences for process-induced uniaxial vs. substrate-induced biaxial stressed Si and Ge channel MOSFETS. In: Tech. Dig. Int. Electron Device Meet., San Francisco, CA, pp 221–224

Thompson SE, Sun GY, Choi YS, Nishida T (2006a) Uniaxial-process-induced strained-Si: extending the CMOS roadmap. IEEE Trans Electron Devices 53:1010–1020

Thompson SE, Suthram S, Sun Y, Sun G, Parthasarathy S, Chu M, Nishida T (2006b) Future of strained Si/Semiconductors in nanoscale MOSFETS. In: Tech. Dig. Int. Electron Device Meet., San Francisco, CA, pp 11–13

Timbrell PY, Baribeau JM, Lockwood DJ, McCaffrey JP (1990) An annealing study of strain relaxation and dislocation generation in $Si_{1-x}Ge_x/Si$ heteroepitaxy. J Appl Phys 67:6292–6300

Uchida K, Krishnamohan T, Saraswat KC, Nishi Y (2005) Physical mechanisms of electron mobility enhancement in uniaxial stressed MOSFETS and impact of uniaxial stress engineering in ballistic regime. In: Tech. Dig. Int. Electron Device Meet., San Francisco, CA, pp 129–132

Ueno Y, Fujii H, Sawano H, Kobayashi K, Hara K, Gomyo A, Endo K (1993) 30-mw 690-nm high-power strained-quantum-well AlGaInp laser. IEEE J Quantum Electron 29:1851–1856

Valster A, van der Poel CJ, Finke MN, Boermans MJB (1992) Effect of strain on the threshold current of GaInP/AlGaInP quantum well lasers emitting at 633 nm. In: Conference Digest 13th IEEE Inter. Semicond. Laser Conf., Takamatsu, Japan, pp 152–153

Vogl P, Hjalmarson HP, Dow J (1983) A semi-empirical tight-binding theory of the electronic structure of semiconductors. J Phys Chem Solids 44:365–378

de Walle CGV (1989) Band lineups and deformation potentials in the model-solid theory. Phys Rev B 39:1871–1883

Wand YP, Wu SL, chang SJ (2005) Optimized Si-cap layer thickness for tensile-strained-Si/compressively strained SiGe dual-channel transistors in 0.13 μm complementary metal-oxide-semiconductor technology. Jpn J Appl Phys 44

Wang YC, Hong M, Kuo JM, Mannaerts JP, Kwo J, Tsai HS, Krajewski JJ, Chen YK, Cho AY (1999) Demonstration of submicron depletion-mode GaAs MOSFETS with negligible drain current drift and hysteresis. IEEE Eletron Dev Lett 20:457–459

Washington L, Nouri F, Thirupapuliyur S, Eneman G, Verheyen P, Moroz V, Smith L, Xu X, Kawaguchi M, Huang T, Ahmed K, Balseanu M, Xia LQ, Shen M, Kim Y, Rooyachers R, Meyer KD, Schreutelkamp R (2006) PMOSFET with 200% mobility enhancement induced by multiple stressors. IEEE Electron Dev Lett 27

Weisstein E (1999–2009) Euler angles. In MathWorld-a Wolfram web resource. http://mathworld.wolfram.com/EulerAngles.html

Welser J, Hoyt JL, Gibbons JF (1992) NMOS and PMOS transistors fabricated in strained silicon/relaxed silicon-germanium structures. In: Tech. Dig. Int. Electron Device Meet., San Francisco, CA, pp 1000–1002

Wiley JD (1971) Polar mobility of holes in iii–v compounds. Phys Rev B 4:2485–2493

Xiang J, Lu W, Hu Y, Wu Y, Yan H, Lieber CM (2006) Ge/Si nanowire heterostructures as highperformance field-effect transistors. Nature 441: 489–493

Xiang Q, Goo JS, Pan J, Yu B, Ahmed S, Zhang J, Lin MR (2003) Strained silicon NMOS with nickel-silicide metal gate. In: Symposium on VLSI Technology Digest of Technical Papers, Kyoto, Japan, pp 101–102

Xie YH, Gilmer GH, Roland C, Silverman PJ, Buratto SK, Cheng JY, Fitzgerald EA, Kortan AR, Schuppler S, Marcus MA, Citrin PH (1994) Semiconductor surface roughness: dependence on sign and magnitude of bulk strain. Phys Rev Lett 73:3006–3009

Yamakawa S, Ueno H, Taniguchi K, Miyatsuji K, Masaki K, Ravaioli U (1998) Electron mobility and Monte Carlo device simulation of MOSFETS. VLSI Design 6:27–30

Yan L, Olsen SH, Escobedo-Cousin E, O'Neill AG (2008) Improved gate oxide integrity of strained Si n-channel metal-oxide-silicon field effect transistors using thin virtual substrates. J Appl Phys 103:094,508-1–094,508-10

Yang HS, Malik R, Narasimha S, Li Y, Divakaruni R, Agnello P, Allen S, Antreasyan A, Arnold JC, Bandy K, Belyansky M, Bonnoit A, Bronner G, Chan V, Chen X, Chen Z, Chidambarrao D, Chou A, Clark W, Crowder SW, Engel B, Harifuchi H, Huang SF, Jagannathan R, Jamin FF, Kohyama Y, Kuroda H, Lai CW, Lee HK, Lee WH, Lim EH, Lai W, Mallikarjunan A, Matsumoto K, McKnight A, Nayak J, Ng HY, Panda S, Rengarajan R, Steigerwalt M, Subbanna S, Subramanian K, Sudijono J, Sudo G, Sun SP, Tessier B, Toyoshima Y, Tran P, Wise R, Wong R, Yang IY, Wann CH, Su LT, Horstmann M, Feudel T, Wei A, Frohberg K, Burbach G, Gerhardt M, Lenski M, Stephan R, Wieczorek K, Schaller M, Salz H, Hohage J, Ruelke H, Klais J, Huebler P, Luning S, van Bentum R, Grasshoff G, Schwan C, Ehrichs E, Goad S, Buller J, Krishnan S, Greenlaw D, Raab M, Kepler N (2004) Dual stress liner for high performance sub-45nm gate length SOI CMOS manufacturing. In: Tech. Dig. Int. Electron Device Meet., San Francisco, CA, pp 1075–1077

Yang X, Lim J, Sun G, Wu K, Thompson SE (2006) Strain-induced changes in the gate tunneling currents in *p*-channel metal-oxide-semiconductor field-effect transistors. Appl Phys Lett 88:052,108-1–052,108-3

Yoshinobu T, Iwamoto A, Iwasaki H (1994) A self-consistent Monte Carlo simulation for two-dimensional electron transport in MOS inversion layer. Jpn J Appl Phys I 26:1447–1452

Yu PY, Cardona M (1996) Fundamentals of semiconductors. Spinger, Berlin

Zawadski W (1974) Electron transport phenomena in small-gap semiconductors. Adv Phys 23:435–522

Zhang D, Nguyen BY, White T, Goolsby B, Nguyen T, Dhandapani V, Hildreth J, Foisy M, Adams V, Shiho Y, Thean A, Theodore D, Canonico M, Zollner S, Bagchi S, Murphy S, Rai R, Jiang J, Jahanbani M, Noble R, Zavala M, Cotton R, Eades D, Parsons S, Montgomery P, Martinez A, Winstead B, Mendicino M, Cheek J, Liu J, Grudowski P, Ranami N, Tomasini P, Arena C, Werkhoven C, Kirby H, Chang CH, Lin CT, Tuan HC, See YC, Venkatesan S, Kolagunta V, Cave N, Mogab J (2005) Embedded SiGe S/D PMOS on thin body SOI substrate with drive current enhancement. Honolulu, HI, pp 26–27

Index

E

F

G

H

I

J

K

L

M

N